TRIBOLOGY SERIES, 16

TRIBOLOGY OF PLASTIC MATERIALS

Their Characteristics and Applications to Sliding Components

Yukisaburo Yamaguchi
Professor Emeritus, Kogakuin University, Tokyo, Japan

ELSEVIER
Amsterdam • Oxford • New York • Tokyo 1990

ELSEVIER SCIENCE PUBLISHERS B.V.
Sara Burgerhartstraat 25
P.O. Box 211, 1000 AE Amsterdam, The Netherlands

Distributors for the U.S.A. and Canada:

ELSEVIER SCIENCE PUBLISHING COMPANY, INC.
655 Avenue of the Americas
New York, N.Y. 10010, U.S.A.

ISBN 0-444-87445-3 (Vol. 16)
ISBN 0-444-41677-3 (Series)

Printed in The Netherlands

PREFACE

Plastic materials excel in lightness, electric and heat insulation, corrosion resistance, absorption of impact and vibration, colourfulness and mouldability. They can be economically produced, and their properties are easily modified by forming composites or blending. Uses in various forms are thus widespread and range from toys and home appliances to industrial tools and machine elements.

In addition to the above-mentioned properties, it is worth noting that many plastic materials have excellent self-lubricating characteristics. Historically, phenolic resin was used for journal bearings and gears because of its ability to operate without conventional lubrication and its vibration-absorbing qualities. More recently, applications of some semi-crystalline plastic materials, especially polytetrafluoroethylene and polyacetal, have been greatly extended to include sliding machine parts, as a direct result of their self-lubricating capabilities.

The friction and wear characteristics of plastic materials have been studied for over 50 years, but generally accepted theories and definitive experimental data have yet to be established because of the proliferation of new materials and the complications of simulating appropriate practical conditions in the laboratory. Nevertheless, when these materials are intended to be used for sliding parts, workable theories and reliable experimental data are required.

During the past 25 years, the author's experiments on the sliding behaviour of plastic materials and their applications in machine elements have led to the accumulation of considerable experimental data and the formulation of practical theories. Most of the data presented in this book were obtained in a single laboratory. Therefore, if these data are to be used in practical situations, caution must be exercised and the conditions carefully analyzed.

This book is a translation, for the most part, of a book entitled *"Lubricity of Plastic Materials"*, which was published originally in Japanese by the Nikkan Kogyo Newspaper Co. after previously appearing as a series of articles in the journal *"Engineering Materials"* during the course of one year.

The book is divided into four parts. Chapters 1 and 2 deal with current theories of friction and wear, and include discussion of various hypotheses based upon experimental studies. Chapter 3 details experiments designed to improve tribological performance via polymer blending and composite production, whilst Chapter 4 explains how the data obtained from these experiments can be applied to sliding machine parts. It is the author's hope that the information may prove useful for the design of plastic materials and components and that it may be a stepping-stone toward future innovations in this field.

I would like to thank Dr. Y. Oyanagi, Mr. S. Amano, Mr. S. Sato, and especially Dr. I. Sekiguchi of the High Polymeric Material Laboratory at Kogakuin University, for their help and assistance throughout the course of this project. I am also grateful to the staff of Nikkan Kogyo Newspaper Co. for their constant support, to Dr. Y. Hazeyama for his help with the translation into English, to Dr. John Lancaster for his final editing of the text, and also to the Oiless Kogyo Co. for their financial assistance. I would also like to take this opportunity to thank Dr. John Lancaster and Dr. Brian Briscoe for being instrumental in arranging for this work to be published in English.

Y. Yamaguchi

FOREWORD

Recent years have seen the publication of several books in English on the subject of Tribology, and indeed the "Elsevier Tribology Series" has contributed significantly to this number. In the particular area of Polymer Tribology there have been two significant Soviet texts as well as at least one important dedicated conference publication. In addition, a recent compilation on "Composite Tribology" is largely devoted to polymeric systems. Many international conferences continue to devote sections to Polymer Tribology and a number of non-tribological texts have reviewed the subject in self-contained chapters. Compared with the situation perhaps twenty years ago, the subject of Polymer Tribology is thus now reasonably well furnished with general introductory material.

The present book naturally contributes directly to this information source, being one of four dedicated textbooks on the subject. The main structure of the book is laid out on classical lines and incorporates many ideas which have evolved in the Western literature. It should be borne in mind, however, that this book is very much a Japanese view of the important elements of the subject and, in detail, tends to concentrate on those topics in which the author and his group have made notable original contributions. In many ways, this is perhaps the main value of the text. The Western Tribologist has now, with this book and the Soviet ones, an overall international view of the way in which Polymer Tribology has developed as a subject. Perhaps the main surprise for these readers will be how similar the development of the subject has been in the three geographical areas. Clearly, this must reflect the many international contacts which occurred during the formative years of the subject. There are obviously differences in emphasis, style and approach, but the basic ingredients of fundamental principles coupled with a desire to develop a reasoned and confident predictive capacity is a common theme.

We, personally, were particularly pleased to be asked to provide the foreword to this text as we were fortunate enough to meet Professor Yamaguchi on a visit to Japan in 1985. At that time, we were able to visit his laboratories, gain a good appreciation of the wide range of his activities in polymer science and technology and also see the present book in the original, Japanese version. Our main overall recollection of that visit to Japan was the stimulating feeling engendered by the discovery that Polymer Tribology was such a strongly developed subject in that country. A similar conclusion could, of course, be drawn from a perusal of the published literature, but personal contacts are naturally more telling. Although

Professor Yamaguchi's book was just one element which contributed to this opinion, we felt then, and indeed also feel now, that this impression deserved a wide audience. Hence our encouragement to the author to undertake the translation of his book into English. The book has its own technical merit, but perhaps its lasting contribution will be to provide a view of the development of Polymer Tribology in Japan. Polymer Tribology has now a reasonably long history and as such deserves to be recorded.

B.J. Briscoe
J.K. Lancaster

CONTENTS

CHAPTER 1

FRICTION

1.1 SLIDING FRICTION; THEORY AND EXPERIMENT

1.1.1 *Theory of sliding friction*

The so-called "adhesion-shearing theory" was advocated originally by Bowden and Tabor [1,2] to account for sliding friction and, more recently, a theory based on surface energy has been presented by Lee [3]. In this book, a theory which is based on the "adhesion-shearing" mechanism is presented and frictional resistance is discussed as it is related to the shearing force required to break the interface of the contacting parts. As shown in Fig. 1.1 the true contact surface area A ($= \int_0^n a$) is far smaller than the apparent contact area A_o. Minute contact areas such as a_1, a_2 shown in Fig. 1.1 adhere to each other under a normal pressure p, and relative sliding motion between A and B is then possible only by destruction of the interface in shear. Accordingly, the frictional resistance F (the resistance to movement along the contact surface) is the sum of the shearing destructive force Fs and the resistance Fd to deform the contact part:

$$F = Fs + Fd \tag{1.1}$$

In reality, $F \simeq Fs$ since Fd is far smaller than Fs, and the following equation may thus be obtained:

$$F = A \cdot \tau \tag{1.2}$$

where A is the true contact area and τ is the shear strength of the contact material. The true contact area A is presented generally as follows:

$$A = kP^m \tag{1.3}$$

where P is the normal load, and k and m are constants, depending on the materials. The value of A may be obtained theoretically by using Hertz's elastic law [4] or Meyer's law [5] on hardness as follows:

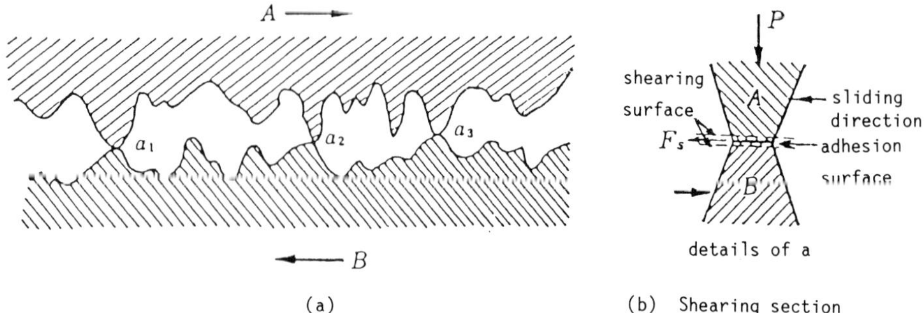

Fig. 1.1 Macroscopic section of sliding contact surfaces

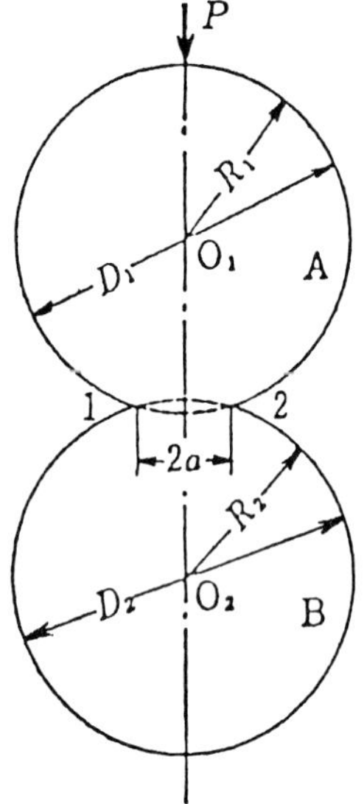

Fig. 1.2 Two spheres in direct contact

(i) *Theory based on Hertz's elastic law*

Two contact cases applying Hertz's law are discussed. One is for two opposing spheres and the other is for a sphere on a plane surface.

(1) *Two opposing spheres.* When two spheres with radii of R_1 and R_2 are in contact under a normal load P, as shown in Fig. 1.2, the radius a of the contact area predicted by Hertz's elastic law [4] is shown in the following equation:

$$a^3 = \frac{3}{4} \cdot \frac{R_1 \cdot R_2}{R_1 + R_2} \cdot \left(\frac{1-\nu_1^2}{E_1} + \frac{1-\nu_2^2}{E_2} \right) \cdot P \tag{1.4}$$

where ν_1 and ν_2 are Poisson's ratios and E_1 and E_2 are Young's moduli of spheres A and B, respectively. The contact area A is then

$$A = \pi a^2 = \pi \left\{ \frac{3}{4} \cdot \frac{R_1 \cdot R_2}{R_1 + R_2} \left(\frac{1-\nu_1^2}{E_1} + \frac{1-\nu_2^2}{E_2} \right) \right\}^{2/3} \cdot P^{2/3} \tag{1.5}$$

For similar materials $E_1 = E_2 = E$, and of $\nu_1 = \nu_2 = 0.4$, then

$$A = 1.16\pi \left\{ \frac{R_1 R_2}{E(R_1 + R_2)} \right\}^{2/3} \cdot P^{2/3} \text{ and} \tag{1.6}$$

$$\mu = \frac{F}{P} = \frac{A\tau}{P} = 1.16\pi \left\{ \frac{R_1 R_2}{E(R_1 + R_2)} \right\}^{2/3} \cdot P^{-1/3} \cdot \tau \tag{1.7}$$

(2) *Sphere on a plane surface.* When a large sphere of radius R having smaller spherical asperities of radius r is in contact with a plane surface xx under a normal load P as shown in Fig. 1.3, the radius a of the circular contact area from Hertz [4] is given by:

$$a = K_1 (PR)^{1/3} \tag{1.8}$$

where K_1 is a constant containing an elastic modulus and other constants. The maximum pressure P_o occurs at the centre and is

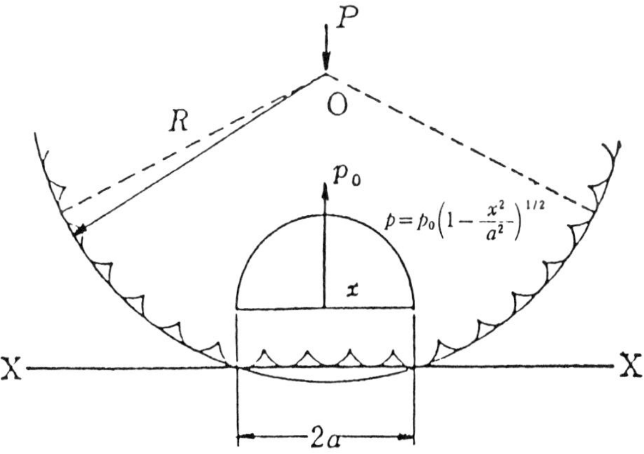

Fig. 1.3 Two surfaces, sphere and plane, in contact at many points

$$P_o = K_2 \cdot P^{1/3} \cdot R^{-2/3} \tag{1.9}$$

and the pressure p at any annulus of radius x is

$$p = p_o \left(1 - \frac{x^2}{a}\right)^{1/2} \tag{1.10}$$

If it is assumed that the asperities have equal radius x and are distributed uniformly as n/cm^2, the minute area dA of a circular ring with a breadth of dx at a radius x is $2\pi x \cdot dx$, and the load w supported by this area is

$$w = \frac{p}{n} \tag{1.11}$$

The area $\bar{a}$ of asperities is then

$$\bar{a} = K_3(w \cdot r)^{2/3} = K_3 \left(p \frac{r}{n}\right)^{2/3} \tag{1.12}$$

and the total true contact area A is

$$A = \int_0^a \cdot 2\pi n K_3 \left(\frac{r}{n}\right)^{2/3} \cdot p^{2/3} \cdot x \cdot dx$$

$$= \int_0^a K_4 r^{2/3} \cdot n^{1/3} \cdot p_o^{2/3} \left(1 - \frac{x^2}{a^2}\right)^{1/3} \cdot x \cdot dx$$

$$= K_5 \cdot r^{2/3} \cdot n^{1/3} \cdot R^{2/9} \cdot P^{8/9} \tag{1.13}$$

$$\mu = \frac{A\tau}{P} = K_5 \cdot r^{2/3} \cdot n^{1/3} \cdot R^{2/9} \cdot P^{-1/9} \cdot \tau \tag{1.14}$$

(ii) *Theory based on Meyer's law*

Pascoe and Tabor [6] have reduced equation (1.19) to find the true contact area and frictional coefficient by applying Meyer's law concerning indentation hardness. According to this, the relationship between the load P, diameter D of the indenter and diameter d of the indentation, as shown in Fig. 1.4, is:

$$P = ad^n \tag{1.15}$$

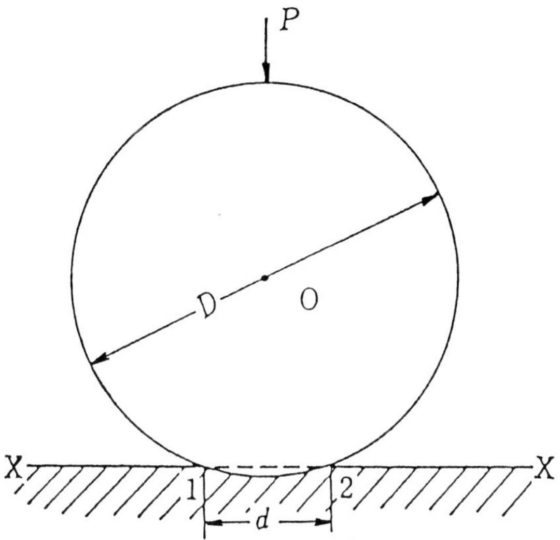

Fig. 1.4 Sphere and penetration of plane (Meyer's law)

$$K = a_1 \cdot D_1{}^{n-2} = a_2 \cdot D_2{}^{n-2} = a_3 \cdot D_3{}^{n-2} \ldots \qquad (1.16)$$

From (1.15) and (1.16)

$$P = \frac{k}{D^{n-2}} \cdot d^n \qquad (1.17)$$

and the true contact area A is

$$A = \frac{\pi}{4} d^2$$

$$= \frac{\pi}{4} \left(\frac{1}{K}\right)^{2/n} \cdot D^{2(n-2)/n} \cdot P^{2/n} \qquad (1.18)$$

The coefficient of friction μ, presented similarly to equation (1.7) is therefore

$$\mu = \frac{A\tau}{P}$$

$$= \frac{\pi}{4} \cdot \tau \left(\frac{1}{K}\right)^{2/n} \cdot D^{2(n-2)/n} \cdot P^{(2/n-1)} \qquad (1.19)$$

When the value of n in equation (1.19) is evaluated with respect to the value of m in equation (1.3), the following values of n are obtained: n is equal to 3 in equation (1.5) and is equal to 2.25 in equation (1.13). It is assumed that the value of n must be 3 when the material is perfectly elastic and 2 when the material is perfectly plastic. In other words, the value of A is proportional to P^m or $P^{2/3\sim 1}$, and the value of μ is proportional to $P^{(m-1)}$ or $P^{-(1/3\sim 0)}$.

1.1.2 *Experimentation based on sliding friction theories*

The relation where μ is proportional to $P^{(m-1)}$ has been previously explained by Shooter [7] and Lincoln [8], and has also been verified experimentally, to some extent, using Meyer's law by Pascoe and Tabor [6]. In this section, a discussion of two experimental cases is presented. One case is that of the contact of a steel sphere with a polymer plane applying Meyer's law and the other is that of contact between two polymer spheres applying Hertz's law.

(i) *Contact between a steel sphere and a polymer plane*

Using a Rockwell hardness tester, a steel sphere of diameter D was indented into a polymer or a steel plane under a load P, and the diameter d of the indentation was measured for each size of the sphere. The constants a and n in equation (1.15) and K in equation (1.16) were obtained and are presented in Table 1.1. Figure 1.5 shows the relationship between the load P and μ obtained from these constants and the shearing strength τ of various plastics from equation (1.19).

TABLE 1.1
Constants in Meyer's Law*

Material	a	n (mean)	K (mean)	μ (for Al)
Phenolics (PF)	6.9-38	2.76	54.8	0.37
Melamine Resin (MF)	9.5-35	2.67	49.1	0.30
PMMA	3.3-5.9	2.79	25.4	0.46
Polystyrene (PS)	4.5-4.8	2.75	23.5	0.36
Polycarbonate (PC)	2.3-5.9	2.75	19.8	0.56
Nylon 6	1.8-5.1	2.84	15.3	0.44
Steel	47-101	2.2	92.8	

*dia. of steel ball: 1/16-1/2", 20°C

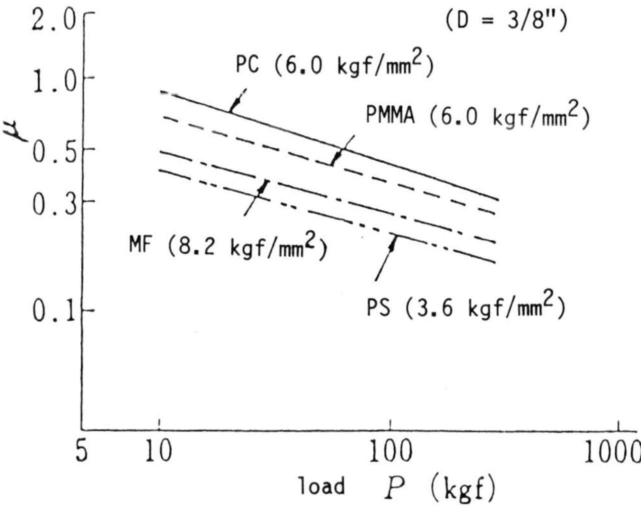

Fig. 1.5 Relationship between μ and P from Equation (1.19) for various polymers at different pressures, (___) slow the value of τ (20°C)

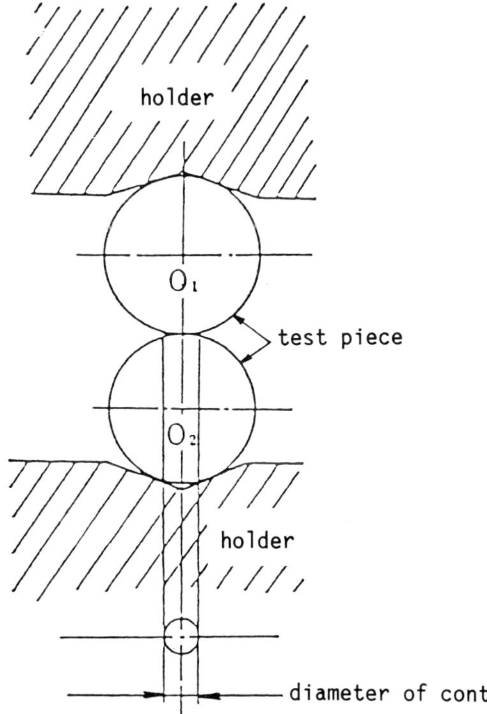

Fig. 1.6 Apparatus for measuring the contact area between two spheres

The experimental values of the static friction coefficient between the plastics and aluminium are also presented in Table 1.1. The theoretical values of μ in Fig. 1.5 are roughly equivalent to the experimental values in Table 1.1, although the former become slightly larger than the latter in the smaller load range.

(ii) *Contact between two polymer spheres*

(1) *Experimental method.* When a load P was applied to two polymer spheres in contact, using the apparatus shown in Fig. 1.6, the diameter d(=2a) was measured from the transverse side with a comparator. In this case, polymer spheres were used of three different diameters (5/16", 1/2" and 3/4") at eight different loads (ranging from 16 to 151 Kgf) and at five different temperatures (0, 20, 40, 50 and 60°C).

(2) *Results and discussion.* Examples of the relationships between the load P and the diameter d of the contact area between two polymer spheres with the same diameter are shown in Fig. 1.7. These relationships are equivalent to Meyer's law.

Meyer's constants, a, n and K in equation (1.7), obtained from the curves in Fig. 1.7, are presented together in Table 1.2.

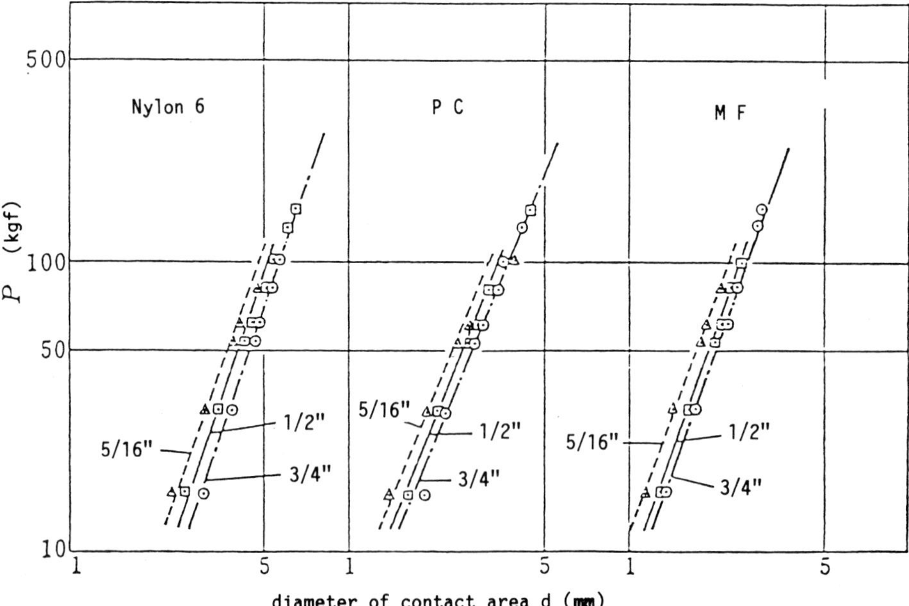

Fig. 1.7 Relationship between diameter of contact area and load (50°C)

TABLE 1.2
Constants in Meyer's Law for each plastic at various temperature

	Constants: a			n				$K = a_1D_1^{n-2} = a_2D_2^{n-2} = \ldots$			
Temp. °C	Dia: 3/4"	1/2"	5/16"	3/4"	1/2"	5/16"	Ave.	3/4"	1/2"	5/16"	Ave.
Nylon 6:											
0	1.64	3.75	5.60	3.08	2.36	2.41	2.62	20.2	18.1	20.2	16.2
20	1.61	2.36	3.35	2.82	2.68	2.48	3.66	11.3	12.6	13.2	12.4
40	0.60	0.80	4.05	3.27	2.05	2.01	2.44	2.2	8.6	10.1	7.0
50	0.73	0.90	1.25	2.90	2.90	2.75	2.85	8.9	7.8	7.2	8.0
60	0.56	0.72	1.12	2.99	2.99	2.82	2.93	8.7	7.6	7.7	8.0
PC:											
0	4.00	5.30	8.00	2.68	2.61	2.48	2.59	22.8	23.7	27.2	24.6
20	4.90	6.10	7.30	2.50	2.52	2.50	2.51	22.0	22.3	21.0	21.8
40	3.20	4.00	5.20	2.75	2.82	2.82	2.80	33.9	30.4	27.4	30.6
50	4.60	5.50	6.70	2.41	2.41	2.41	2.41	15.4	15.6	15.7	15.6
60	4.45	5.90	8.00	2.36	2.32	2.25	2.31	13.1	13.0	15.2	13.8
MF:											
0	8.30	12.40	18.00	2.89	2.91	2.98	2.93	128.5	131.6	123.8	123.0
20	8.80	10.90	13.00	2.88	2.85	2.88	2.88	117.2	101.8	80.6	99.9
40	7.40	9.20	11.20	2.99	2.99	2.99	2.99	137.0	114.1	87.0	112.7
50	7.50	8.60	11.90	2.85	2.85	2.85	2.85	92.3	74.4	69.3	73.7
60	4.90	6.30	8.40	2.99	2.90	2.93	2.94	78.4	68.7	58.8	63.6

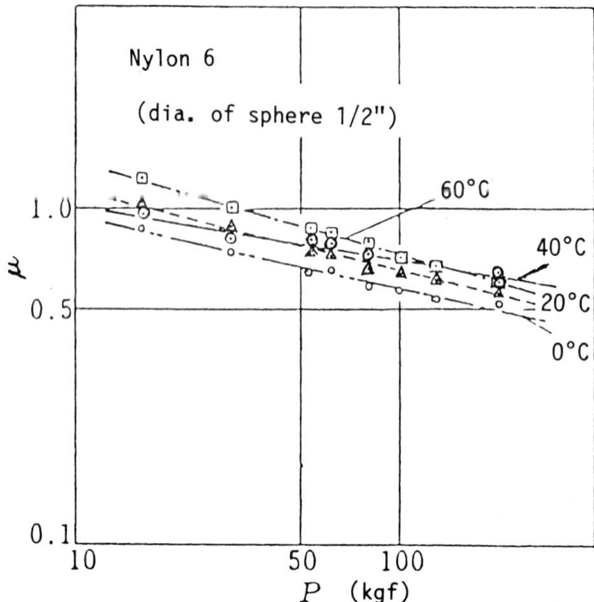

Fig. 1.8 Theoretical μ-P diagram for Nylon 6

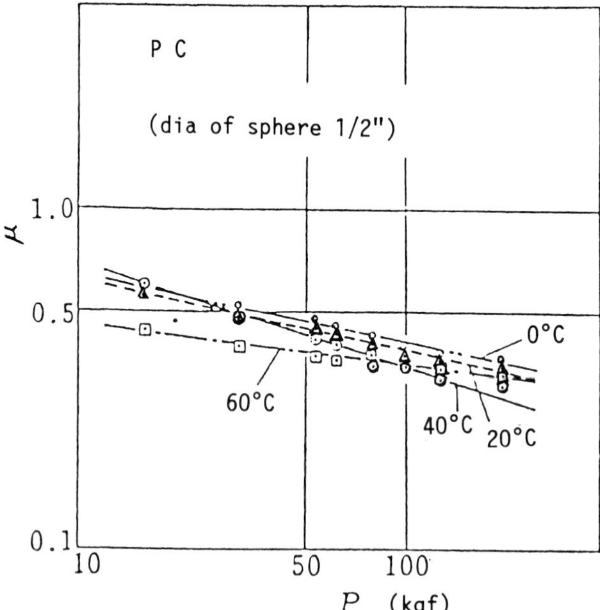

Fig. 1.9 Theoretical μ-P diagram for PC

TABLE 1.3
Diameter of contact area and theoretical frictional coefficient for each load (d - 1/2", 20°C)

Load	Nylon 6				Polycarbonate				Melamine Resin			
P, kgf	Experiment 2a,mm	Hertz's Two Sphere $2a_H$,mm	μ_H	μ_M	Experiment 2a,mm	Hertz's Two Sphere $2a_H$,mm	μ_H	μ_M	Experiment 2a,mm	Hertz's Two Sphere $2a_H$,mm	μ_H	μ_M
16	2.16	2.06	1.108	1.025	1.48	1.69	0.750	0.579	1.14	1.03	0.424	0.538
31	2.71	2.69	0.894	0.881	1.83	2.20	0.655	0.503	1.29	1.34	0.375	0.426
53	3.63	3.33	0.800	0.766	2.44	2.73	0.585	0.488	1.76	1.66	0.334	0.370
61	3.84	3.52	0.781	0.741	2.64	2.89	0.571	0.438	1.82	1.76	0.326	0.358
81	4.25	4.00	0.760	0.689	2.35	3.24	0.541	0.411	2.03	1.97	0.309	0.326
101	4.67	4.33	0.707	0.651	3.17	3.54	0.517	0.392	2.29	2.15	0.295	0.304
131	5.23	4.77	0.667	0.615	3.63	3.91	0.487	0.373	2.99	2.38	0.278	0.283
151	5.62	5.08	0.656	0.590	3.85	4.17	0.480	0.348		2.54	0.274	0.271
	E = 100 kgf/mm^2 τ = 4.88 kgf/mm^2				E = 200 kgf/mm^2 τ = 5.37 kgf/mm^2				E = 900 kgf/mm^2 τ = 8.2 kgf/mm^2			

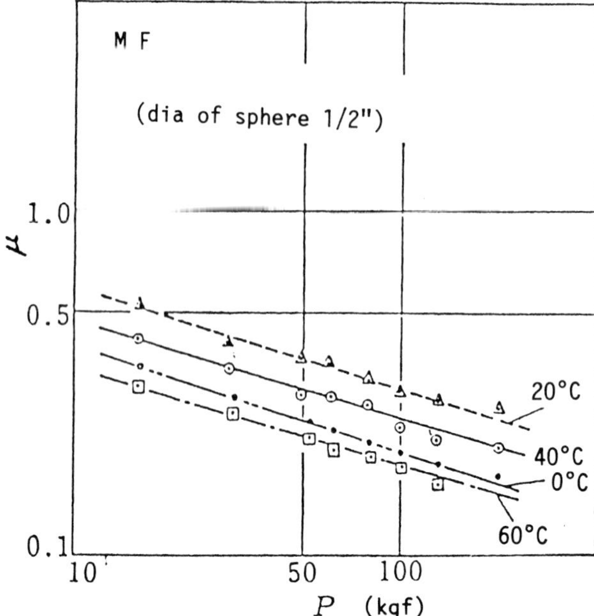

Fig. 1.10 Theoretical μ-P diagram for MF

The relationships at various temperatures between the load P and the values of μ obtained from equation (1.19) are shown for Nylon 6, polycarbonate and Melamine resin in Figs. 1.8, 1.9 and 1.10, respectively. For each plastic, Table 1.3 presents the experimental values of d(=2a) of the contact area for each load P at 20°C and D=1/2"; the diameter $2a_H$ of the contact area between spheres obtained from Hertz's equation (1.4); the magnitude of μ_H obtained from the Young's modulus E, the shear strength τ of each plastic and equation (1.7); and the magnitude of μ_M obtained from the above-mentioned Meyer's constants and equation (1.19). As can be seen from this table, a comparison between the experimental value of the diameter 2a and the theoretical value of $2a_H$ based on Hertz's analysis shows that the former is slightly larger than the latter for Nylon 6 and the melamine resin, but is slightly smaller for the polycarbonate. It is also clear from Table 1.3 that μ_M is smaller than μ_H for polycarbonate ($2a<2a_H$), μ_M is larger than μ_H for melamine resin ($2a>2a_H$) and μ_M is almost equal to μ_H for nylon.

Figure 1.12 shows the relationships between the contact pressure p and the value of μ measured with a sliding friction test apparatus, shown in Fig. 1.11, for each of the above plastics. The relationships between the load P and the theoretical value of μ_M based on Meyer's equation (1.19) are also presented. It is clear from these diagrams that the theoretical value of μ is greatest for nylon but the experimental value is greatest for polycarbonate.

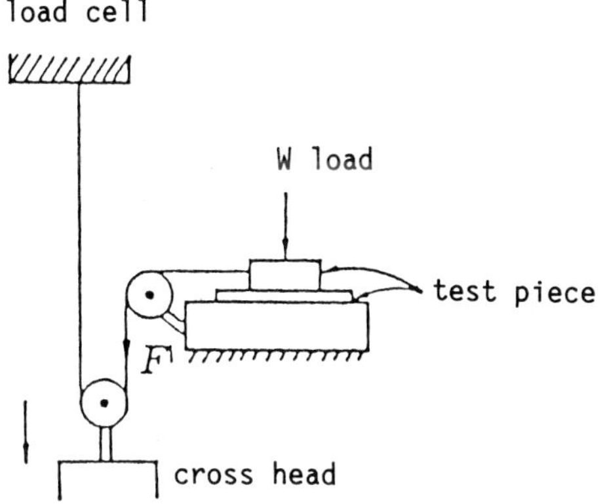

Fig. 1.11 Apparatus for measuring μ

Also, the theoretical values μ_H and μ_M are generally larger than the experimental values. The trend of decreasing μ with increasing load P is similar in both the theoretical and experimental cases. From the comparison of 2a with $2a_H$, μ_H with μ_M in Fig. 1.5, the theoretical values of μ with the experimental ones μ_M in Fig. 1.12 and the trends shown in Table 1.1 and Fig. 1.5, together with the fact that there can be considerable scatter in the measurement of friction coefficients, it may be concluded that the experimental values of μ are closely related to the theoretical values. The "adhesion-shearing theory" thus appears to be applicable, with some allowances, as a basis for explaining the sliding friction phenomena of plastics.

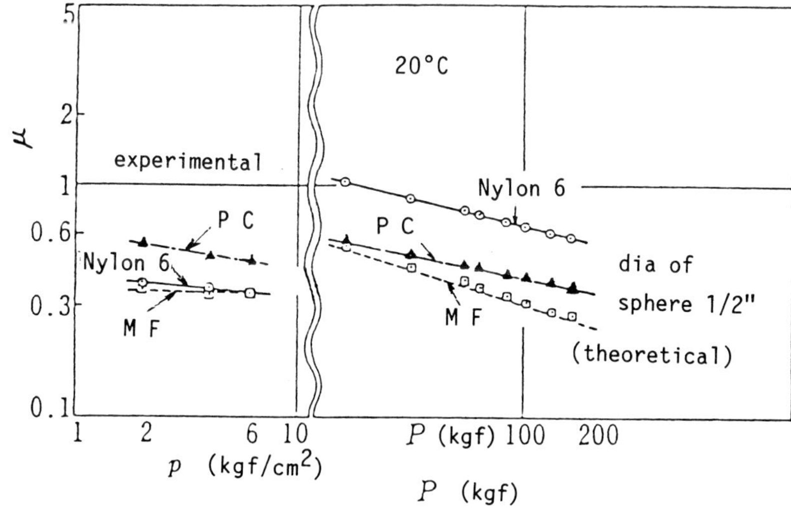

Fig. 1.12 Relationships between the theoretical or experimental values of μ and load for various plastics (20°C)

1.2 THEORY OF ROLLING FRICTION [12].

Much theoretical research on the friction of metallic materials rolling over plane surfaces has been conducted since Reynolds [13]. The rolling friction of plastics, however, is a more recent field of research and has been investigated by Tabor [16], Flom [11-19], May [20], Norman [21] and Takemura [22,23]. Other contributions include a review by Johnson [24], experimental studies by Cremer [25], Tanaka [26] and Matsubara [27]. In this section, the major theories on rolling friction in plastics are introduced and discussed for the combination of a rigid rolling body and a viscoelastic plane surface. The relevance to rolling element bearings will be given later in the applications section [4.1.3].

1.2.1 *Rolling friction theory based on hysteresis loss*

The rolling friction theory by Tabor [16] is based on the idea that when a body rolls over a viscoelastic surface, such as rubber or plastics, there is a hysteresis loss resulting from the deformation of the surface under load and its subsequent recovery after unloading. In other words, when the elastic deformation energy per unit rolling length is ϕ, and the hysteresis loss following recovery of this deformation energy is α, then the rolling friction resistance F is

$$F = \alpha\phi \tag{1.20}$$

As shown in Fig. 1.13(a), when the normal load per unit length is $\overline{W}$ and the roller O rolls on a plastic plane, the work done by the roller per unit length, that is the total deformation energy, ϕ_1 may be obtained as follows.

According to Hertz's equation [4], the pressure p at a point x far from the centre O_1 of the contact area is $p = p_o(1-x^2/a^2)^{1/2}$, and the pressure p_0 at the centre O_1 is $p_0 = 2\overline{W}/\pi a$. The total couple C exerted by all the elements on the front half of the band of contact is

$$C = \int_0^a px \cdot dx = \frac{2\overline{W}\cdot a}{3\pi} \tag{1.21}$$

and $\phi_1 = C/R$ where R is the radius of the roller. Thus,

$$\phi_1 = \frac{2}{3\pi} \cdot \frac{\overline{W}\cdot a}{R} \tag{1.22}$$

The value of a is given by the following equation (from Hertz) where E is the elastic modulus, ν is Poisson's ratio and the roller is rigid

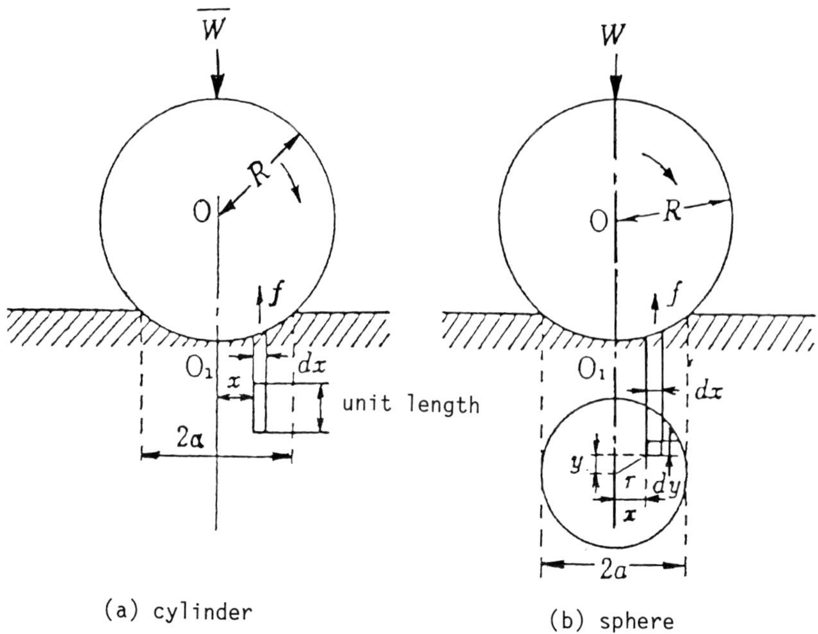

Fig. 1.13 Change in shape of the surface of a plastic material caused by a rolling body

$$a = \frac{2}{\sqrt{\pi}} \left\{ \overline{W} \cdot R \left(\frac{1-\nu^2}{E} \right) \right\}^{1/2} \tag{1.23}$$

When a ball O with radius R is rolling under a load W, as shown in Fig. 1.13(b), the pressure p at a point r far from the centre O_1 is

$$p = \frac{3W}{2\pi a^2} \left(1 - \frac{r^2}{a^2} \right)^{1/2} \tag{1.24}$$

The normal force P over an elementary area (dx·dy) is p·dx·dy and $r^2=x^2+y^2$, and so the total vertical force f on the whole elementary strip is

$$f = \frac{3W}{4a^3} (a^2-x^2) \, dx \tag{1.25}$$

Since the couple exerted on the ball is fx, the total couple C exerted by all elements on the front half of the circle is

$$C = \frac{3W}{4a^3} \int_0^a (a^2-x^2)x \cdot dx = \frac{3Wa}{16} \tag{1.26}$$

Hence, the elastic work done ϕ_2(=C/R) rolling forward a unit distance is

$$\phi_2 = \frac{3}{16} \cdot \frac{Wa}{R} \tag{1.27}$$

From Hertz,

$$a = \left(\frac{3}{4} \cdot W \cdot R \cdot \frac{1-\nu^2}{E}\right)^{1/3} \tag{1.28}$$

If α is the hysteresis loss in simple tension or compression, and γ is the correction factor due to slippage, the actual energy loss, α_e is $\gamma\cdot\alpha$, and the frictional resistance F = $\alpha_e\cdot\phi$. Since μ = F/W, the coefficients of rolling friction μ_1 (roller) and μ_2 (ball) are given by the following equations

$$\mu_1 = \gamma_1\alpha\phi_1/\overline{W} = \alpha_{e1}\phi_1/\overline{W} = \alpha_{e1}\left(\frac{2}{3\pi} \cdot \frac{a}{R}\right) \tag{1.29}$$

$$\mu_2 = \gamma_2\alpha\phi_2/W = \alpha_{e2}\left(\frac{3}{16} \cdot \frac{a}{R}\right) \tag{1.30}$$

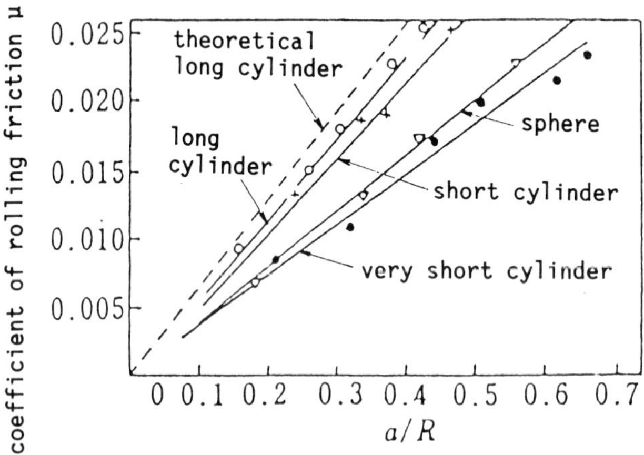

Fig. 1.14 Relationship between the rolling friction coefficient μ and a/R for a rigid rolling body on rubber

For simple tension or compression in rubber, α = 0.08, Fig. 1.14 shows the theoretical and experimental relationships between the coefficient of rolling friction μ of a rigid rolling body on rubber and a/R. The value of γ is experimentally equivalent to the following:

γ_1 = 3.3α for a long cylinder
γ_1 = 2.9α for a short cylinder
γ_1 = 2α for a very short cylinder
γ_2 = 2.2α for a sphere

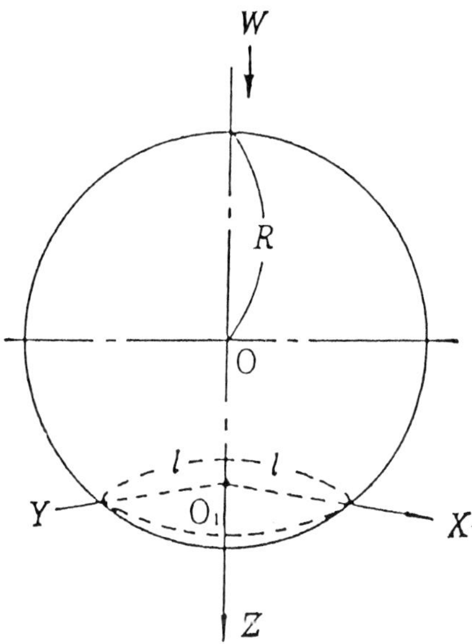

Fig. 1.15 Deformation of the base material by a hard sphere in rolling contact

1.2.2 *Viscoelastic theory and experiment*

Deformation of plastic materials due to loading or unloading is influenced by the duration of loading because of their viscoelastic properties. A theoretical approach to rolling friction of plastic materials considering these properties was made by Flom and Bueche [17] and is as follows. Referring to Fig. 1.15, if p is the axial pressure in the Z direction on a sphere of radius R, η is the coefficient of viscosity and G is the elastic modulus of the plastic material in a Voight's mechanical model, then

$$p = kGz + k\eta(dz/dt) \tag{1.31}$$

where z is the indentation depth on the base plastic material and k is a constant with units of mm^{-1}.

From Fig. 1.15, $Z=(R^2-x^2-y^2)^{1/2}-(R^2\ \ell^2)^{1/2}$, and when the base material moves to the -x direction at a speed s, the velocity of the contact point dz/dt is

$$dz/dt = x\cdot S(R^2-x^2-y^2)^{-1/2} \tag{1.32}$$

The retardation time τ in Voight's model is $\tau=\eta/G$, and when $\beta=\tau S/R$, $\phi=1/R$, equation (1.31) becomes

$$p = \frac{1}{2} kGR \left\{ \phi^2 - \frac{x^2}{R^2} - \frac{y^2}{R^2} + 2\beta \frac{x}{R}\left(1 + \frac{1}{2}\cdot\frac{x^2}{R_2} + \frac{1}{2}\cdot\frac{y^2}{R_2}\right) \right\} \tag{1.33}$$

Therefore, the total force W in the z direction must be

$$W = \int_{y=-\ell}^{y=+\ell} \int_{x=R\xi}^{x=(\ell^2-y^2)^{1/2}} \cdot p\cdot dx\cdot dy \tag{1.34}$$

and $$R\xi = R\beta - R\left\{ \beta^2 + \phi^2 - \left(\frac{y^2}{R^2}\right) \right\}^{1/2}$$

$$\therefore W = kGR^3\left\{\left(\frac{\pi}{8}\phi^4 + \frac{\pi}{24}\phi^6 + \frac{1}{4}\beta\phi^2 - \beta^3\phi\ \ \frac{7}{6}\beta^3\phi^3\right)\right.$$
$$\left. + \left(\frac{1}{2}\beta^2\phi^2 + \frac{1}{4}\phi^4 + \frac{1}{4}\beta^4 + \beta^6 + \frac{3}{2}\beta^4\phi^2 + \frac{1}{2}\beta^2\phi^4\right)\cdot \sin^{-1}\frac{\phi}{(\beta^2+\phi^2)^{1/2}} \right\} \tag{1.35}$$

Experimentally, W, G and R are usually constant. Therefore, equation (1.35) shows that ϕ varies with β.

At either a low rolling speed or low retardation time such that $\beta \ll \phi$, $\sin^{-1}\{\phi/(\beta^2+\phi^2)^{1/2}\}$ approaches the value $\pi/2$, and equation (1.35) reduces approximately to

$$W = \frac{\pi}{4} kGR^3\phi^4 \tag{1.36}$$

The value of k is approximately $4.85/\ell$, from Hertz's analysis in static conditions. The moment M opposing steady rolling of the sphere is

$$M = \int_{y=-\ell}^{y=+\ell} \int_{x=R\xi}^{x=(\ell^2-y^2)^{1/2}} p\cdot x\cdot dx\cdot dy \tag{1.37}$$

Therefore, the frictional force F is

$$F = \frac{M}{R} = kGR^3 \left\{ -\frac{5}{12}\beta^2\phi^3 - \frac{1}{4}\beta^4\phi - \frac{35}{24}\beta^6\phi - \frac{145}{72}\beta^4\phi^3 \right.$$

$$- \frac{7}{12}\beta^2\phi^5 + \frac{\pi}{8}\beta\phi^4 + \frac{\pi}{24}\beta\phi^6 + (\beta^2+\phi^2)\cdot\left(\frac{1}{4}\beta\phi^2 + \frac{1}{4}\beta^3 + \frac{35}{24}\beta^5\right.$$

$$\left.\left. + \frac{25}{24}\beta^3\phi^2 + \frac{1}{12}\beta\phi^4\right) \cdot \sin^{-1} \cdot \frac{\phi}{(\beta^2+\phi^2)^{1/2}} \right\} \quad (1.38)$$

From this, the coefficient of rolling friction $\mu(=F/W)$ may be obtained. However, it is distributed over three regions according to the value of β. That is, region (1), $\beta \ll \phi$; region (3), $\beta \gg \phi$; and region (2), $\beta=\phi\sim5\phi$, having a maximum value of μ. The value of F in region (1) is

$$F = \frac{\pi}{4} kGR^3\beta\phi^4 \quad (1.39)$$

therefore

$$\mu = \frac{F}{W} = \beta = \tau \cdot \frac{S}{R}$$

$$= 2\pi\tau n \quad (1.40)$$

where n is the speed of revolution (s^{-1}) of a sphere.

The value of μ in region (3) ($\beta \gg \phi$) is

$$\mu = \frac{3\pi}{16R}\left(\frac{3}{9.7} \cdot \frac{W}{G\cdot\beta}\right)^{1/2} = \frac{0.327}{R}\left(\frac{W}{G\cdot\beta}\right)^{1/2} \quad (1.41)$$

The values of W, F and μ in region (2) where $\beta=m\phi$ (m=1~5) are presented in Table 1.4. The relationships having β as a horizontal axis are shown in Fig. 1.16. These theoretical relationships were examined experimentally by Flom using elastomers [18] and plastics [19].

Figure 1.17 shows the apparatus for testing rolling friction and Fig. 1.18 shows the effect of speed on the coefficient of rolling friction μ for each different size of steel ball on polybutyl elastomer. It is clear that the value of μ decreases with increasing ball size and with decreasing speed in region (2) where μ is proportional to $W^{1/2}$. Figure 1.19 shows the effect of speed on the coefficient of rolling friction μ for steel balls on a PMMA plate.

TABLE 1.4
Values of W, F and μ for each region

β	Region	W	F	$\mu(=F/W)$
$\beta<<\phi$	1	3.81 $GR^2\phi^3$	3.81 $GR^2\phi^3\beta$	$\beta(=\tau S/R)$
ϕ		5.71 "	2.48 $GR^2\phi^4$	0.43ϕ -$0.243(W/GR^2)^{1/3}$
2ϕ		8.68 "	4.43 "	$0.51\phi=0.248($ " $)$
3ϕ	2	11.8 "	6.35 "	$0.537\phi=0.236($ " $)$
4ϕ		15.0 "	8.26 "	$0.551\phi=0.223($ " $)$
5ϕ		18.2 "	10.2 "	$0.559\phi=0.212($ " $)$
$\beta>>\phi$	3	3.23 $\beta GR^2\phi^2$	1.91 $\beta GR^2\phi^3$	$0.590\phi \frac{0.327}{R}\left(\frac{W}{G\beta}\right)^{1/2}$

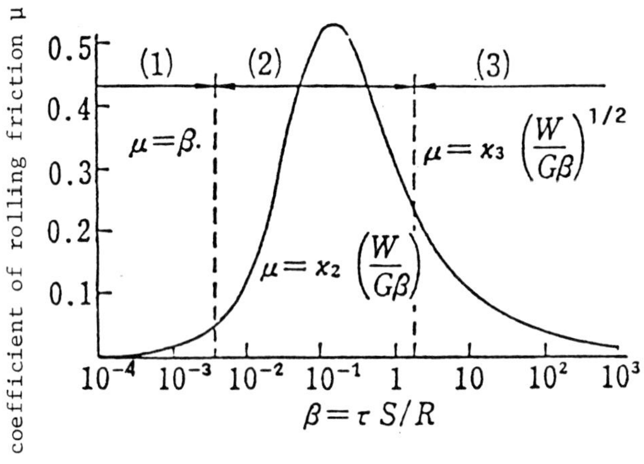

Fig. 1.16 Relationship between μ and β (W=100 gf, G=10^6 gf/cm^2, R=0.1 cm)

The coefficient of rolling friction μ may be greatly affected near the glass transition temperature where the mechanical loss (tan δ) becomes much larger because μ is directly proportional to the retardation time which is greatly influenced by temperature. This relationship is presented in Fig. 1.20, which shows a comparison between the friction of steel balls (1/2" dia.) on Nylon 6 and mechanical loss in nylon. It is clear that the values of μ and tan δ both increase appreciably near the glass transition temperature of 50°C and follow the trend predicted for region (2).

May [20] has also discussed the role of viscoelasticity in rolling friction between a cylinder and a plastic plate using Maxwell's model with a relaxation time.

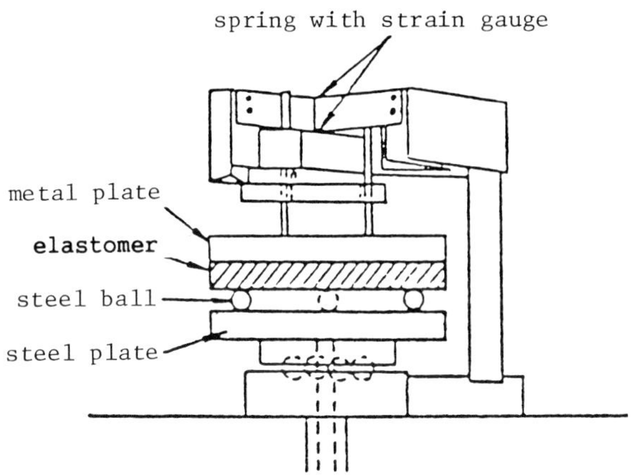

Fig. 1.17 Rolling friction test apparatus

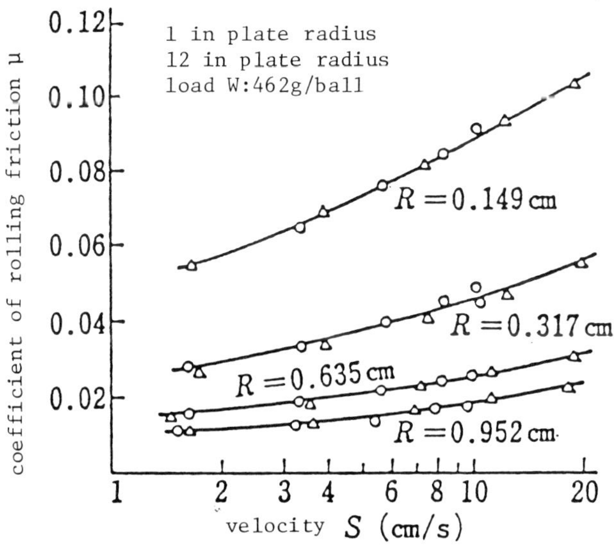

Fig. 1.18 Effect of speed on the rolling friction of a steel ball on polybutyl elastomer

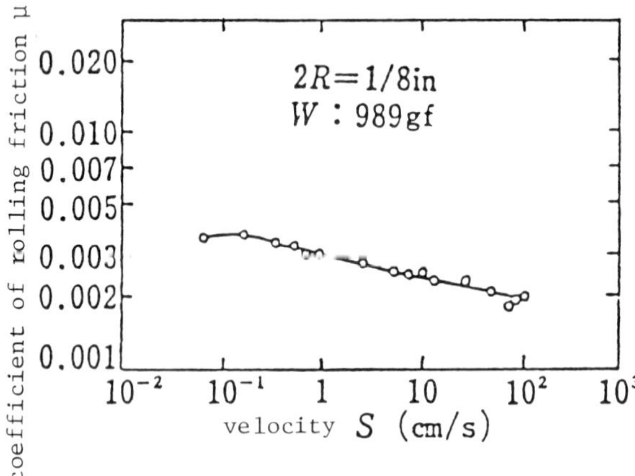

Fig. 1.19 Effect of speed on the rolling friction of a steel ball on a PMMA plate (989 gf/3 balls, 25°C)

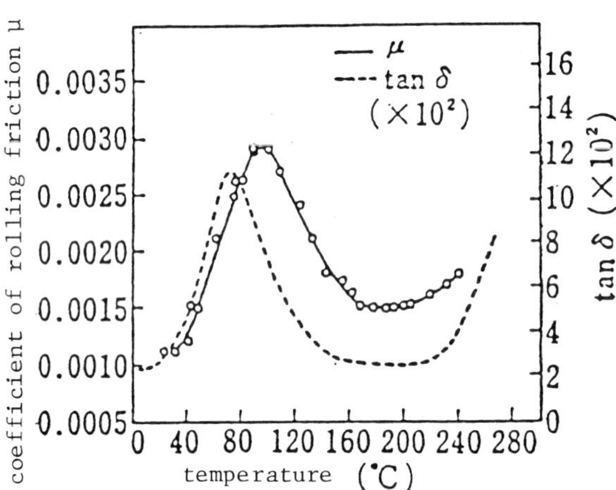

Fig. 1.20 Comparison of the rolling friction of steel balls (1/2" dia) on nylon with mechanical loss (W=1516 gf/3 balls)

1.3 METHODS OF MEASURING FRICTION

There are many methods to measure the coefficient of friction between plastics or plastics and other materials. Those used most often are outlined in this section.

1.3.1 *The coefficient of static friction*

The methods shown in Fig. 1.21 have previously been used to measure the coefficient of dry static friction between solids. Figure 1.21(a) shows the apparatus used to measure the angle of inclination θ at which sliding motion between A and B begins; the coefficient of static friction μ_s between A and B is then obtained from $\mu_s=\tan\theta$. Figure 1.21(b) shows the apparatus used to measure the weight W which exerts a horizontal force through a fine wire S to initiate sliding between A and B under a load P. The coefficient of static friction μ_s is obtained from the equation $\mu_s=W/P$.

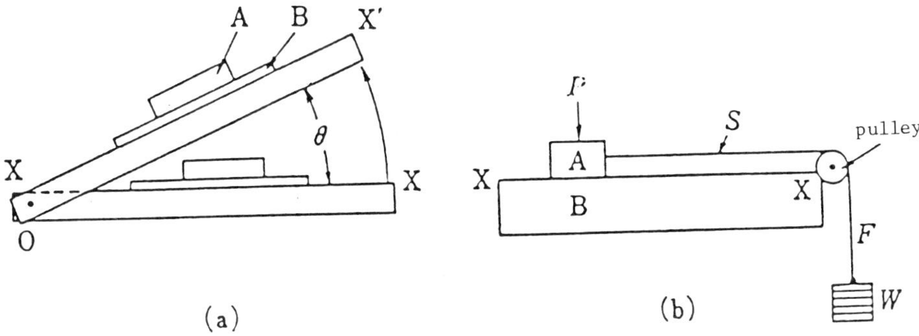

Fig. 1.21 Apparatus for measuring static friction

1.3.2 *The coefficient of kinetic friction*

(i) *Between the end surfaces of two cylinders*

When sliding occurs between two cylinder ends, aa of A and B with an inner diameter r_1 and outer diameter r_2, as shown in Fig. 1.22, the coefficient of kinetic friction μ_k between them is obtained from the following equation by measuring the frictional resistance couple $F\ell$ at a distance ℓ from the centre O.

$$F_R R = 2F\ell = \int_{r_1}^{r_2} 2\pi f r^2 \cdot dr$$

$$= \frac{2}{3}\pi f(r_2^3 - r_1^3)$$

where f is the frictional resistance per unit area and F_r is the representative force at the representative radius R.

$$f = \frac{F_R}{\pi(r_2^2 - r_1^2)}$$

therefore

$$R = \frac{2}{3} \cdot \frac{r_2^3 - r_1^3}{r_2^2 - r_1^2}$$

$$F_R = \frac{2F\ell}{R}$$

therefore

$$\mu_k = \frac{F_R}{P} = \frac{3F\ell(r_2^2 - r_1^2)}{P(r_2^3 - r_1^3)} \tag{1.42}$$

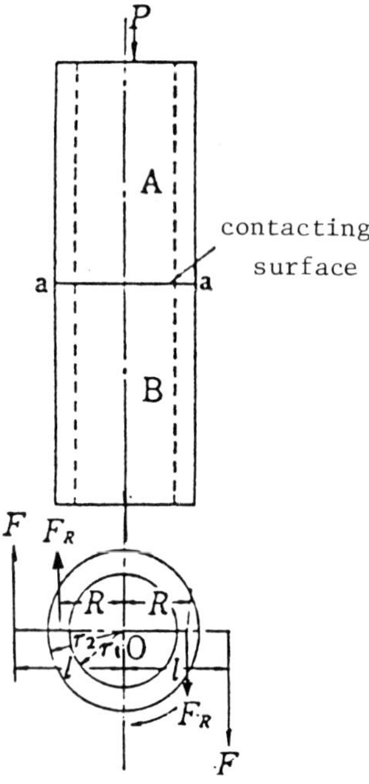

Fig. 1.22 Contact between cylinder ends

(ii) *Between a cylinder and a plane plate*

The coefficient of kinetic friction μ_k is obtained from μ_k=F/P, by measuring the tangential frictional force F when a plane plate A is loaded with a vertical load P against the surface of a revolving cylinder B, as shown in Fig. 1.23.

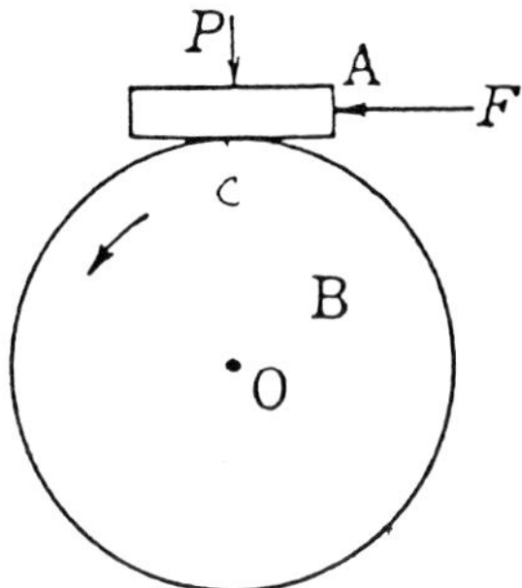

Fig. 1.23 Contact between a cylinder and a plate

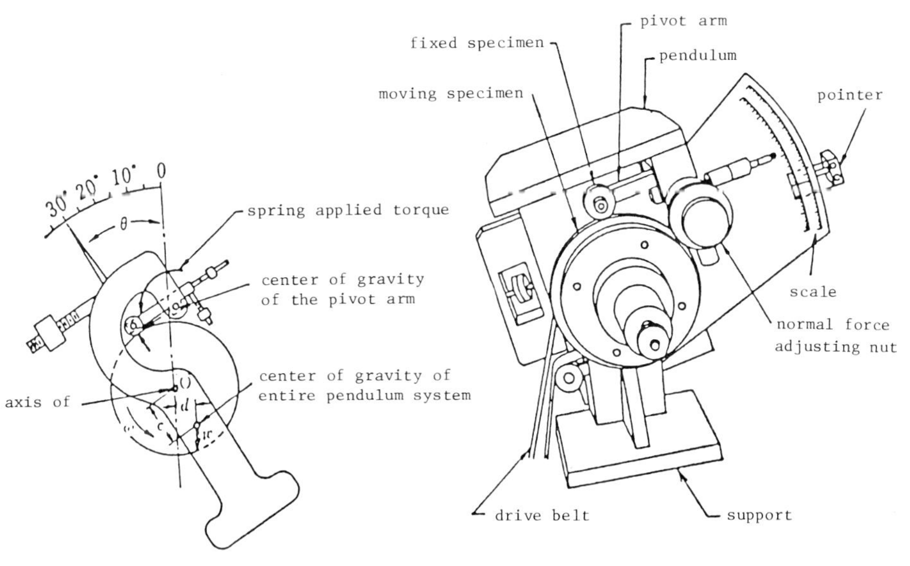

(a) Schematic friction meter components (b) Instrument configuration

Fig. 1.24 Friction meter components (ASTM-D-3028)

Examples of this method are used by ASTM-D 3208 and FT-51-107. Figure 1.24 shows the apparatus of ASTM-D 3208. When sliding friction occurs between a fixed specimen and a revolving cylinder under a normal load n (N), the angular pendulum displacement θ (deg.) changes with the frictional force F (N) as shown in the schematic diagram (a) of Fig. 1.24. The coefficient of kinetic friction μ_k is obtained from the following equation.

$$\mu_k = \frac{M.\sin\theta}{0.05n} \tag{1.43}$$

where M is the maximum pendulum moment of the instrument, N.m.

(iii) *Between two plain surfaces*

In this method, as shown in Fig. 1.21, the tensile force F through a fine wire S is measured during horizontal sliding between A and B under a vertical load P, and the coefficient of kinetic friction μ_k is obtained from μ_k=F/P. This method is usually used to measure μ_k between plastic films, as described in ASTM-D-1894 and DIN 53375.

TABLE 1.5
Standard measuring coefficients of friction in various nations

Standard	Characteristics to measure	Combination	Test conditions
ASTM-D 1894-72	μ_s, μ_k	film on film other material on film	v:0.25 cm/s P:sled weight
ASTM-D 3028-72	μ_k	plastics on plastics other material on plastics	v:10-300 cm/s p:0.55 kgf/cm^2
BS-2782-70	μ_s, μ_k	film on film	v:1.33 cm/s p:0.01-0.50 kgf/cm^2
DIN-53375-72	μ_s, μ_k		v:25 cm/s p:0.01 kgf/cm^2
FT-51-107-71	V_w, μ_k	revolving disc on plate	v:63 cm/s P:1 kgf (0.25-4)
USSR, ГОСТ 11629-65	μ_k	plastics on steel plate	v:30 cm/s p:0.3 kgf/cm^2

1.3.3 *Standardization of testing methods*

No method of measuring the coefficient of friction of plastics has yet been standardized in JIS (Japan Industrial Standards). However, there are some standards in other countries for measuring the coefficient of friction, as shown in Table 1.5, where μ_s is the coefficient of static friction, μ_k is the coefficient of kinetic friction, and V_w is the wear rate.

1.4 EFFECT OF THE INTERNAL STRUCTURE OF MATERIALS

This section discusses the effects of the internal structure of plastics such as molecular structure, super structures, crystallization and direction of molecular chain orientation, on frictional characteristics.

1.4.1 *Effect of molecular structure*

(i) *Arrangement of atoms*

The theoretical basis of sliding friction has already been discussed in terms of the "adhesion-shearing theory", that is, frictional resistance is equivalent to the force needed to break the true contact area in shear. Figure 1.25 shows the microscopic aspects of contact and friction between two solid surfaces using polyethylene (PE) and steel as an example. In this case, relative sliding occurs between molecules A and B because the shear strength between molecule A of the PE and the steel surface to which it adheres is greater than that between the PE molecules. The resistance to motion between molecules A and B is thus equivalent to the molecular bonding energy between them.

The concept of the shearing resistance between the two molecules A and B can also be considered from a morphological standpoint disregarding bonding energy. For example, the atomic arrangements of PE and polystyrene (PS) are shown in Fig. 1.26.

The side atoms of hydrogen (H) in the PE molecule are located symmetrically along the axis of the main carbon (C) chain, and the outside line of H atoms is parallel to the axis and smooth. In PS, however, the side chain of H atoms and benzene rings are situated asymmetrically along the main chain, so the outside profiles of the side chains are irregular along the axis. It would be expected that the shearing resistance between molecules is likely to be much larger for an asymmetrical arrangement of atoms than for a symmetrical one. Therefore, in considering the friction coefficient of plastic materials, the degree of symmetry of the molecular structure must be taken into account, in addition to the shearing resistance needed to break the molecular bonds.

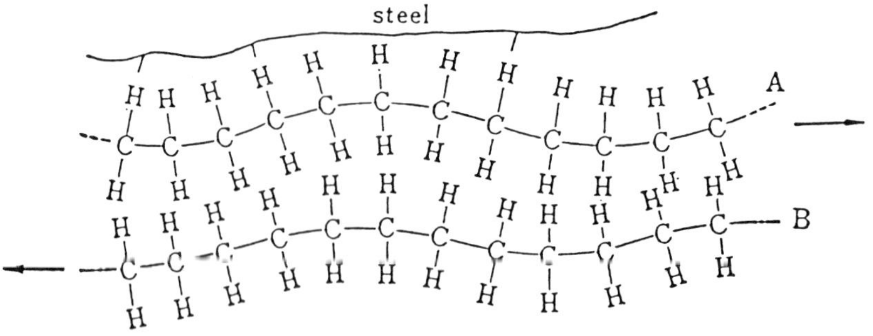

Fig. 1.25 Model of sliding between molecular chains

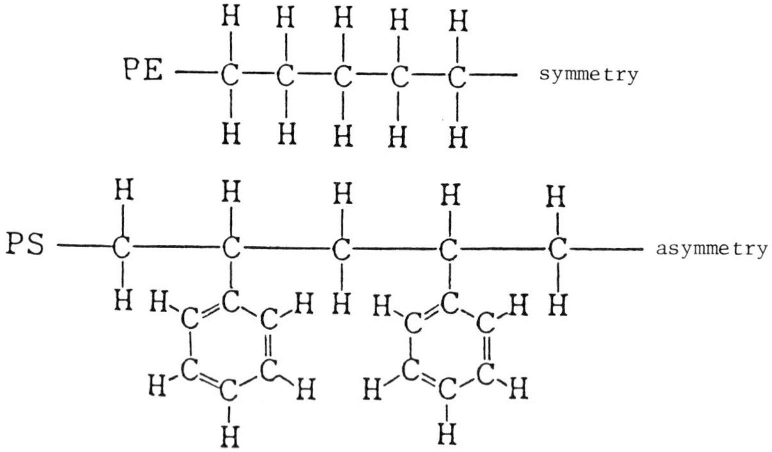

Fig. 1.26 Chemical structures of symmetry (PE) and asymmetry (PS)

As another example, the molecular structures of graphite and molybdenum disulfide (MoS_2), which are widely used as excellent solid lubricants, are shown in Figs. 1.27 and 1.28. As can be seen in Fig. 1.27, which illustrates the structure of graphite, when a shearing force is applied transversely between layers A and B or B and C, easy shearing will occur parallel to those layers because the distance between the layers is larger than the distance between the atoms in each layer. In Fig. 1.28, which illustrates the structure of MoS_2, it can be seen that the distance between C and B or that between A and C is much shorter than that between A and B; easy shear thus takes place transversely between A and B because of the weaker bonding between them.

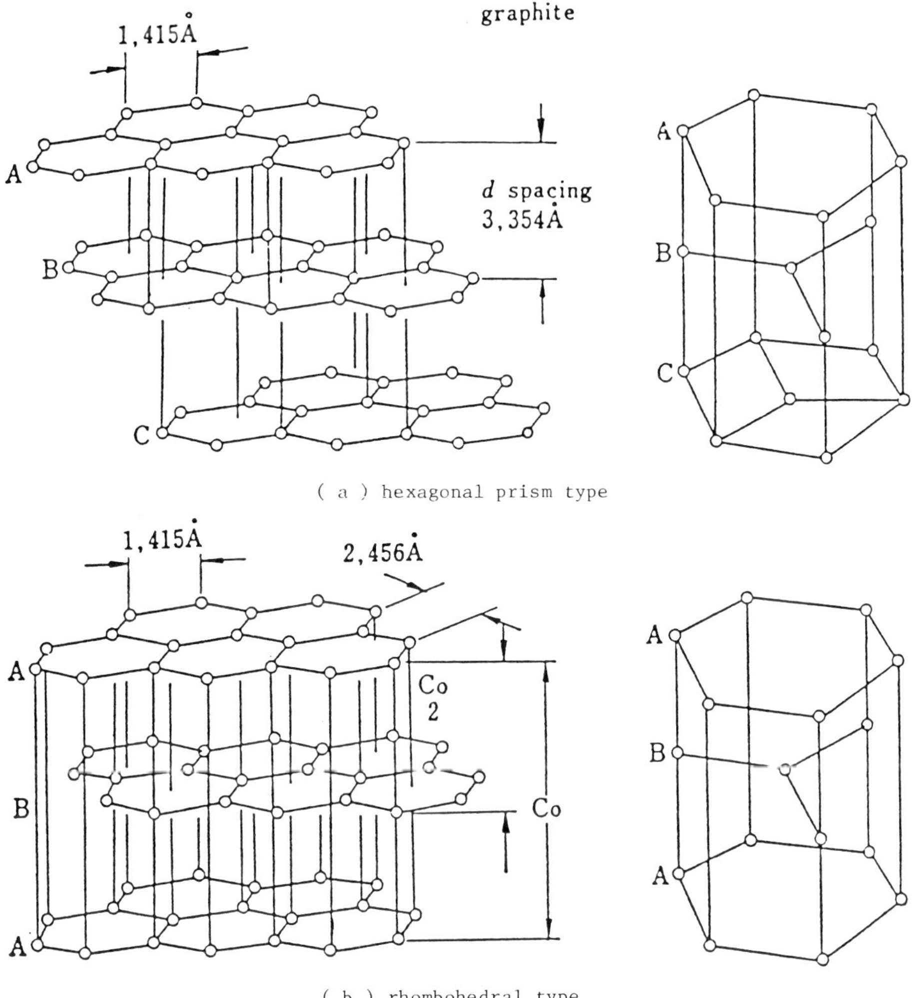

Fig. 1.27 Crystal structures of graphite

Figure 1.29 presents photographs of the end surface of a polyacetal (POM) cylinder subjected to sliding friction against steel; (a) shows the surface prior to the test and (b) is that of a surface which has been partially welded under high pressure and temperature. Figure 1.30 shows the X-ray diffraction patterns of (a) and (b). The Debye-Scherrer ring in (a')

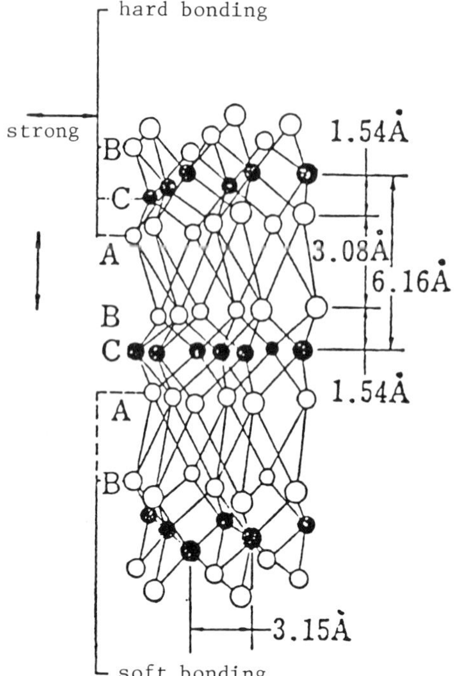

Fig. 1.28 Crystal lattice of MoS_2

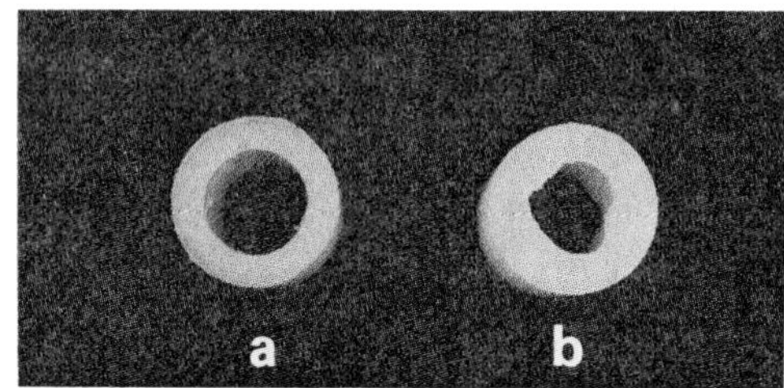

Fig. 1.29 Frictional surfaces of POM

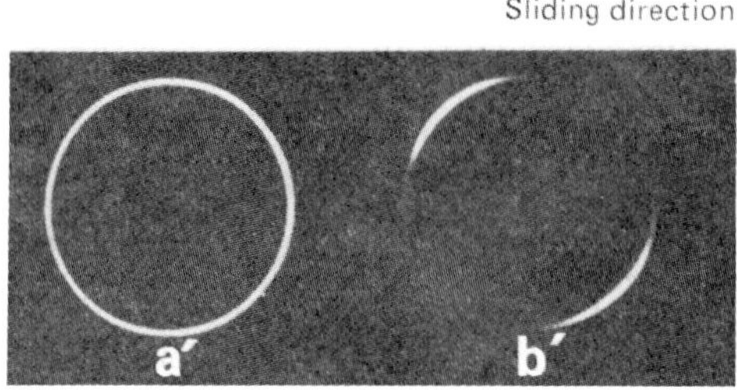

Fig. 1.30 X-ray diffraction patterns

shows a random molecular arrangement, but (b') shows some molecular orientation parallel to the frictional direction. This means that forces exerted upon the molecules of the polymer have changed their orientation to minimize the frictional resistance. This phenomenon is apparent in Fig. 1.31 which shows that the value of μ_k in the starting period of friction is larger than that during the later period; this example is for polyimide at a high temperature for which μ_k is much reduced.

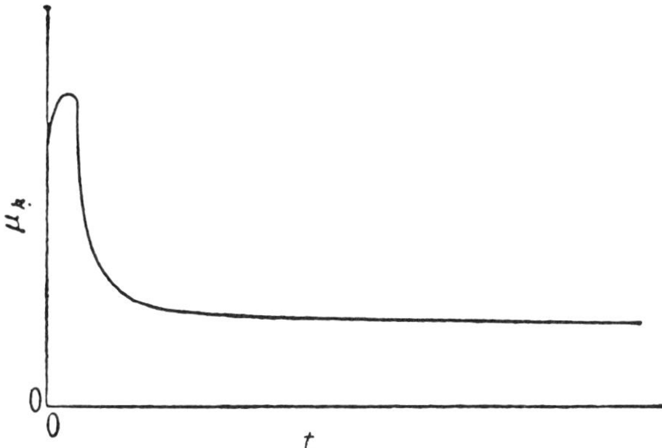

Fig. 1.31 Relationship between μ_k and the duration of sliding t

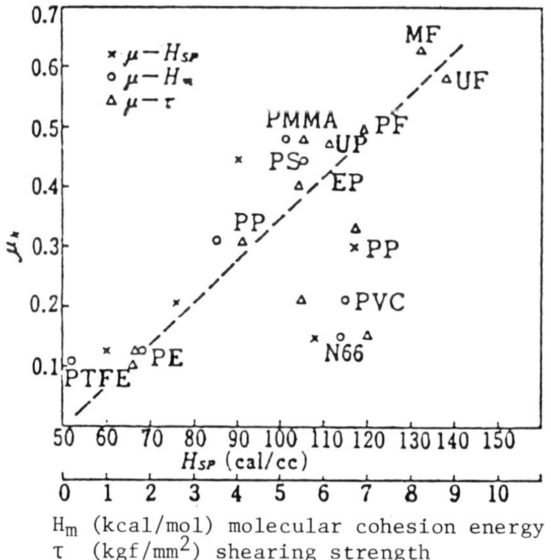

Fig. 1.32 Relationship between μ_k and molecular cohesive energy (H_m or H_{sp})

The relationship between frictional resistance and the intermolecular binding energy (or Van der Waal's force) is as follows. Figure 1.32 shows the coefficient of sliding friction μ_k of plastics against steel plotted against the specific cohesive energy H_{sp} [34] (represented by x), the molecular cohesive energy H_m [35] which represents the cohesive bonding force between molecules (represented by o), and the shear strength, τ, of the plastics (represented by Δ). It becomes clear from this figure that for many plastics μ_k is directly proportional to H_{sp}, H_m or τ. However, some materials, such as PVC, N66 and PS, are anomalous and it cannot therefore always be assumed that the frictional resistance of plastics is represented solely by the constants H_{sp}, H_m or τ.

The relationship between μ_k and the arrangement of atoms in a molecule, or the symmetry of the atomic arrangement is as follows. Table 1.6 shows individual and average values of μ_k for various types of plastics which have been measured under constant conditions of v = 6.2 cm/s, p = .083 Kgf/cm^2. A sliding friction tester was used to test the cylinder-end surfaces of nineteen materials, including six kinds of thermosetting plastics and thirteen kinds of thermoplastics, which were either amorphous or semicrystalline. S/P indicates that steel was on the top during testing and the plastic on the bottom, whilst P/S indicates the inverse configuration.

These plastics may be classified from a morphological standpoint by their atomic symmetry within a molecule into the following classes. The cross-linked thermosetting plastics are block polymers and are classified as asymmetric. Amorphous high polymers are generally asymmetric and are classified into two groups, high asymmetry and asymmetry. The semi-crystalline high polymers are classified into three groups, low symmetry, quasisymmetry and perfect symmetry, respectively, as shown in Table 1.6.

According to the relationship between μ_k and symmetry as shown in Table 1.6, the average μ_k value of thermosetting plastics is generally large at around 0.5 because of asymmetry, especially in those containing cellulose. In the thermoplastics, it is clear that the value of μ_k decreases with increasing symmetry. It may therefore be concluded that the μ_k of plastics against steel is not only influenced by their molecular cohesive energy or their shear strength, but is also affected greatly by the morphological structure of their molecules in regard to symmetry.

(ii) *Molecular weight*

The effect of the molecular weight of high polymers on the coefficient of friction has been studied experimentally for high density polyethylene PE as follows. The specimens examined were three types of "Sholex" manufactured by the Showadenko Company:

1. 6120V: molecular weight 60,000
2. 6040V: molecular weight 90,000
3. 6002B: molecular weight 150,000

TABLE 1.6

Coefficient of kinetic friction between steel and various plastics and their chemical structure, v=6.2cm/s, p=0.83kgf/cm^2

Type of Plastic	Coefficient of Kinetic Friction μ_k S/P	P/S	Average	Average	Average	Chemical Structure of Monomer	Note
Phenolics (PF)	0.468	0.524	0.496			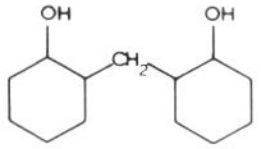	
Urea resin (UF)	0.453	0.711	0.582	0.568		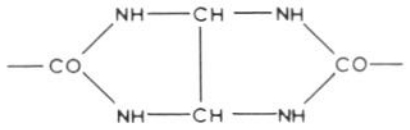	
Melamine resin (MF)	0.567	0.686	0.626		0.504	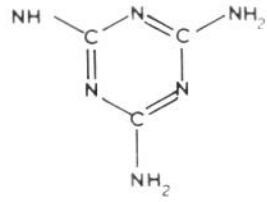	
Unsaturated polyester (UP)	0.48	0.46	0.47			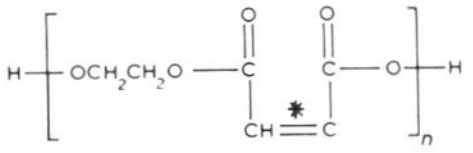	
Epoxy resin (EP)	0.43	0.37	0.40	0.446		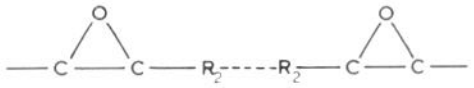	
Polyimide (PI)	0.47	-	0.47			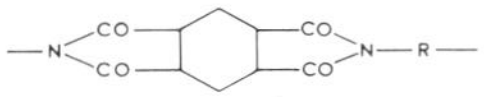	v=41cm/s p=8.5kgf/cm^2

TABLE 1.6 CONTINUED
Coefficient of kinetic friction between steel and various plastics and their chemical structure, v=6.2cm/s, p=0.83kgf/cm²

Type of Plastic	Coefficient of Kinetic Friction μ_k S/P	P/S	Average	Average	Average	Chemical Structure of Monomer	Note
Polymethyl-methacrylate (PMMA)	0.568	0.385	0.476				
Polystyrene (PS)	0.368	0.517	0.442	0.429 high asymmetry			
ABS resin (ABS)	0.366	0.376	0.371				
					0.352		
Polyvinyl chloride (PVC)	0.219	0.216	0.217				
Polyether-sulfone (PES)	0.22	0.45	0.33	0.328 asym-metry			v=62cm/s p=32kgf/cm²
Polyallylate (U Polymer)	-	0.41	0.41				v=35cm/s p=14kgf/cm²

TABLE 1.6 CONTINUED

Coefficient of kinetic friction between steel and various plastics and their chemical structure, v=6.2cm/s, p=0.83kgf/cm^2

Type of Plastic	Coefficient of Kinetic Friction μ_k S/P	P/S	Average	Average	Average	Chemical Structure of Monomer	Note
Polycarbonate (PC)	0.302	0.362	0.331				
Polyphyenylene-sulfide (PPS)	0.216	-	0.216	0.285 low symmetry			v=38cm/s p=2.9kgf/cm^2
Polypropy-lene (PP)	0.300	0.316	0.308				
Nylon 6 (PA)	0.192	0.129	0.148	0.151 quasi-symmetry	0.197		
Polyacetal (POM)	0.129	0.180	0.154				
Polyethylene (PE)	0.139	0.109	0.124	0.116 perfect symmetry			
Polytetra-fluoroethylene (PTFE)	0.117	0.100	0.108				

S/P top:steel; bottom:plastics; P/S its reciprocal

The coefficient of sliding friction μ_k between steel and the plastic specimens made by injection moulding were measured using a Suzuki-type friction tester, as shown in Fig. 1.22, and a FT-type tester composed of a steel disk and a plastic plate. The former were tested at 5.69 kgf/cm^2 pressure and a speed of 60 cm/s, whilst the latter were tested at 7 kgf normal load and 60 cm/s. The experimental relationships between μ_k and molecular weight W are shown in Fig. 1.33. It is clear that when using the Suzuki-type tester, the μ_k value increases from 0.29 to 0.43 with increasing molecular weight. However, the absolute values of μ_k are not the same for the two types of test arrangements.

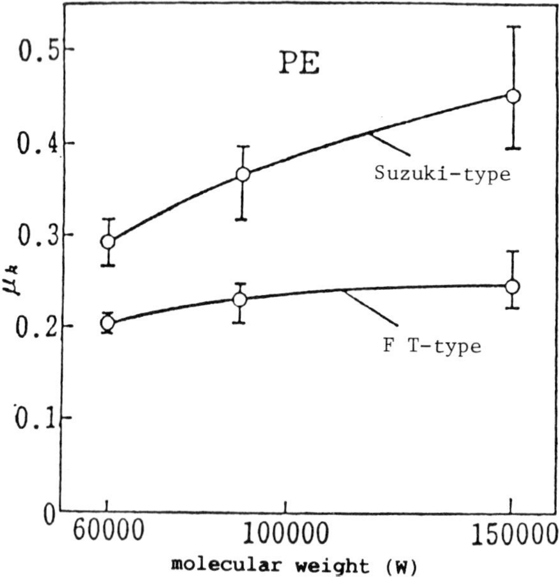

Fig. 1.33 Effect of molecular weight on kinetic friction μ_k of PE

1.4.2 *Crystal structure and molecular orientation*

(i) *Crystal structure [37]*

It has already been shown that the values of μ_k for semicrystalline plastics are generally small. The influence of the degree of crystallization on the μ_k of these plastics is now discussed.

Experiments were made with two types of polyethylenterephthalate (PET) with different degrees of crystallization, made by injection moulding either with rapid cooling at a mould temperature of 60°C or by slow cooling at a mould temperature of 140°C. X-ray diffraction patterns of the specimens are shown in Figs. 1.34(a) and 1.34(b). The degree of crystallization of the materials was measured from these diffraction patterns and found to be 30% and 45%, respectively. The extent of chain orientation and the degree of crystallization of polyethylene (PE) mouldings are changed by stretching

100-500% at a temperature of 100°C or 60°C. The diffraction patterns of these specimens before or after stretching are shown in Figs. 1.34 and 1.35, and the degree of crystallization of PET was from 37% to 40% and of PE was from 38% to 56%.

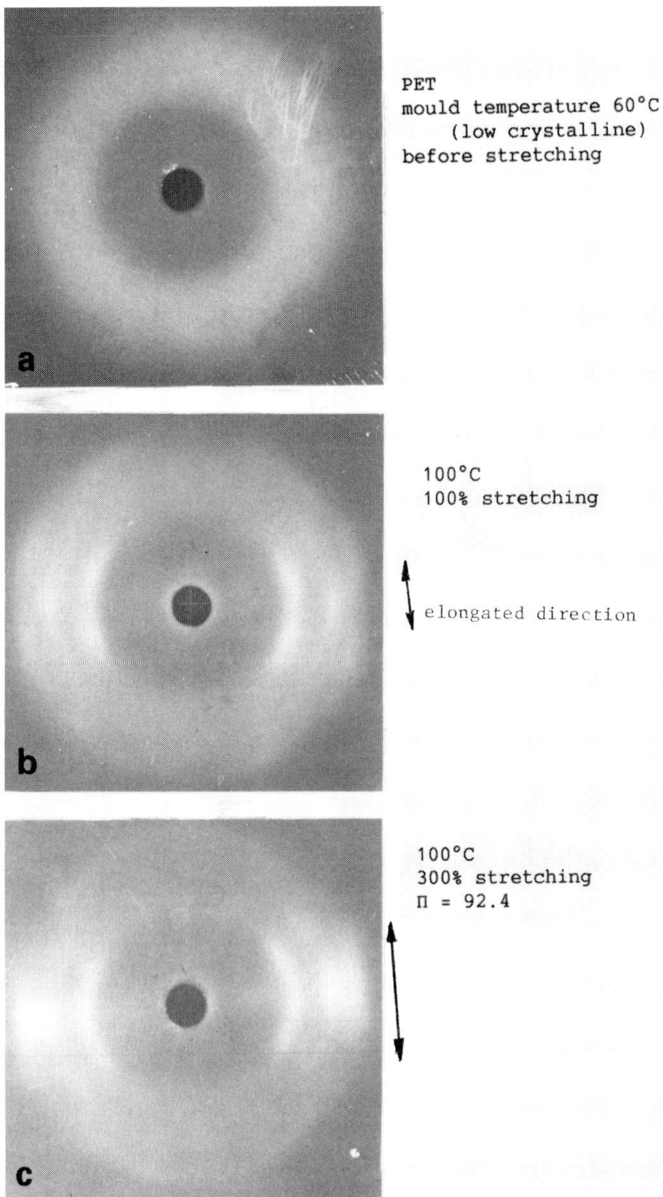

Fig. 1.34 X-ray diffraction patterns of low-crystalline PET before and after stretching

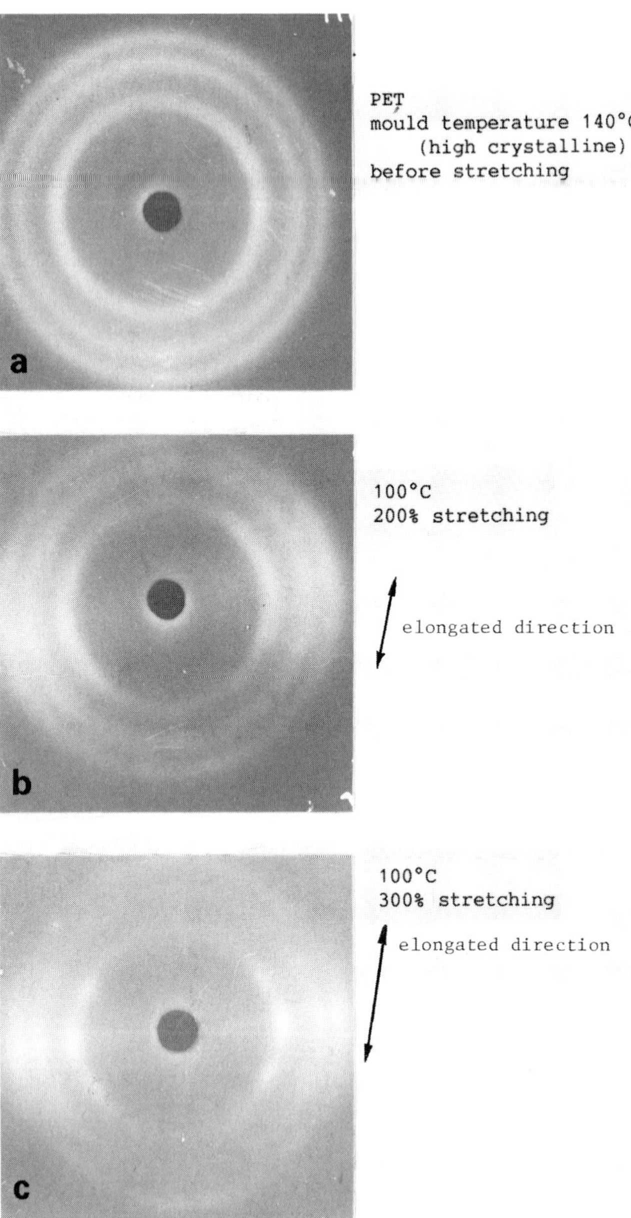

Fig. 1.35 X-ray diffraction patterns of high-crystalline PET before and after stretching

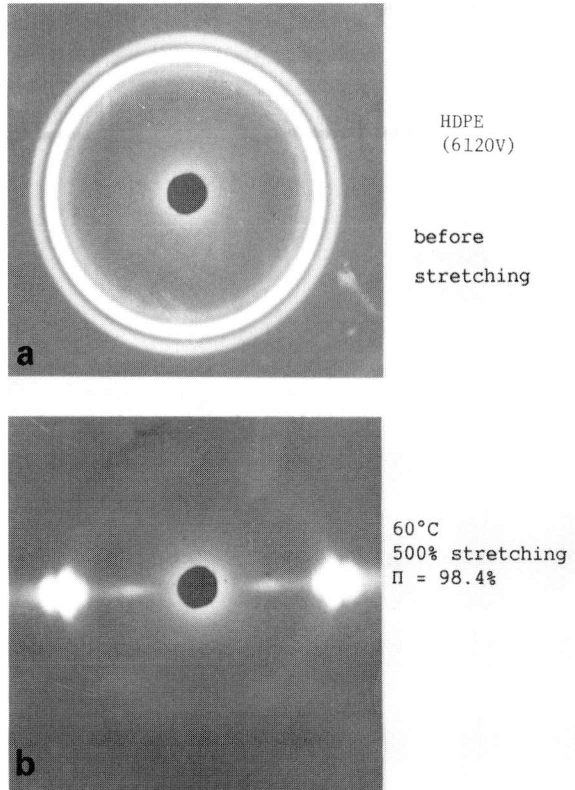

Fig. 1.36 X-ray diffraction patterns of PE before and after stretching

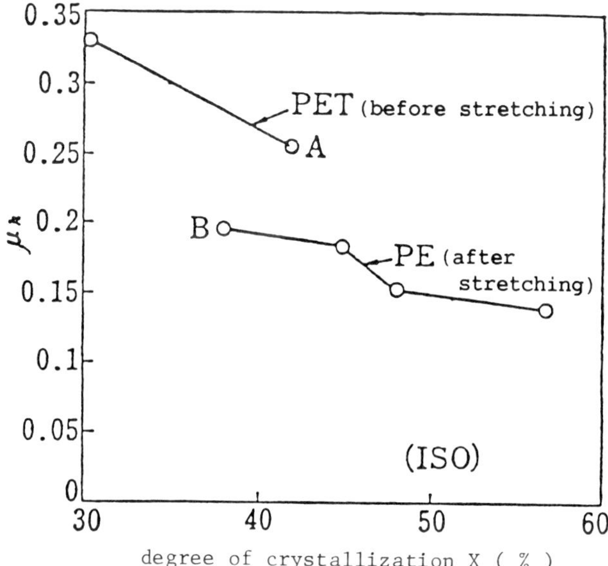

Fig. 1.37 Relationship between μ_k and the degree of crystallization

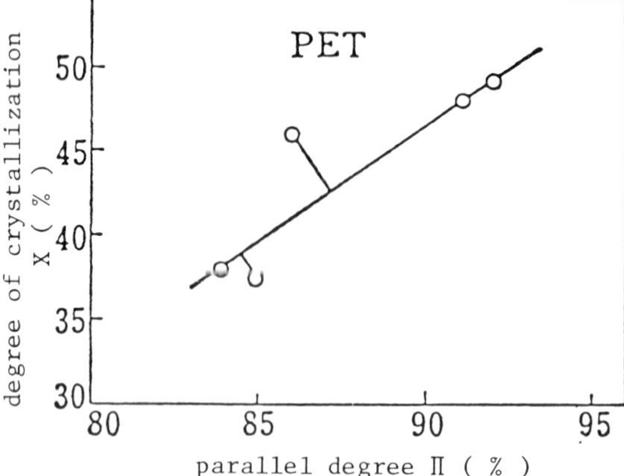

Fig. 1.38 Relationship between the degree of crystallization and the extent of parallel orientation for PET

The coefficient of sliding friction μ_k between steel and plastic specimens with different degrees of crystallization were obtained using an FT-type friction tester and the relationship between μ_k and the degree of crystallization is presented in lines A and B of Fig. 1.37. Line A shows the relationship for unstretched PET with different degrees of crystallization and line B shows the corresponding relationship for stretched PE. Both the amount of chain orientation and the degree of crystallization arc generally increased by stretching, as shown in Figs. 1.36 and 1.38, but line B in Fig. 1.37 shows that the μ_k along the direction of stretching for PE decreased in both of these areas after stretching. From these results it is clear that the μ_k value of unstretched plastics decreases with an increasing degree of crystallization.

TABLE 1.7
Effect of cooling methods on PE and PET

Specimen	Cooling Method	Degree of Crystallization (%)	Specific Gravity
PE	annealing for 22 hrs at 129°C	53	0.977
	cooling in hearth for 2 hrs at 120°C	54	0.981
	air cooling	50	0.961
	water cooling (0°C)	47	0.950
PET	annealing for 22hrs at 315°C	50	2.172
	air cooling	46	2.157
	water cooling (0°C)	42	2.135

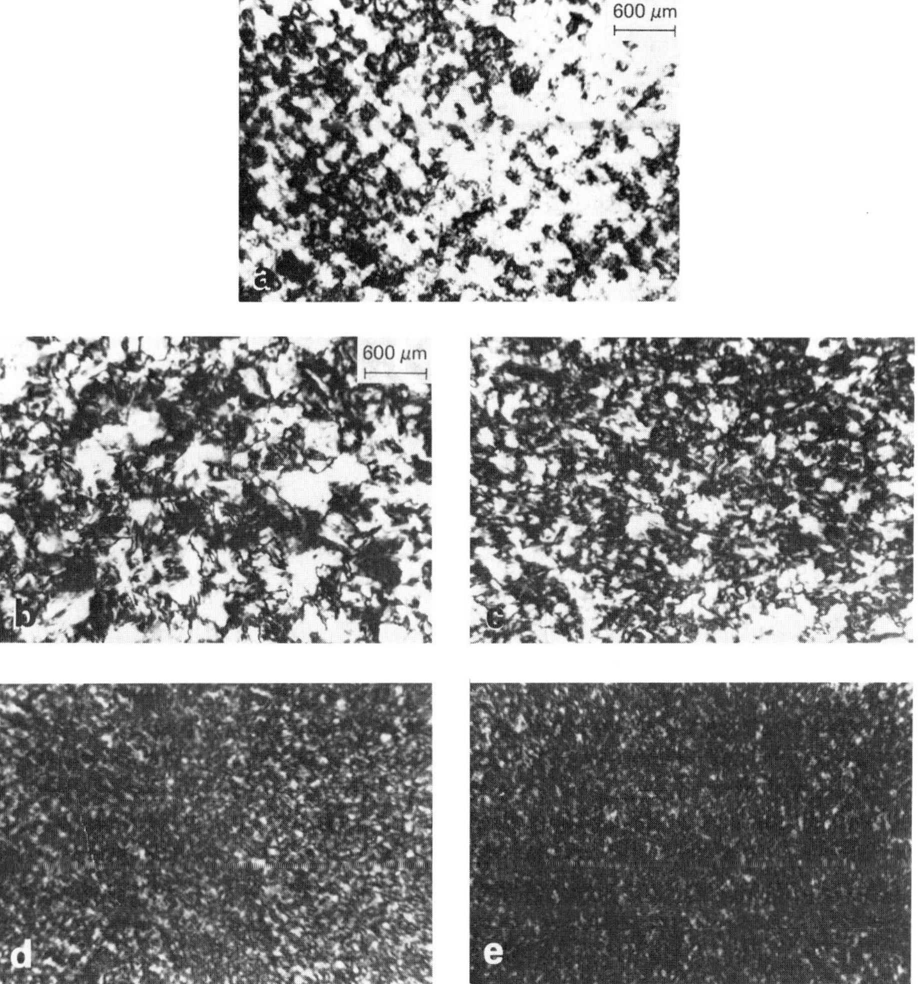

Fig. 1.39 Microstructures under polarized light (x 600) of PE treated by various methods. (a) 129°C annealing; (b) hearth cooling (120°C); (c) air cooling; (d) water cooling; (e) water cooling at 0°C from melting.

As shown in Table 1.7, the degree of crystallization of PE and PET changes from 47% to 54% for the former and from 42% to 50% for the latter, depending on whether they are cooled or heat treated. Also, the microstructures of PE under polarized light are presented in Fig. 1.39 [38] and those of PET in Fig. 1.40 [39].

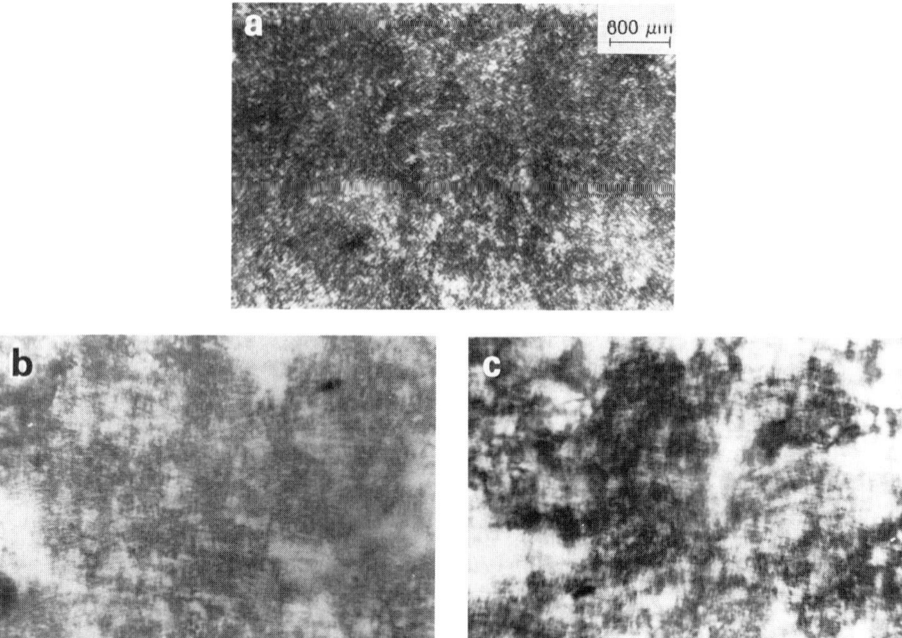

Fig. 1.40 Microstructures under polarized light (x 600) of PTFE treated by various methods. (a) annealing (315°C); (b) air cooling; (c) water cooling (0°C).

The relationships between μ_k and the distance of sliding for PE treated by each method are shown in Fig. 1.41, and it appears that the value of μ_k becomes large for the specimen with a low degree of crystallization induced by rapid cooling.

Figures 1.42 and 1.43 show the relationships between either the coefficient of friction μ_k or the specific adhesive wear rate V_s, and the degree of crystallization for PE and PTFE, respectively. It is clear from these figures that for both polymers the value of μ_k decreases with increasing degree of crystallization and V_s only decreases with increasing crystallization for PE; but for PTFE, V_s increases.

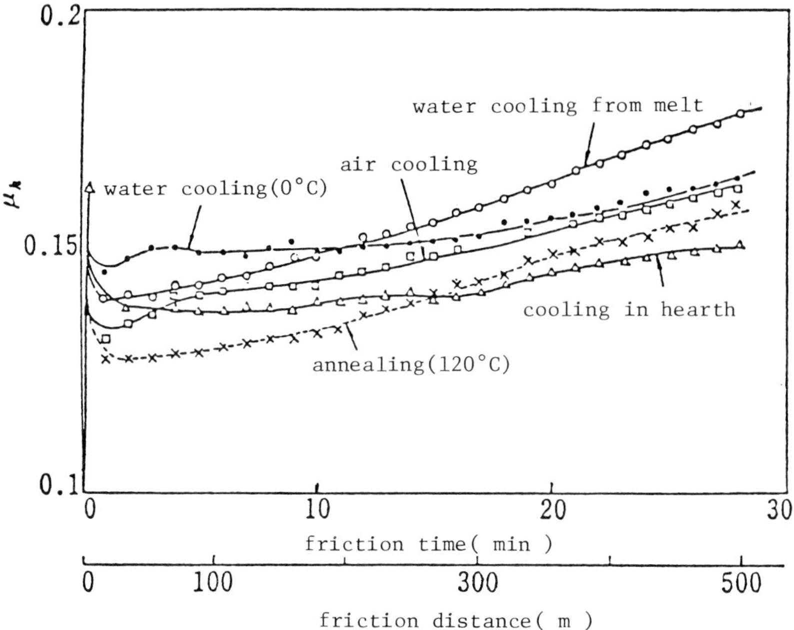

Fig. 1.41 Relationship between μ_k and duration, or distance, of sliding for PE treated in various ways

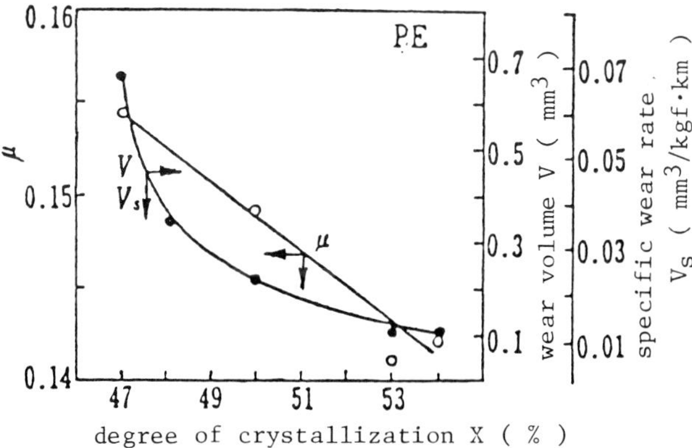

Fig. 1.42 Relationship between μ_k, Vs and the degree of crystallization x for PE

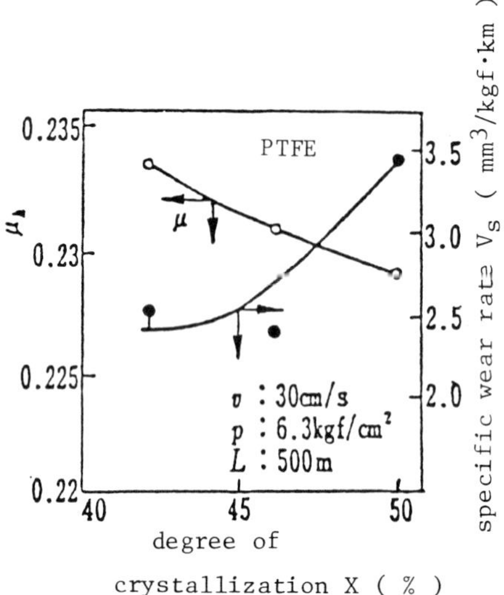

Fig. 1.43 Relationship between μ_k, V_s and the degree of crystallization x for PTFE

(ii) *Molecular orientation*

To obtain samples with different molecular orientations, a specimen similar to that shown in Fig. 1.44(a) was stretched parallel to direction YY in the range from 100% to 600% elongation at a moderate temperature. Specimen A elongated parallel to its axis, and specimen B, with its axis at right-angles, were then made, as shown in Fig. 1.44(b). The amount of elongation d(%) is given by the following equation, when ℓ_o and ℓ' are the lengths of the specimen before and after stretching

$$d = \frac{\ell' - \ell_o}{\ell_o} \times 100\% \qquad (1.44)$$

The degree of orientation π(%) of the molecular chains was obtained by measuring the black density curve obtained from the half Debye-Scherer patterns of the specimens before and after stretching, using a micro-photometer.

$$\pi = \frac{180^\circ - \theta^\circ}{180^\circ} \times 100\% \qquad (1.45)$$

where θ is the breadth (in degrees) at the half height in the black density curve.

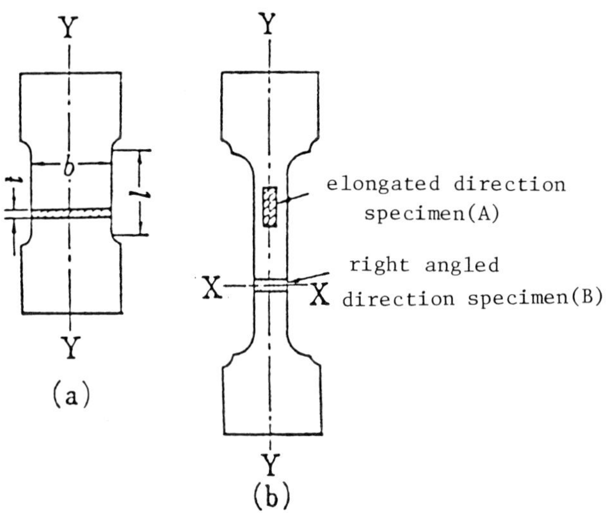

Fig. 1.44 Specimen before and after stretching

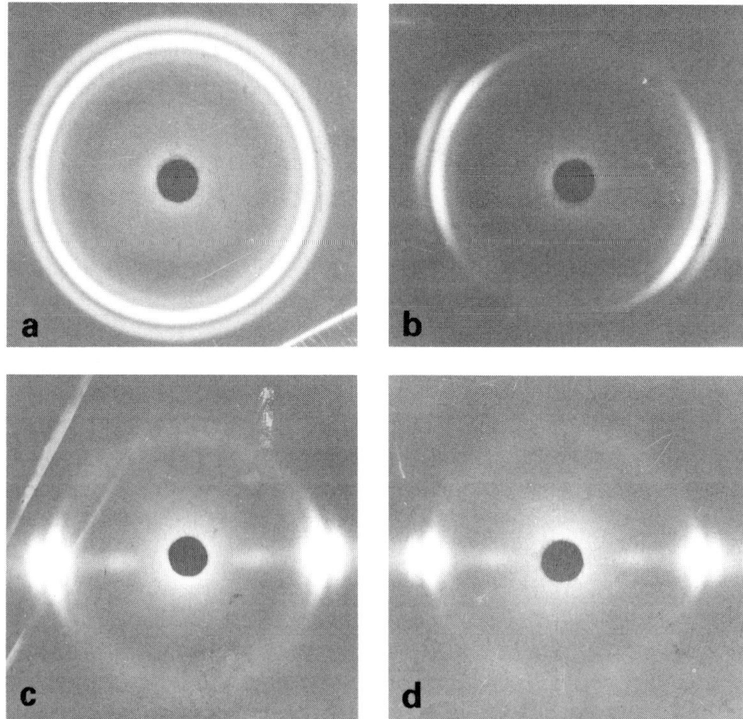

Fig. 1.45 X-ray diffraction patterns of POM before and after stretching. (a) before stretching; (b) 100% stretching (120°C) π:52%; (c) 200% stretching, π:82%; (d) 300% stretching (120°C) π:84%

Figures 1.34 and 1.35 show X-ray diffraction patterns before and after stretching, and the degree of chain orientation π obtained for each PET of low or high crystallization. Figure 1.36 shows similar data for PE. Table 1.8 [39] shows the degree of crystallization x and the degree of chain orientation π for four types of plastics after stretching: polyacetal (POM), polyethylenterephthalate (PET), polycarbonate (PC), and Nylon 6 (PA). Figure 1.45 shows the X-ray diffraction patterns and degree of chain orientation π obtained from stretched polyoxymethylene (POM). Figure 1.46 shows the microstructure of Nylon 6 (PA) under polarized light before and after stretching and sliding relative to the direction of stretching.

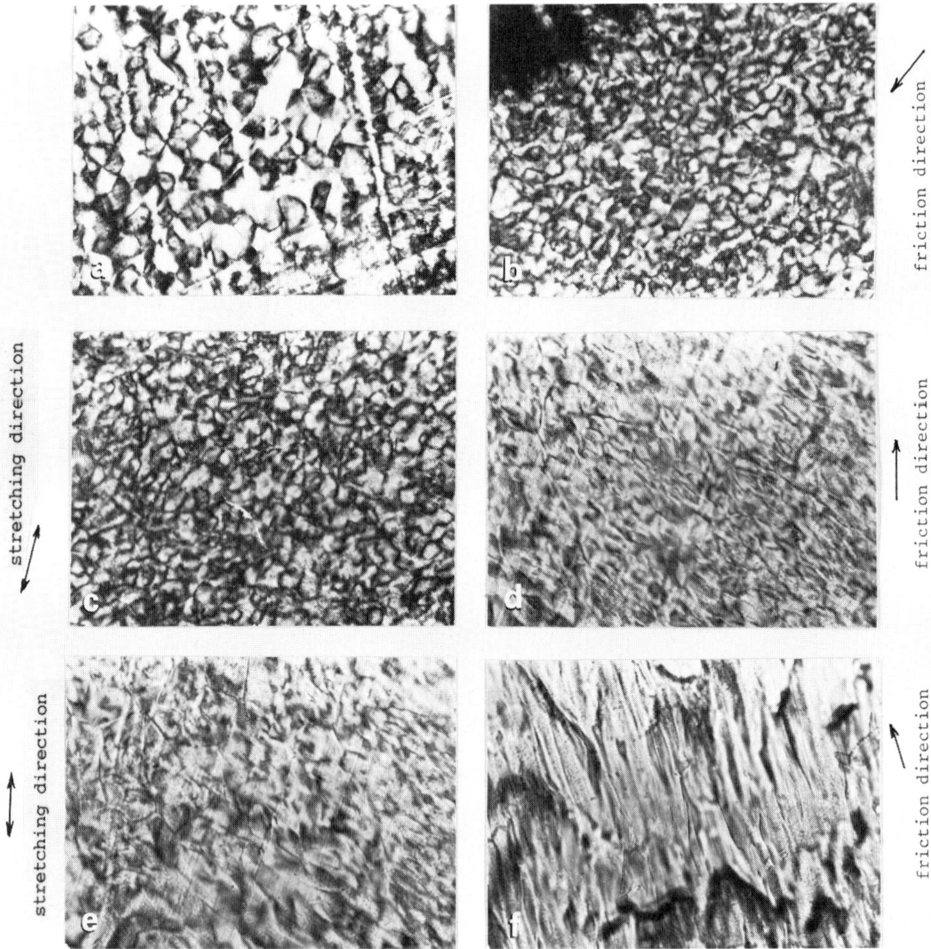

Fig. 1.46 Microstructures under polarized light of PA before and after stretching and friction test. (a) no elongation, before friction; (b) no elongation, after friction; (c) 100% elongation, before friction; (d) 100% elongation, after friction R.D.; (e) 100% elongation, before friction; (f) 100% elongation, after friction D.D.

TABLE 1.8
Effect of stretching on plastic specimens

Specimen	Stretch (%)	Direction	Degree of Crystallization (%)	Degree of chain Orientation
POM	0[1]		68	0
	100 (120°C)	P[2]	70	52
		R[3]		
	200 (120°C)	P	74	82
		R		
	300 (120°C)	P	80	84
		R		
PET	0		48	0
	100 (100°C)	P	51	71
		R		
	200 (100°C)	P	54	79
		R		
PC	0		35	0
	100 (140°C)	P	38	57
		R		
PA	0		38	0
	100 (22°C)	P	45	71
		R		

[1] before stretching
[2] parallel to the stretch direction
[3] at right angles to the stretch direction

Figures 1.47 to 1.54 give the relationships obtained experimentally between the coefficient of friction μ_k and the amount of stretching d, the degree of chain orientation π and the degree of crystallization x. Figure 1.47 shows the relationship between μ_k and d for stretching direction (P) or right-angled direction (R) in polyethylene, and it appears that the value of μ_k decreases with an increase of stretch (d) for stretching direction but increases slightly for right-angled direction.

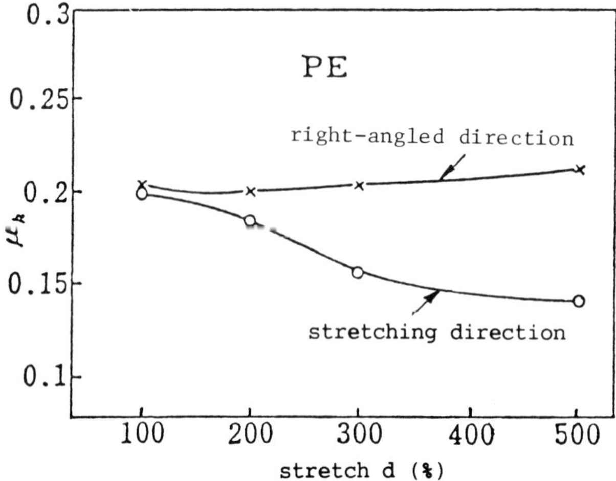

Fig. 1.47 Relationship between μ_k and the amount of stretching of PE

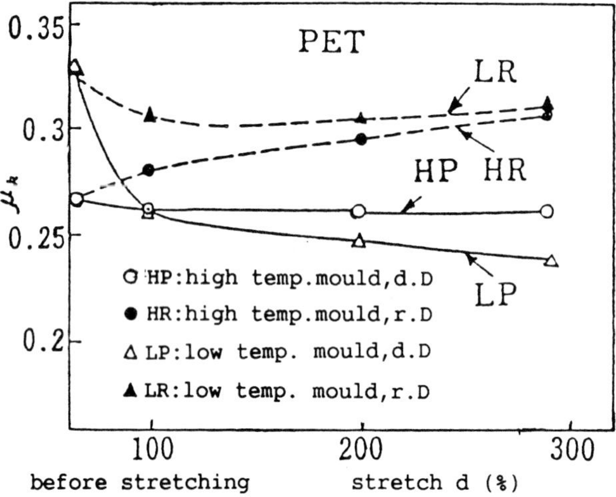

Fig. 1.48 Relationship between μ_k of PET for different mould temperatures and stretching direction

Figure 1.48 shows a similar relationship for PET moulded by high temperature mould (high degree of crystallization) or low temperature mould (low degree of crystallization) and it is clear that the value of μ_k decreases with increase of d for stretching direction, but does not change or only increases slightly for right-angled direction. Both the parallel degree and degree of crystallization increase by stretching, and it is difficult to distinguish which has an effect on μ_k.

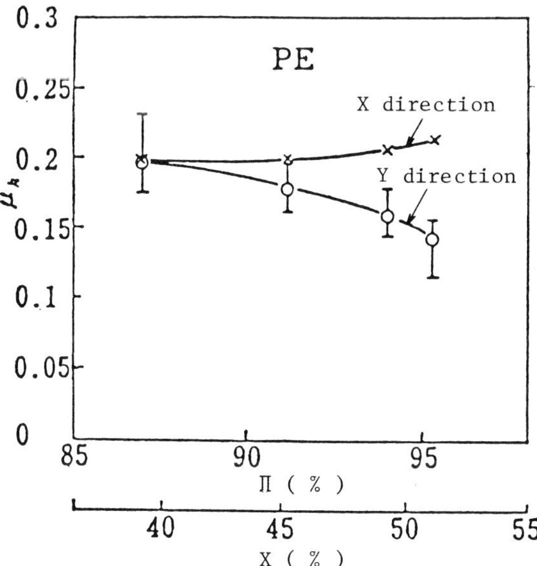

Fig. 1.49 Relationship between μ_k and the degree of parallel orientation for PE after stretching

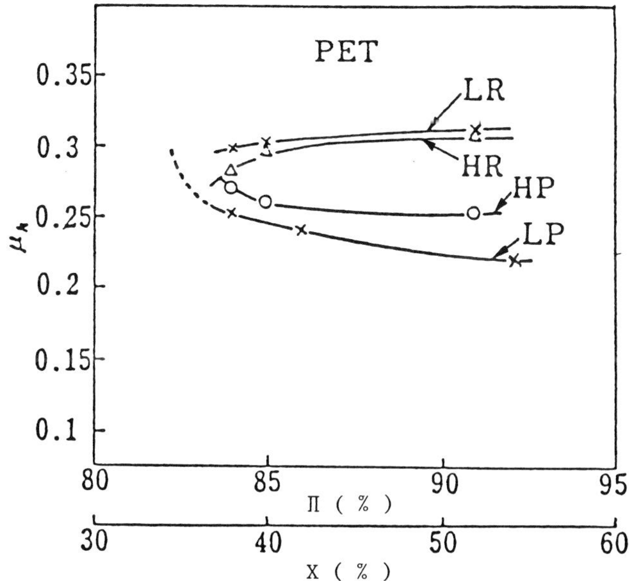

Fig. 1.50 Relationship between μ_k and the degree of parallel orientation for PET after stretching

Figure 1.49 shows the relationship between μ_k and parallel degree or degree of crystallization increased by stretch for both stretching and right-angled directions in PE, and Fig. 1.50 shows a similar relationship for high or low degree of crystallization of PE. It is clear from this figure that the μ_k value for the stretching direction of PE having a 95% parallel degree becomes 0.13, which is about half that before stretching, and the μ_k of PET for stretching direction decreases to 0.23 from 0.34, which is one of random direction.

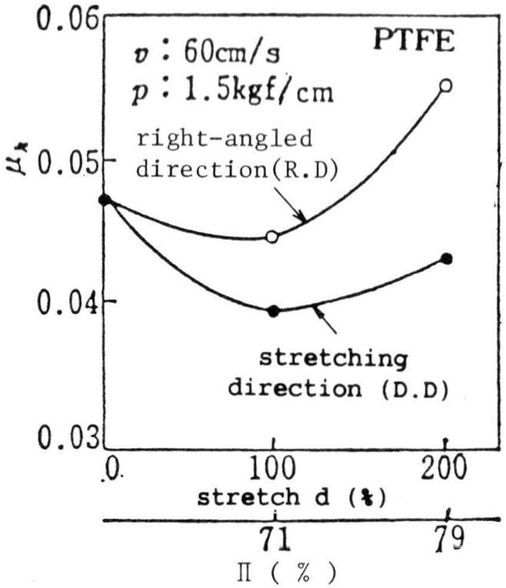

Fig. 1.51 Relationship between μ_k and stretching d or π for PTFE

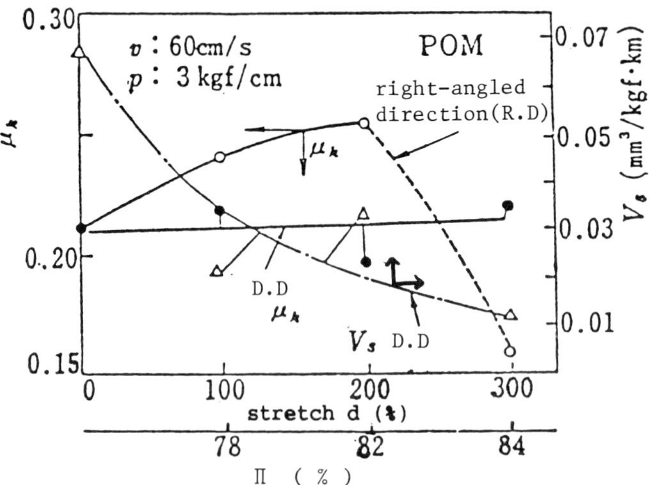

Fig. 1.52 Relationship between μ_k or V_s and stretching or π for POM

Figure 1.51 shows the relationship between μ_k and stretch or parallel degree for PET, and Fig. 1.52 shows the relationship between μ_k or specific wear rate V_s and stretch or parallel degree for POM. The relationship for PET is similar to that of PE as presented above, and μ_k value for the stretching direction does not change but the V_s value for the stretching direction greatly decreases with increasing stretch or parallel degree. Figure 1.53 shows the effect of stretch on μ_k value for each direction of stretching and at right-angles for polycarbonate and Nylon 6 (PA); it appears that the μ_k value for PC is generally larger than that of PA and decreases for stretching direction with increasing stretch but does not change for right-angled direction. Figure 1.54 shows the relationship between the V_s value and stretch d or parallel degree π. V_s value for stretching direction increases with increasing stretch, and this tendency is opposite that of POM.

Previous research has been conducted on the effects of molecular weight on friction by Tanaka [41] and Pratt [42], and on the effects of the degrees of crystallization by Pratt [42], Lontz [43], Bely [44] and Tanaka [45].

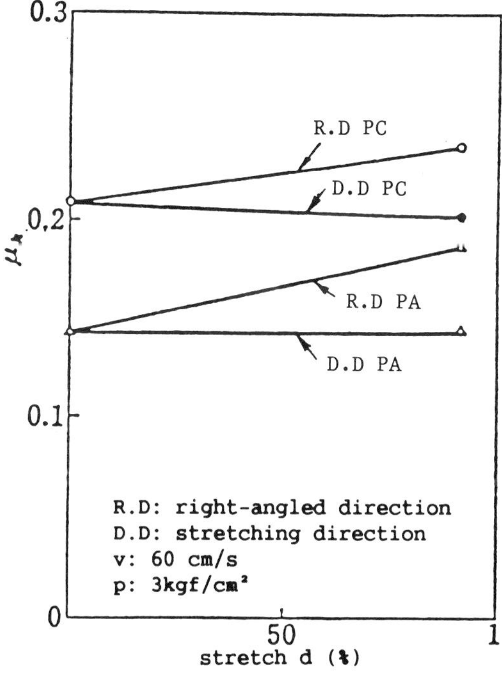

Fig. 1.53 Effect of stretch on μ_k for PC and PA (Nylon 6)

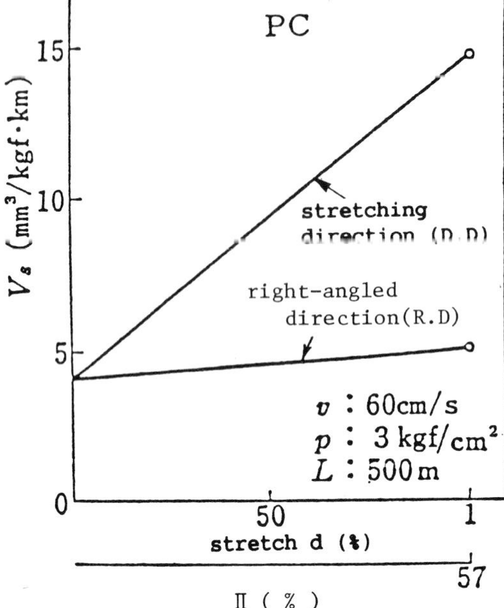

Fig. 1.54 Relationship between V_s and stretch d or π for PC

1.5 PRACTICAL ASPECTS OF FRICTION

There are many kinds of friction which can affect plastic materials, such as sliding friction, rolling friction, lubricated friction, dry or solid friction, static friction and kinetic friction. In this section, dry sliding friction - the most common type - will be discussed. Rolling friction will also be discussed in relation to rolling bearings, and lubricated friction will be discussed as applied to sliding bearings and brake shoes.

Dry friction will be discussed by dividing it into the two categories of static and kinetic friction.

1.5.1 *Coefficient of static friction*

The coefficient of static friction is the ratio of the tangential resisting force needed to initiate relative motion between two contacting solid bodies to the normal force applied between the contacting surfaces. While the coefficient of static friction is not a general characteristic like the coefficient of kinetic friction, the former is not affected as much as the latter by conditions such as sliding speed, temperature rise and the progress of wear, so its characteristic value is usually more stable than the latter. Previous research has been reported by Shooter [16], Tabor [47] and

Rabinowicz [48] concerning the coefficient of friction between plastics and metals, and by Matsubara [49], Sato [50], Tanaka [51], Flom [52] and Sietgoff and Kucsma [53] for plastic/plastic combinations. Only experimental data obtained in the author's laboratory will be discussed in this section.

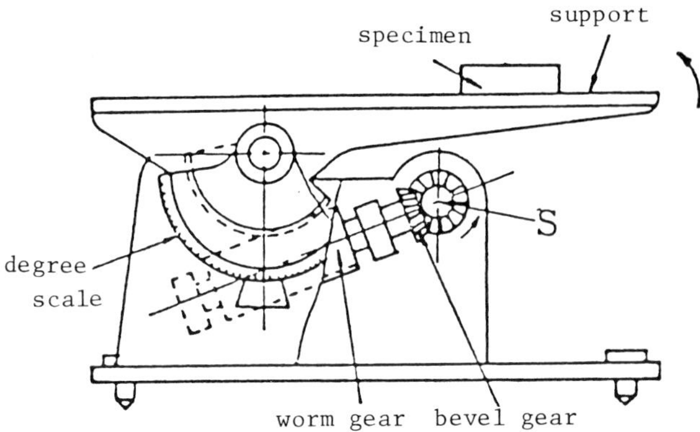

Fig. 1.55 Measuring apparatus for coefficient of static friction

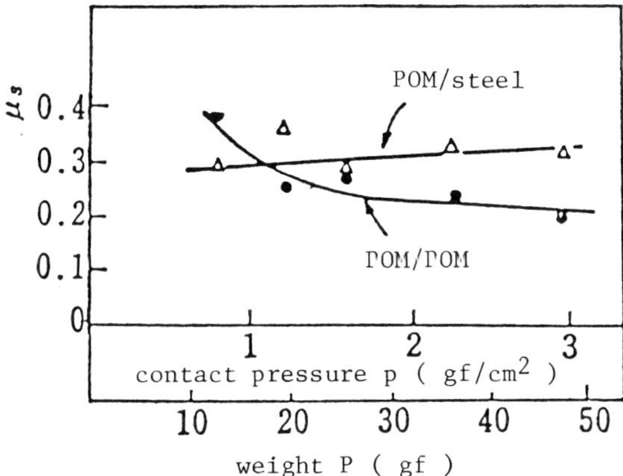

Fig. 1.56 Relationship between μ_s and contact pressure or load

(i) *Effect of surface contact pressure and contact area*

Figure 1.55 shows the experimental apparatus. The coefficients of static friction μ_s between various materials at 20°C and 60% R.H. were obtained from the equation μ_s=tanθ, where θ is the angle of inclination to initiate sliding between the contacting materials when increasing the inclination at a rate of 1.13°/s. The relationship between μ_s and the vertical contact force P is shown by Fig. 1.56, when the vertical load P or contact pressure p changes from 10 gf to 50 gf or from 0.75 gf/cm² to 3.0 gf/cm² over a constant apparent contact area of 16 cm². It is clear from the figure that the value of μ_s between polyacetal (POM) and steel is almost constant regardless of p, but the value of μ_s for POM against itself is almost constant above about 1.5 gf/cm² and increases with decreasing p below 1.5 gf/cm². These relationships agree with the results reported by Shooter and Tabor [47] and Flom [52].

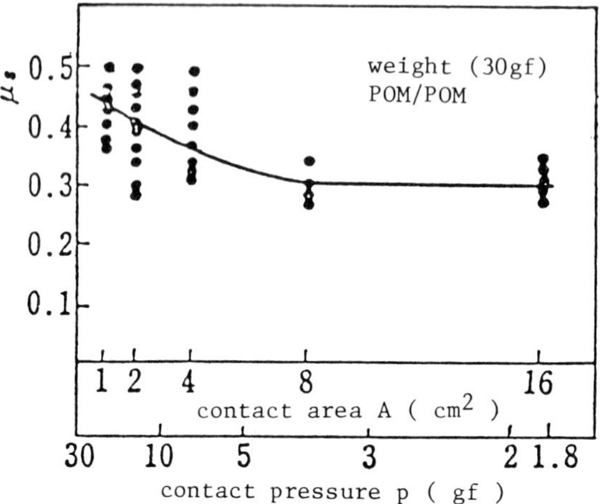

Fig. 1.57 Relationship between μ_s and contact area or contact pressure

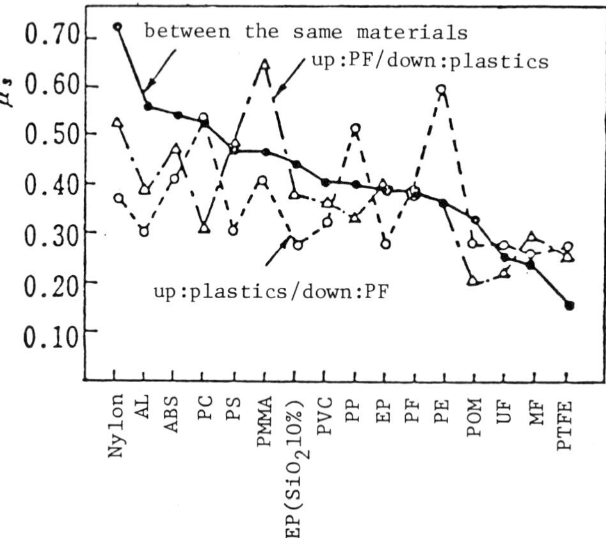

Fig. 1.58 μ_s for combinations of PF and various materials

Figure 1.57 shows the relationship between μ_s and the apparent contact area A under a constant normal load P of 30 gf over a range of A from 1 to 16 cm^2. It can be seen that for over half of this range, the results agree with Coulomb's law whereby μ_s is independent of A.

(ii) *Between various plastics or between plastics and metal*

The results of measurements of μ_s between seventeen kinds of materials, including five thermosetting plastics, ten thermoplastics, aluminium and steel, are shown in Table 1.9. Table 1.10 shows the static coefficient of friction for materials positioned either on the top or the bottom during testing. These results were obtained using a contact area of 16 cm^2 (4 x 4 cm) and a vertical load of 30 gf or contact pressure of 1.875 gf/cm^2; they are averages of fifteen tests under these constant conditions.

TABLE 1.9
Plastic specimen

Resin type	Manufacturer	Trade name	Moulding condition
Phenolic	Sumimoto Bake	Sumikon PM40	160°C, 200 kgf/cm^2 5 min
Melamine	Sumimoto Bake	Sumikon MM54	160°C, 200 kgf/cm^2 5 min
Urea	Toyo Koatsu	Yulight	160°C, 200 kgf/cm^2 5 min
Epoxy unfilled	Chiba Co.	Araldite B Hardener; HT901	Cured at 130°C for 15 hours
Epoxy filled with 10% SiO_2		Araldite B Hardener; HT901	
PMMA	Mitsubishi Rayon	Acrypet M100	
Polystyrene	Asahi Daw	Stylon 666	
PVC	Takiron kagaku	Takilon	
ABS	Asahi Daw	Stylack 200	
Polyethylene	Mitsui Sekiyu	Hizex 5000	Extruded sheet
Polypropylene	Nippon Chisso	β1012	
Polycarbonate	Teijin	Panlite	
Polyacetal (POM)	du Pont	Delrin	
Nylon 6 (PA)	Toyo Rayon	CM 1031	
PTFE	du Pont	Teflon	
Metal, Aluminium or Steel (S45C)			

TABLE 1.10

Static coefficient of friction of plastics (A = 16 cm^2, P = 30 gf)

Specimen (top) / Plate (bottom)	Al	POM	Nylon 6	PMMA	Poly-carbo-nate	ABS resin	PVC	Poly-prop-ylene	Poly-ethy-lene	Phen-olics	Urea resin	Mela-mine resin	Poly-sty-rene	Epoxy resin (no filler)	Epoxy resin (10% SiO_2)	PTFE	Mean value
Al	0.550	0.432	0.441	0.466	0.565	0.491	0.383	0.448	0.424	0.390	0.350	0.308	0.367	0.491	0.441	0.350	0.521
POM	0.246	0.318	0.227	0.283	0.227	0.308	0.328	0.383	0.243	0.206	0.237	0.283	0.752	0.194	0.230	0.200	0.252
Nylon 6	0.596	0.334	0.731	0.505	0.484	0.524	0.459	0.445	0.407	0.524	0.344	0.350	0.484	0.581	0.387	0.380	0.455
PMMA	0.704	0.452	0.441	0.462	0.637	0.434	0.418	0.484	0.459	0.653	0.469	0.502	0.476	0.491	0.494	0.357	0.484
Polycarbonate	0.546	0.308	0.441	0.498	0.524	0.604	0.407	0.452	0.558	0.302	0.441	0.380	0.441	0.427	0.360	0.414	0.449
ABS resin	0.509	0.311	0.407	0.484	0.498	0.539	0.441	0.469	0.438	0.473	0.400	0.364	0.577	0.621	0.500	0.337	0.449
PVC	0.535	0.328	0.448	0.370	0.484	0.466	0.397	0.498	0.704	0.363	0.363	0.344	0.341	0.395	0.321	0.387	0.409
Polypropylene	0.383	0.305	0.383	0.321	0.360	0.373	0.367	0.393	0.347	0.391	0.286	0.299	0.404	0.350	0.367	0.224	0.388
Polyethylene	0.243	0.261	0.215	0.315	0.243	0.261	0.209	0.277	0.354	0.360	0.233	0.249	0.271	0.289	0.367	0.200	0.267
Phenolics	0.296	0.280	0.377	0.407	0.531	0.407	0.321	0.513	0.596	0.367	0.274	0.252	0.305	0.277	0.277	0.274	0.350
Urea resin	0.321	0.200	0.312	0.344	0.448	0.457	0.308	0.360	0.410	0.233	0.240	0.243	0.367	0.264	0.230	0.255	0.303
Melamine resin	0.277	0.206	0.299	0.331	0.354	0.387	0.347	0.360	0.387	0.299	0.286	0.229	0.334	0.308	0.230	0.277	0.296
Polystyrene	0.438	0.261	0.328	0.390	0.459	0.581	0.370	0.494	0.349	0.484	0.455	0.448	0.462	0.431	0.367	0.383	0.437
Epoxy resin (no filler)	0.400	0.315	0.462	0.360	0.328	0.271	0.370	0.434	0.542	0.400	0.397	0.348	0.441	0.377	0.427	0.783	0.372
Epoxy resin (10% SiO_2)	0.535	0.293	0.462	0.490	0.527	0.434	0.347	0.539	0.491	0.330	0.410	0.370	0.452	0.330	0.431	0.407	0.417
PTFE	0.312	0.274	0.283	0.331	0.173	0.249	0.182	0.252	0.237	0.255	0.373	0.305	0.312	0.246	0.237	0.134	0.253
Mean value	0.417	0.304	0.391	0.382	0.420	0.417	0.356	0.417	0.462	0.366	0.340	0.332	0.389	0.375	0.348	0.302	0.381 0.376

Figures 1.58 to 1.61 show graphs constructed from the results shown in Table 1.10 and will be used to discuss examples of the characteristics of μ_s between the same plastics; the lowest value of μ_s is 0.134 for polytetrafluoroethylene (PTFE) and the highest μ_s is 0.731 for Nylon 6. Figurc 1.58 shows the μ_s values for phenolformaldehyde resin (PF), that is, those positioning it down side and others to up side are represented by an open circle and its reciprocal by an open triangle.

Similarly, Fig. 1.59 shows μ_s values for polyethylene (PE), Fig. 1.60 for polyacetal (POM), and Fig. 1.61 for polytetrafluoroethylene (PTFE). Table 1.11 lists those combinations for which μ_s is less than 0.25, and it appears that the values for combinations including PTFE are the lowest and those including POM are next to the lowest. Table 1.12 shows those combinations for which μ_s is larger than 0.6. The values of μ_s may vary widely, depending upon the measuring method and the test conditions, and Table 1.13 [46, 55, 56] shows a comparison of the present results with those from other research. However it can be seen that, with the exception of Nylon 6, there is generally good agreement.

TABLE 1.11
Material combinations with μ_s below 0.25

Between the same plastic	PTFE
MF and other plastics	POM
PF and other plastics	UF, MF, PTFE
POM and other plastics	UF, MF, PTFE
PTFE and other plastics	PE, PP, POM, PC

TABLE 1.12
Material combinations with μ_s above 0.5

Between the same plastic	ABS, Nylon, PC
PF and other plastics	PMMA, PE
Nylon and other plastics	EP
PS and other plastics	PE, PVC, PF
PC and other plastics	ABS, PMMA, Al

TABLE 1.13
Comparison of μ_s values with those of other researchers

	Between the Same Plastic	
	μ_s of other researchers	μ_s of author
Polytetrafluoroethylene (PTFE)	0.04 [ref. 46] 0.10 [ref. 56]	0.13
Polyethylene (PE)	0.10 [ref. 46] 0.25 [ref. 55] 0.33 [ref. 56]	0.35
Nylon 6	0.30 [ref. 55]	0.72
Polyvinylchloride (PVC)	0.45 [refs. 46,56]	0.54
Polystyrene (PS)	0.4-0.5 [ref. 55] 0.50 [ref. 46]	0.46
Polymethylmethacrylate (PMMA)	0.4-0.6 [ref. 55] 0.80 [ref. 46]	0.46

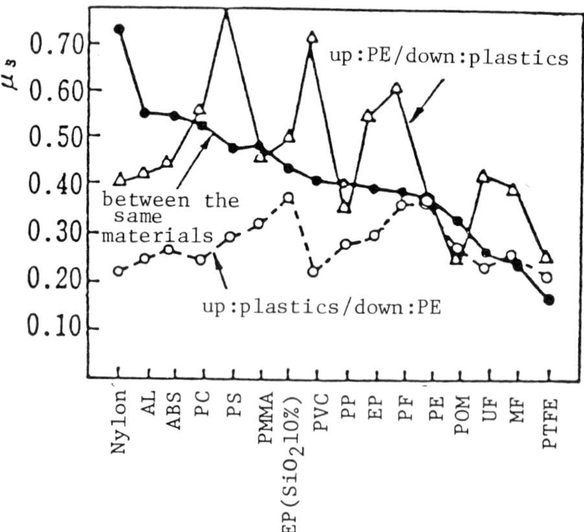

Fig. 1.59 μ_s for combinations of PE and various materials

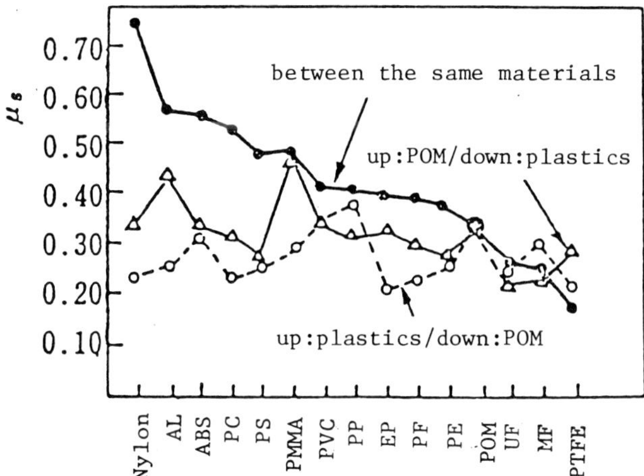

Fig. 1.60 μ_s for combinations of POM and various materials

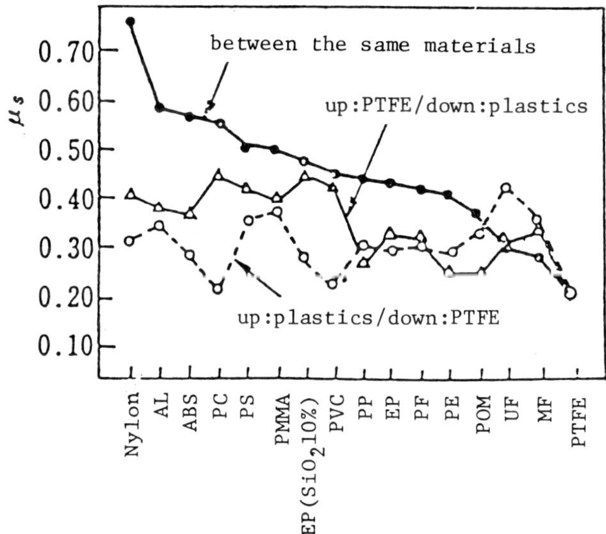

Fig. 1.61 μ_s for combinations of PTFE and various materials

1.5.2 *Coefficient of kinetic friction*

The coefficient of kinetic friction during dry (unlubricated) conditions of sliding is affected by more parameters than the coefficient of static friction and has little meaning without a precise definition of sliding conditions.

The main parameters of relevance are:

(1) Internal structure of the plastics
(2) Surface contact pressure
(3) Sliding speed
(4) Temperature and humidity
(5) Surface roughness
(6) Material of the opposite surface

The effect of internal structure (1) has already been discussed, and temperature and humidity dependence (4) will be covered in section 1.6.2. In all the tests to be described, the surface roughness was kept constant. The effects of (2), (3) and (6) will now be discussed.

(i) *Effect of surface contact pressure*

The influence of contact pressure on the coefficient of kinetic friction μ_k during dry sliding is shown theoretically by equations (1.7), (1.14) and (1.19), as described earlier. In these equations, according to Hertz's law, μ_k is proportional to $p^{-1/3}$ for two contacting spheres and to $p^{-1/9}$ for a sphere with many small asperities contacting a plain surface. According to Meyer's law, μ_k is proportional to $p^{(2/n-1)}$. Thus, in all cases, μ_k is proportional to $p^o \sim p^{-1/3}$, because the value of n is 3 for a perfectly elastic material and 2 for a perfectly plastic material. It follows that μ_k should decrease slightly with increasing contact load P.

Figures 1.62 and 1.63 show experimental determinations of the variations in μ_k with load. Each curve in Fig. 1.62 is similar to the above theoretical relationship. However, in Fig. 1.63, the value of μ_k reaches a maximum at a particular load P_o, and the trend only becomes similar to that predicted theoretically above this load. The critical load P_o decreases with increasing sliding speed. The reason for this is one example of temperature dependence, since the glass transition temperature of Nylon 6 is about 50°C. These phenomena will be discussed in more detail later.

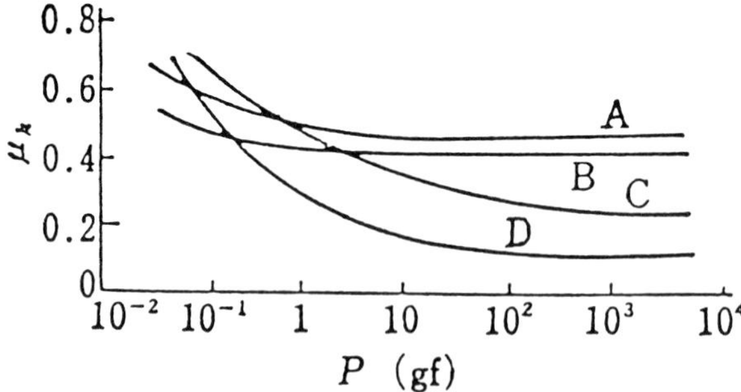

Fig. 1.62 Dependence of μ_k on contact force (for steel, A:PMMA, B:PVC, C:PE, D:PTFE)

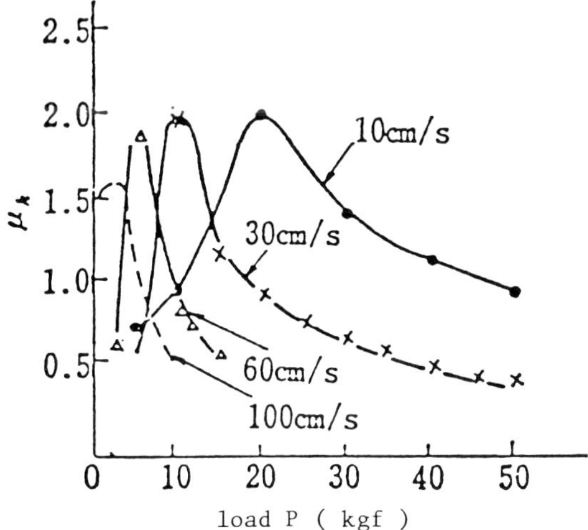

Fig. 1.63 Relationship between μ_k and contact load P for various sliding speeds (Nylon on steel)

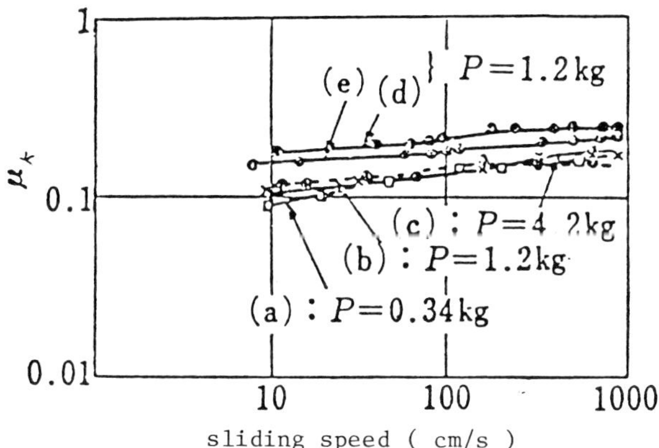

Fig. 1.64 Relationship between μ_k and sliding speed for Nylon (against steel)

(ii) *Effect of sliding speed*

Figure 1.64 shows an example of the effect of speed on the coefficient of kinetic friction μ_k for some thermoplastics, according to Matsubara [49] who suggests the following relationship:

$$\mu_k = cv^{\beta} \tag{1.46}$$

where c and β are constants and the value of β is assumed to be greater than 0 because the tensile strength of a viscoelastic material increases with increasing elongation speed. The value of β is 0.18 for Nylon 6, 0.13 for PTFE and 0.036 for HDPE (high density polyethylene). Recently, Matsubara [58] has explained these relationships with reference to Grosch's [59] paper in which the relationship between μ_k and speed is presented, relative to temperature, in a master curve for ANB rubber as shown in Fig. 1.65.

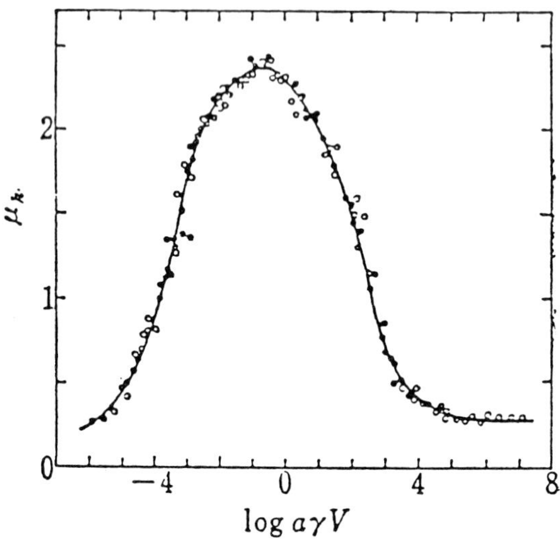

Fig. 1.65 Master curve of the relationship between μ_k and sliding speed v

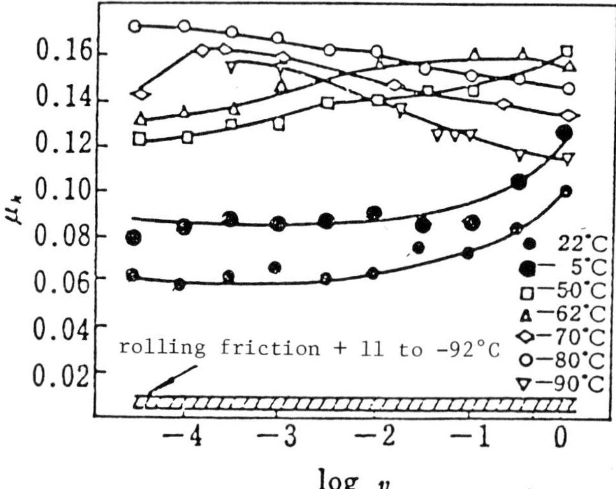

Fig. 1.66 Relationship between μ_k and sliding speed v for PTFE (against a steel ball)

In addition, Ludema [60] has attempted to derive a similar relationship for combinations of soft steel and PTFE or Nylon and LDPE (low density polyethylene) as shown in Figs. 1.66, 1.67 and 1.68. However, it appears from this work that for plastics the general relationship between μ_k and sliding speed is not simple to explain because the effect of speed is overlapped by the effect of temperature due to frictional heating with increased sliding speed, especially near the glass transition temperature at which discontinuities occur.

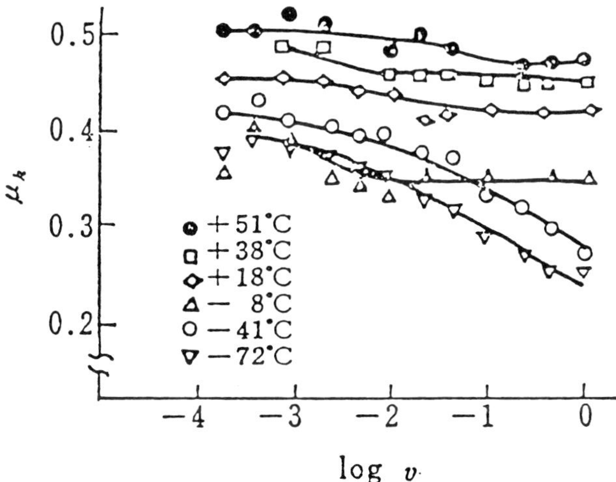

Fig. 1.67 Relationship between μ_k and sliding speed v for Nylon (against a steel ball)

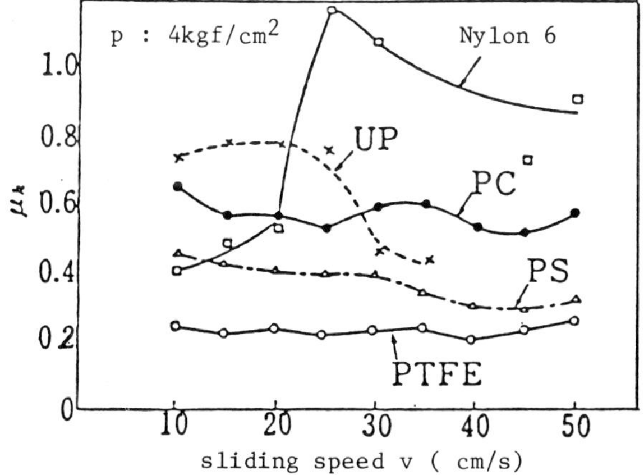

Fig. 1.68 Relationship between μ_k and v for Nylon, PTFE, PC, PS and UP

(iii) *Between various plastics or between plastics and metal*

The coefficient of kinetic friction μ_k can be measured as described in the following equation, by using strain gauges k_1 and k_2 to measure the frictional resistance F between opposing cylinder ends with an inside diameter of r_1 and an outside diameter of r_2 under a load P, as shown in Fig. 1.69(a).

$$\mu_k = \frac{2F_x}{P} = \frac{3FR(r_2^2-r_1^2)}{P(r_2^3-r_1^3)} \qquad (1.42)$$

Figure 1.69(b) shows the shape and dimensions of the test specimens used for these tests. The upper specimen is static and the lower is rotating.

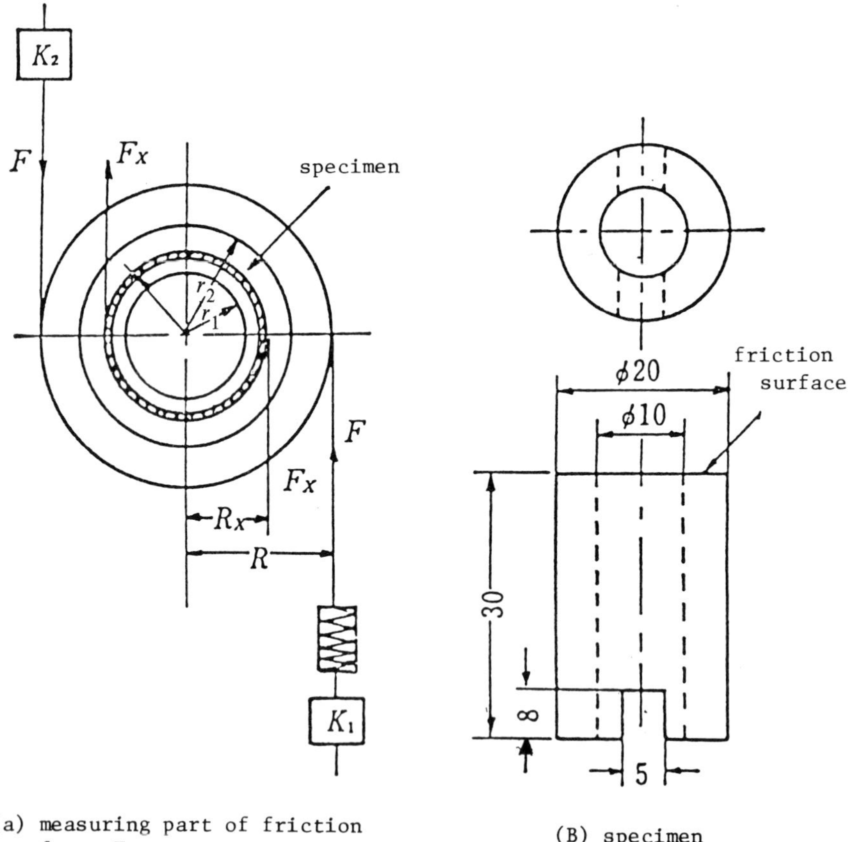

(a) measuring part of friction force F

(B) specimen

Fig. 1.69 Apparatus for measuring the coefficient of kinetic friction

Table 1.14 shows the values of μ_k for various combinations of materials, including thirteen types of plastics and steel, at a contact pressure of 0.83 kgf/cm^2, a sliding speed of 6.2 cm/s, an environmental temperature of 20°C and a relative humidity of 60%. The values of μ_k will now be discussed and compared with μ_s for various combinations: the same materials, steel and plastics, and different types of plastics, such as a thermosetting one, an amorphous one and a semicrystalline one.

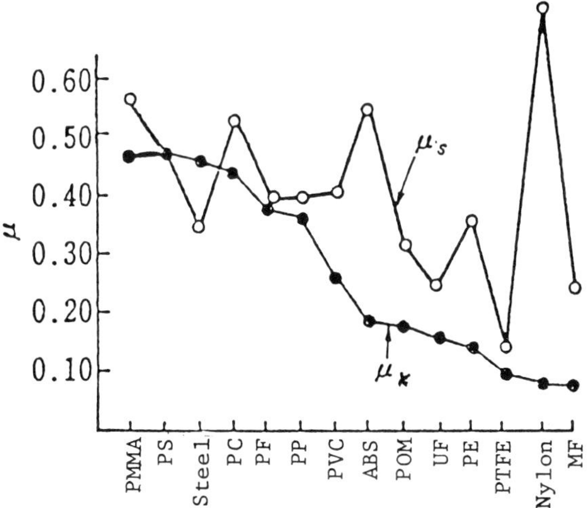

Fig. 1.70 Comparison of μ_s and μ_k between the same materials

Figure 1.70 shows the values of μ_k and μ_s between the same materials, and it appears that μ_k is generally smaller than μ_s, ranging from 0.08 and 0.55. Figure 1.71 shows a comparison of μ_k between steel and various plastics when the material combinations were tested in both the top and bottom positions. It appears that there are some differences depending on the relative positions of the materials. Figure 1.72 shows the values of μ_k and μ_s between three types of thermosetting plastics and various materials, the open circles indicating values between the same materials. Figure 1.73 shows the values of μ_k and μ_s between three types of amorphous thermoplastics and various materials, and Fig. 1.74 shows the values between three types of semicrystalline thermoplastics and various materials. It is clear from these figures that, in general, the values of μ_k for the thermosetting and amorphous thermoplastics are greatly influenced by the type of opposing material. However, the values for the semicrystalline thermoplastics are small, from 0.1 to 0.25, and are not much influenced by the type of opposing material. These relationships agree well with the internal molecular structures, as discussed in a previous section (1.4.1).

TABLE 1.14
Values of μ_k between various combinations of materials (0.83 kgf/cm^2, 6.2 cm/sec)

Top / Bottom	Steel	PF	MF	UF	PC	Nylon	POM	PMMA	PVC	ABS	PP	PS	PE	PFTE	average
Steel	0.468	0.524	0.686	0.711	0.362	0.104	0.180	0.385	0.216	0.376	0.316	0.517	0.109	0.100	0.359
Thermosetting Plastics:															
PF	0.468	0.373	0.083	0.495	0.418	0.514	0.112	0.308	0.200	0.195	0.271	0.503	0.074	0.100	0.261
MF	0.567	0.397	0.071	0.076	0.028	0.054	0.067	0.260	0.101	0.158	0.065	0.273	0.025	0.082	0.158
UF	0.453	0.067	0.089	0.153	0.058	0.087	0.071	0.070	0.071	0.282	0.352	0.127	0.075	0.092	0.146
Thermo-plastics:															
PC	0.302	0.429	0.286	0.486	0.429	0.100	0.195	0.549	0.044	0.487	0.478	0.479	0.088	0.092	0.344
Nylon	0.192	0.152	0.073	0.101	0.129	0.070	0.074	0.088	0.076	0.191	0.075	0.099	0.066	0.099	0.105
POM	0.129	0.190	0.090	0.136	0.142	0.092	0.177	0.091	0.124	0.190	0.180	0.161	0.092	0.095	0.134
PMMA	0.568	0.464	0.470	0.395	0.418	0.168	0.109	0.551	0.386	0.177	0.472	0.452	0.123	0.099	0.436
PVC	0.219	0.256	0.087	0.110	0.222	0.112	0.143	0.313	0.250	0.216	0.317	0.391	0.088	0.128	0.202
ABS	0.366	0.229	0.087	0.125	0.269	0.126	0.167	0.185	0.176	0.180	0.213	0.138	0.096	0.100	0.175
PP	0.300	0.314	0.139	0.308	0.326	0.124	0.188	0.079	0.249	0.316	0.350	0.292	0.133	0.112	0.259
PS	0.368	0.392	0.310	0.438	0.375	0.171	0.053	0.345	0.333	0.263	0.246	0.467	0.156	0.108	0.274
PE	0.139	0.147	0.130	0.092	0.090	0.079	0.086	0.068	0.102	0.127	0.122	0.160	0.141	0.106	0.113
PFTE	0.117		0.075	0.101	0.105	0.094	0.104	0.108	0.097	0.093	0.111	0.106	0.095	0.083	0.092
Average	0.331	0.302	0.191	0.264	0.240	0.108	0.190	0.271	0.201	0.232	0.254	0.290	0.104	0.099	0.218 0.215

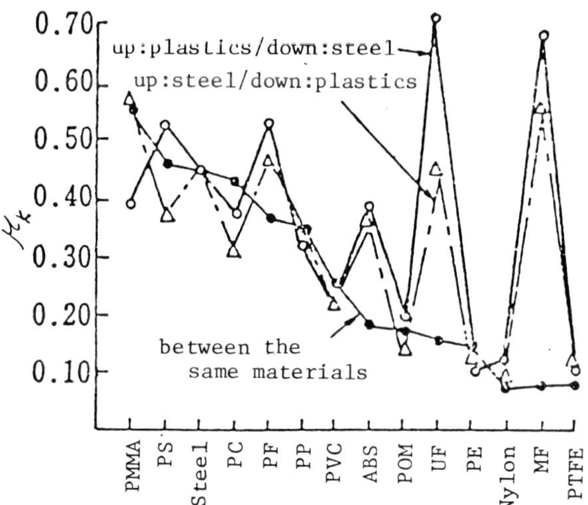

Fig. 1.71 Comparison of μ_k values with the plastic specimen located in either the top or the bottom position during testing (steel and plastic)

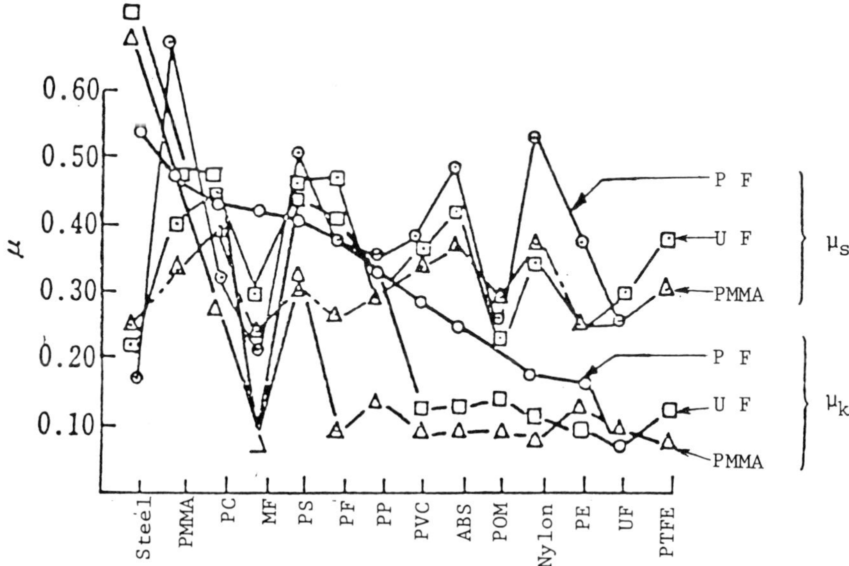

Fig. 1.72 Coefficients of friction between thermosetting plastics and various materials

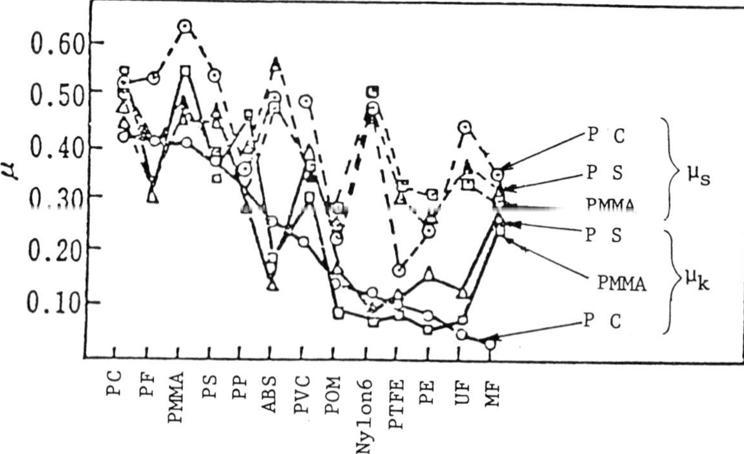

Fig. 1.73 Coefficients of friction between amorphous plastics and various various plastics

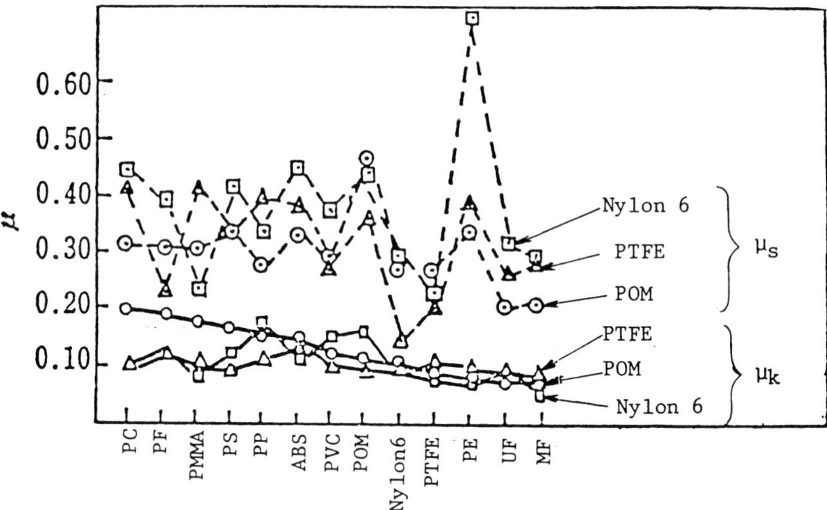

Fig. 1.74 Coefficient of friction between semicrystalline plastics and various other plastics

Table 1.15 shows combinations with values of μ_k under 0.1, and Table 1.16 shows combinations with values of μ_k over 0.5. Although the values of μ_k are greatly influenced by the type of test specimen and the test conditions, which are not always steady and constant, a comparison of these results with those from other work is shown in Table 1.17 [46,48,55,62]. It can be seen from this table that, except for PVC, there is not a great difference in the μ_k values between those reported here and those obtained in other research.

TABLE 1.15
Material combinations with μ_k values below 0.1

Between the same plastics	Nylon, MF, PTFE
Steel and plastic	Nylon, PTFE
POM and other plastics	PE, Nylon, PTFE
Nylon and other plastics	PE, MF, PTFE
PE and other plastics	UF, Nylon, POM, PC
PTFE and other plastics	MF, UF, Nylon, POM, PC, ABS
PC and other plastics	PE, PTFE
MF and other plastics	UF, Nylon, PTFE

TABLE 1.16
Material combinations with μ_k values above 0.5

Between the same plastics	PMMA, PS, PC
Between steel and plastics	PMMA, MF, UF, PF

TABLE 1.17
Comparison of μ_k with values obtained in other research

Material		Top Specimen: Steel Bottom Specimen: Plastic	Plastic Plastic	Plastic Steel
Polyvinylchloride (PVC)	C	0.4-0.45	0.4 -0.45	0.35-0.4
	B	0.35-0.45		
	D	0.45	0.45-0.55	0.45-0.50
	*	0.22	0.22	0.22
Polystyrene (PS)	C	0.4-0.5	0.4-0.5	0.4-0.5
	A	0.35	0.5	0.3
	*	0.34	0.47	0.52
Polyethylene (PE)	C	0.3	0.25	0.25
	D	0.08-0.12	0.11	
	*	0.12	0.14	0.11
Polyacetal (POM)	D	0.13		
	*	0.13	0.18	0.10
Polymethylmethacrylate (PMMA)	C	0.45-0.05	0.4-0.6	0.5
	B	0.45	0.8	0.5
	*	0.57	0.55	0.39
Polyamide (Nylon 6)	C	0.4	0.3	0.25
	D	0.34		0.35
	*	0.19	0.08	0.10
Polytetrafluoroethylene (PTFE)	A	0.10	0.04	0.04
	D	0.05	0.04	
	*	0.12	0.08	0.10
Polycarbonate (PC)	D	0.53		
	*	0.30	0.43	0.36

*: Author's measurements; A: K.V. Shooter [47]; B: K.V. Shooter and D. Tabor [48]; C: K.V. Shooter [63]; D: R.C. Bower [55]

1.6 LIMITING pv VALUE/TEMPERATURE AND HUMIDITY DEPENDENCE

1.6.1 *Limiting pv value*

(i) *Significance*

Frictional heat is produced when continuous sliding occurs between solid bodies. In particular, because the heat liberated is equivalent to $pv\mu_k a$, where a is the contact area, the temperature can greatly increase when the contact pressure p and the sliding speed v become large. Since the heat endurance temperature of plastic materials is much lower than that of metals, due to their lower melting or heat decomposition temperatures, normal sliding frictional motion in plastics cannot continue indefinitely as the temperature rises. The minimum pv beyond which a plastic is unable to continue normal sliding is designated the "limiting pv" or the "pv limit".

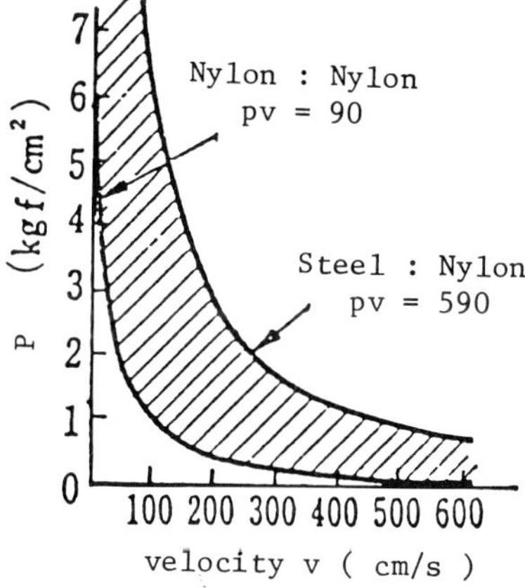

Fig. 1.75 Melting limit of Nylon in pv coordinates

Figure 1.75 shows an example of the limiting pv for plastic materials, as reported by Matsubara [49], and offers an experimental equation [49] as follows:

$$pv^{\gamma} = K \tag{1.47}$$

where γ and K are constants which depend upon the material, and whose values for the combinations in Figure 1.75 are:

$\gamma = 1.1$, K = 90 (Kgf/cm²) for Nylon 6/Nylon 6
$\gamma = 1.1$, K = 590 (Kgf/cm²) for Nylon 6/steel

Research papers concerning the limiting pv of plastic materials have been published by Lomax [63], Awaya [64], Tanaka [65], Lewis [66], Yamaguchi [67] and Miyake [68].

The limiting pv value is affected not only by the conditions of heat generation but also by the extent of heat loss by radiation. In the following section, the limiting pv of plastics is discussed taking radiation conditions into consideration.

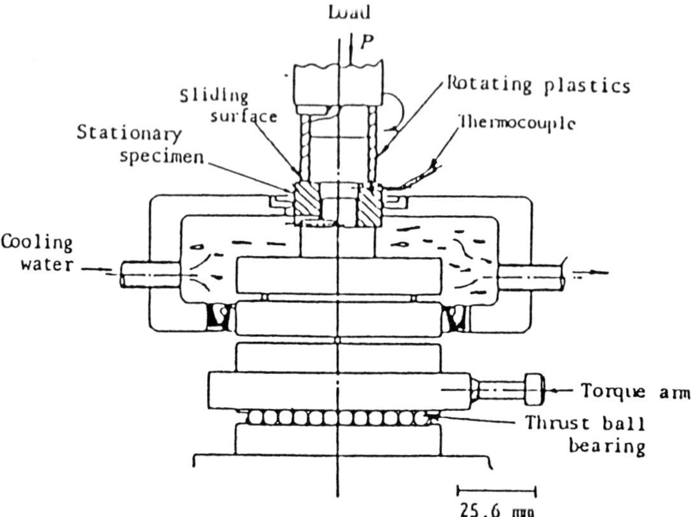

Fig. 1.76 Apparatus

(ii) *Theoretical considerations*

Consider unlubricated continuous sliding, as in Fig. 1.76, where the end surface of two cylindrical specimens are under constant load with an average sliding velocity v, a friction coefficient μ, and a sliding area of a (cm^2). The amount of heat generated per second becomes pvμak (cal), where k is the mechanical equivalent of heat 0.0234 cal/kgf·cm. Let the temperature at the sliding surface be τ_b (°C), the environmental or cooling temperature τ_a (°C), the heat-radiating area A (cm^2) and the average overall heat transfer rate H(cal/cm^2·°C·s). The amount of heat radiated per second then becomes $H\cdot A(\tau_b-\tau_a)$ (cal). When thermal equilibrium is reached,

$$pv\mu ak = H\cdot A(\tau_b-\tau_a)$$

$$pv = \frac{H\cdot A}{\mu ak}(\tau_b-\tau_a)$$

Let A/a = n; then

$$pv = \frac{H \cdot n}{\mu k} (\tau_b - \tau_a) \tag{1.48}$$

Here, a pv coefficient C is defined by putting $H \cdot n/\mu k$ = C ($Kgf/cm^2 \cdot cm/s/°C$). Equation (1.48) then reduces to

$$pv = C(\tau_b - \tau_a) \tag{1.49}$$

The process leading to equilibrium is as follows. As shown in Fig. 1.77, the temperature τ_b at the sliding surface becomes constant after a certain time, corresponding to a specific pv value such as pv_1, pv_2, etc. Once the pv exceeds a critical value, however, a sharp temperature rise will occur after a time lapse, leading to melting, burning or degradation. The minimum pv and the temperature at which these phenomena occur is defined as the limiting pv (pv_{max}) and the limiting temperature (τ_{max}), respectively.

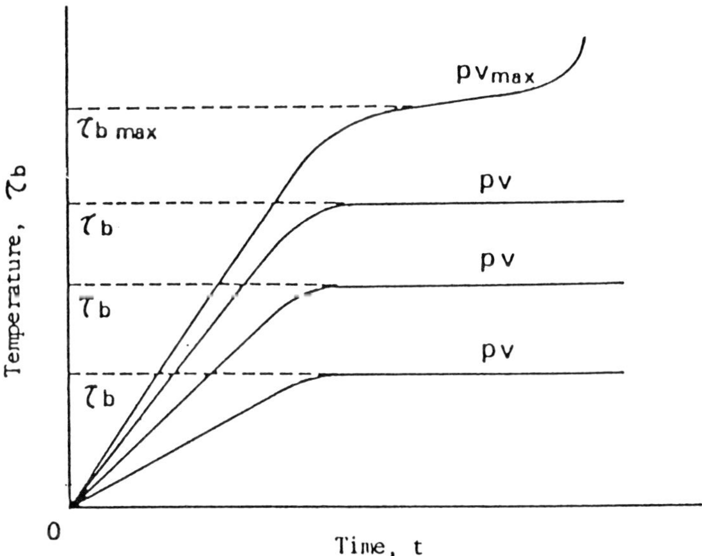

Fig. 1.77 Relationship between temperature and sliding duration at different pv values

This behaviour, expressed in terms of the relationship between pv and τ_b or (τ_b-τ_a), is shown in Fig. 1.78. The vertical axis indicates temperature, generally τ_b, or in some cases, τ_{ar}, the room temperature and τ_{ao}, that of cooling water. If heat is allowed to be radiated at a room temperature of τ_{ar} and with the pv coefficient $H \cdot n/\mu K = C_2$, the pv~$(\tau_b-\tau_a)$ relationship is indicated by AC_2. The length of the intersection B_2 of the horizontal line as the limiting temperature τ_{max} and AC_2 gives the limiting pv ($=pv_{2max}$). With a smaller H and n, or with a larger μ, C becomes C_1, which is smaller than C_2. In this case, the above-mentioned relationship is indicated by the line AC_1. With a larger cooling coefficient, H resulting from the cooling water (whose temperature is lower than the room temperature), C becomes C_3, larger than C_2. In this case, we obtain a limiting pv of pv_{3max}.

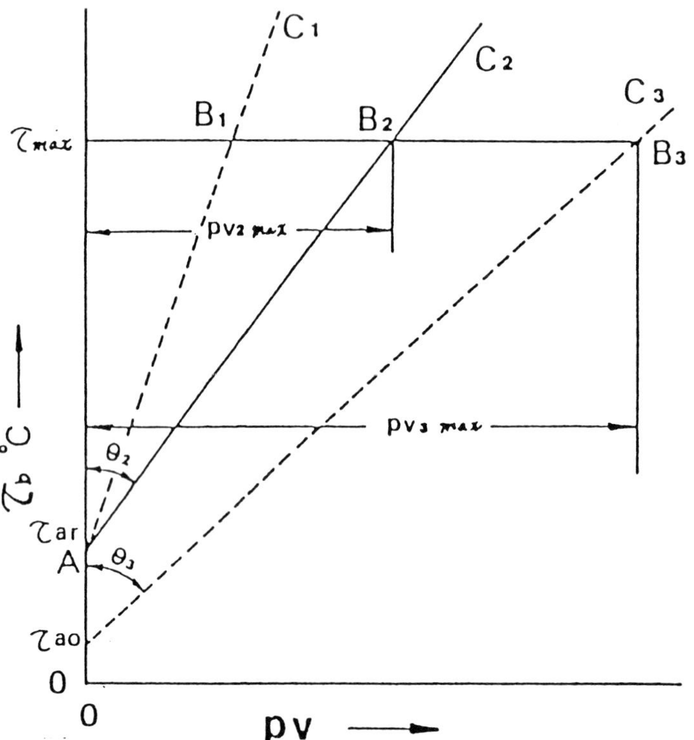

Fig. 1.78 Relationship between the measured temperature τ_b and pv

(iii) *Experimentation*

(1) *Method.* The friction apparatus is shown in Fig. 1.76 and the dimensions of the specimens in Fig. 1.79. The specimens were subjected to unlubricated or "dry" continuous sliding under constant load.

In order to vary the heat radiation conditions, cooling water at various temperatures was used, in addition to various materials with different thermal conductivities. The temperature was measured 1.5 mm from the sliding surface of the stationary specimen with an embedded thermocouple. Although these temperature readings will be somewhat lower than at the actual sliding surface due to heat conductivity of the material, they are nevertheless regarded as τ_b.

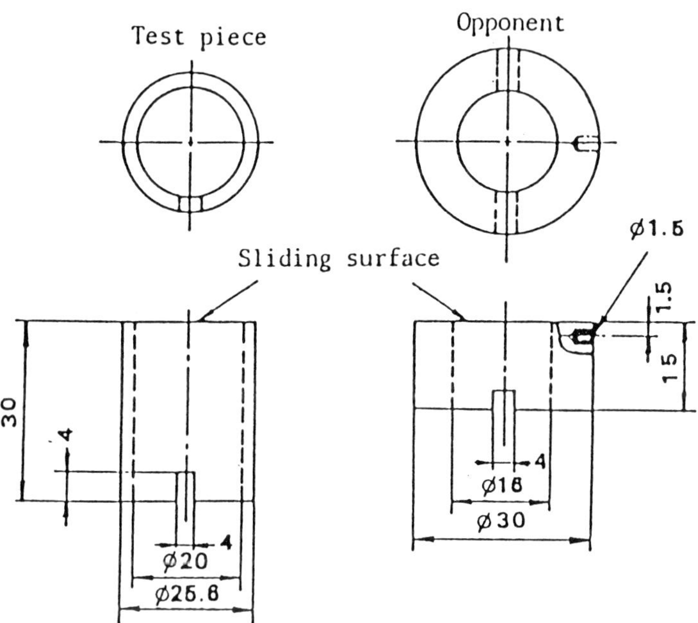

Fig. 1.79 Shape and dimensions of test pieces (in mm)

The properties of the plastic materials and the metals are listed in Tables 1.18 and 1.19, respectively. The sliding surfaces were cleaned with a dry tissue after being polished with 600 grade abrasive paper.

Tests began with a relatively low contact pressure p and velocity v, that is, at a low pv, and continued with increasingly higher values of pv. First, temperatures τ_{b1} and τ_{b2}, which are constant values at constant pv, were measured as in Fig. 1.77. Then, the relationship between pv and τ_b was obtained in terms of line AC as shown in Fig. 1.78. Finally, the limiting pv_{max} and τ_{bmax} were obtained. Tests were made either at room temperature of 23°C without cooling water, or with cooling water at temperatures of 2°C, 16°C and 50°C.

TABLE 1.18
Properties of plastics

Properties	PF (cloth filled)	ABS	PC	POM	PTFE
Trade name	Fudolite PMFR-30	Stylac-200	Panlite	Delrin	Teflon
Melting temp. (τ_m °C)			240	175	327
Moulding condition or shape	p: 200 kg/cm^2		Circular bar		

TABLE 1.19
Properties of metals

Properties	Materials Steel	Cu	Zn
Melting temp. (τ_m °C)	1390 - 1420	1083	419.46
α: thermal conductivity {cal/s·cm^2(°C/cm)}	0.134 - 0.086	0.94	0.27

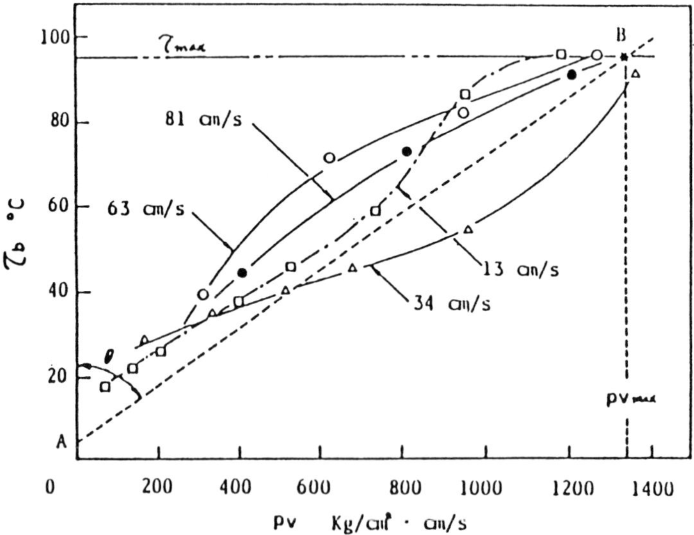

Fig. 1.80 τ_b vs. pv for POM on steel (S45C) at different sliding speeds (AB: average line)

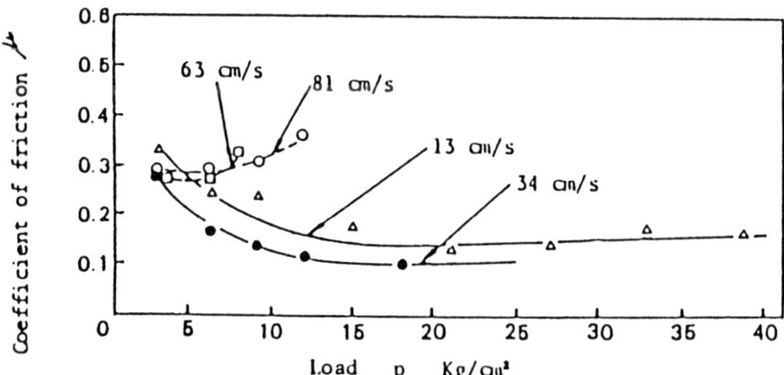

Fig. 1.81 μ vs. p for POM on steel (S45C) at different sliding speeds

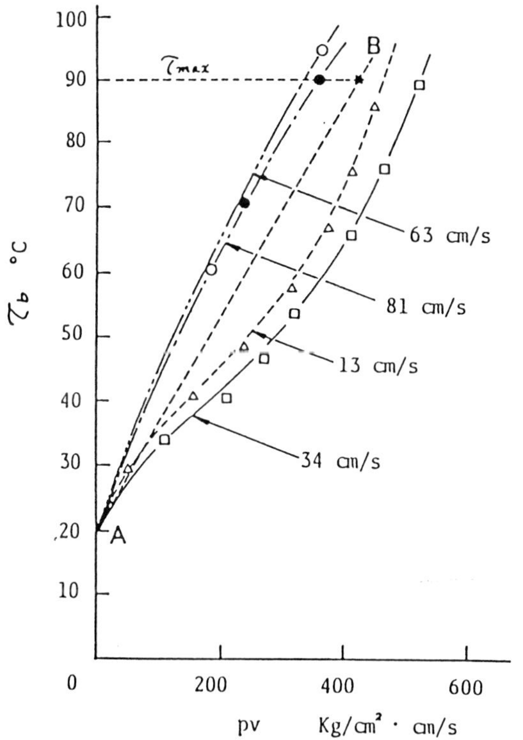

Fig. 1.82 τ_b vs. pv for ABS on Cu at different sliding speeds. AB: average line

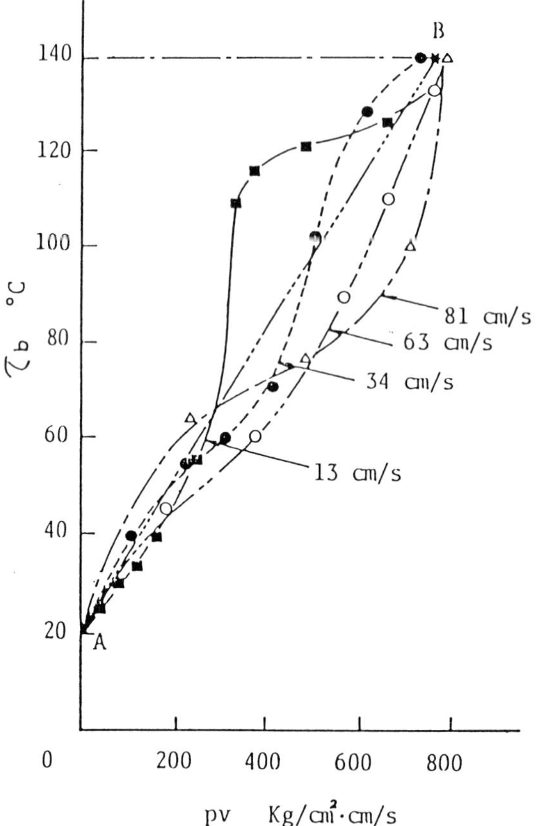

Fig. 1.83 τ_b vs. pv for PC on steel (S45C) at room temperature and different sliding speeds. AB: average line

(2) *Results and discussion.* The equilibrium temperatures τ_b at each pv, obtained under a given cooling condition and for given combinations as shown in Fig. 1.77, and the pv-τ_b relationship corresponding to each specific velocity, obtained as in Fig. 1.78, are shown in Figs. 1.80, 1.82 and 1.83. In Fig. 1.80, the pv-τ_b relationships are given for the combination of POM and steel (S45C), cooled with water at 2°C, and for velocities of 13, 34, 63 and 81 cm/s. The resulting curves are not identical with each other but depend on the velocity v. As can be clearly seen from Fig. 1.81, the greater the velocity, the larger the friction coefficient μ, even when the conditions are the same as in Fig. 1.80. An increased μ causes pv to decrease for the same τ_b, as Fig. 1.77 shows, resulting in a smaller C, which in turn shifts the curve to a lower position. The average relationship for all these pv-τ_b curves is represented approximately by the dotted line

AB. The curves in Fig. 1.82 show the pv-τ_b relationships for the combination of ABS and Cu at a room temperature of 23°C and at the same four different velocities. These curves are also displaced similar to those in Fig. 1.80. In thc vicinity of τ_{max}, however, the curves almost coincide, giving a common τ_{max} of nearly 90°C. The average of the four curves is represented by the dotted straight line AB, where point B represents both pv_{max} and τ_{max}.

Corresponding results for the combination PC and steel (S45C), again at room temperature, are shown in Fig. 1.83. Although the curves alter with v, pv_{max} and τ_{max} are represented by almost the same point B. From the pv-τ_b relationships found in these three examples, equations (1.48) and (1.49) can be considered to be valid in practice. Figure 1.84 shows some examples of the relationships between p and v when pv_{max} is reached for the combinations of PTFE/PTFE and PTFE/S45C at room temperature. It can be seen from the figure that pv_{max}, the pv value under abnormal friction (G.P.S. in Fig. 1.84) is somewhat smaller outside of than in continuous generation of powders (C.G.P. in Fig. 1.84). For both pv values, however, Matsubara's equation pv^{γ} = K [47] holds, where γ ranges from 0.85 to 1.1 with an average of approximately 1.

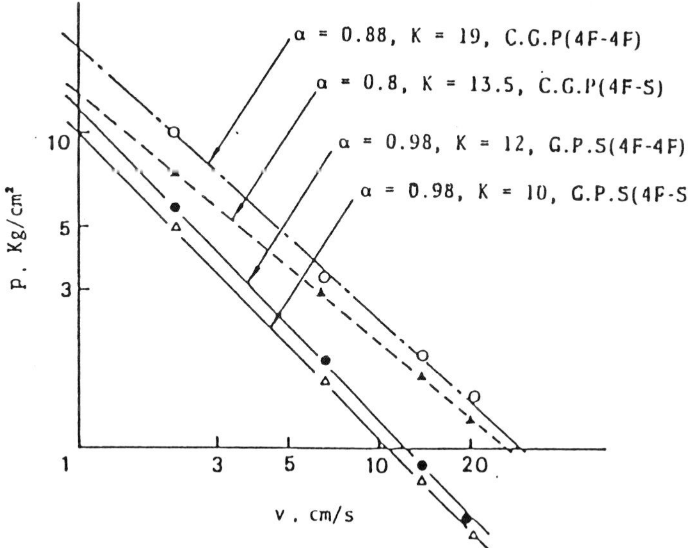

Fig. 1.84 pv vs. v for critical conditions between PTFE and PTFE or steel (S45C) (C.G.P.: continuous generation of power; G.P.S.: generation of pf power at starting period)

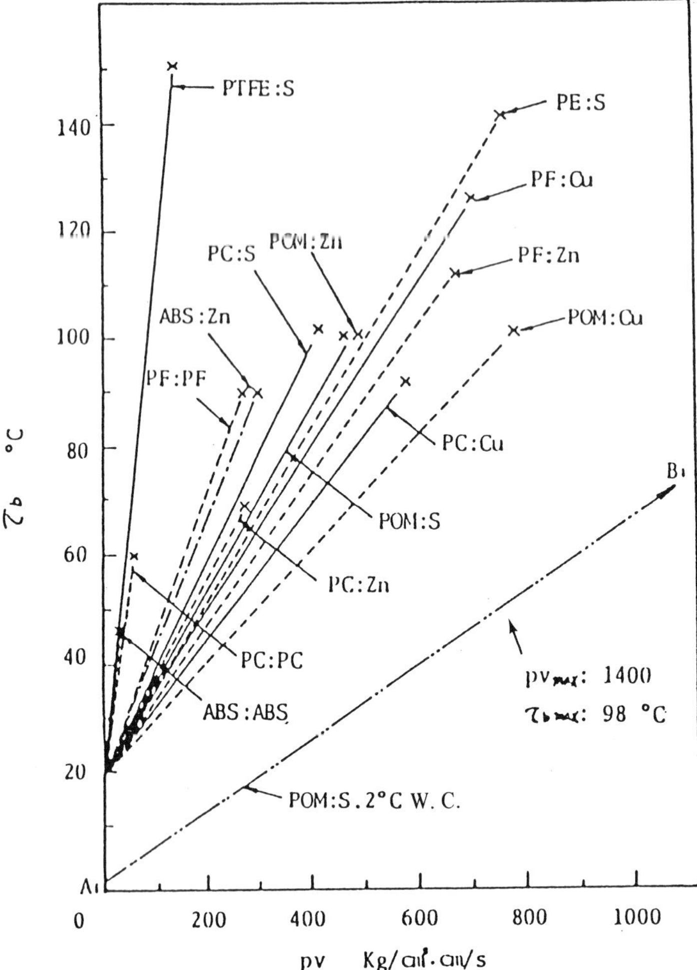

Fig. 1.85 Relationships between τ_b and mean pv for various material combinations at room temperature; A_1B_1: water cooling at 2°C

pv-τ_b Curves for each combination of five plastic materials and three metals at various sliding velocities were obtained by the method shown in Fig. 1.78. The curves in Fig. 1.85 show the average pv-τ_b relationships for each combination under spontaneous cooling. The curve A_1B_1 is added for reference, representing the pv-τ_b relationship for the combination of POM and S45C cooled with water at 2°C. The crosses, x, correspond to each pv_{max} and τ_{max}.

Table 1.20 presents the frictional coefficient μ_k, τ_{max} (apparent) and pv_{max} for combinations of five plastics against themselves and against metals when cooled with room temperature air or with water at 2°C, 16°C and, exceptionally, at 50°C. As Fig. 1.85 and Table 1.20 together show, all the combinations of POM and the three metals have almost the same τ_{max} of 100°C. This result conforms to the theory in Fig. 1.78. In materials such as PF, PC and ABS, on the other hand, τ_{max} is not constant. When plastics slide against themselves, the apparent τ_{max} is very low because of the poor heat conductivity. The pv_{max} is also small because of the lower heat radiation rate.

The pv coefficient C obtained from the data in Fig. 1.85 and Table 1.20, following the equation $C = pv_{max}/(\tau_{bmax}-\tau_a)$, is 1~10{(Kg/ cm$_2$·cm/s)/°C} when cooling at room temperature. The plastics POM, PC, PF, PTFE and ABS have decreasing values of C, in that order. When sliding against the metal counterfaces, Cu, Zn and S45C, the C values are higher because of the higher heat conductivity. As Fig. 1.85 shows, when cooled with water at 2°C, the combination of POM and S45C has a C value of about 14{(Kg/cm^2·cm/s)/°C} and this value is about 2.3 times that obtained with normal cooling, 6{(Kg/cm^2·cm/s)/°C}. Also, the pv_{max} is markedly improved. The thermosetting resin PF has a relatively large pv_{max} because of its higher τ_{max}, although its C value is not particularly high.

1.6.2 *Temperature and humidity dependence*

(i) *Reason for temperature dependence*

As explained previously, the frictional resistance F of plastics consists almost wholly of the product of the true contact area and its shear strength. The true contact area A is proportional to $E^{-\alpha}$ (α=2/3) according to equation (1.5). It is assumed that F generally increases with a rise in temperature, because A increases due to the decrease in E with increasing temperature, as shown by line ab in Fig. 1.86. On the other hand, the value of F can decrease due to a rise in temperature because of the decreasing shear strength of τ_o, as shown by line cd in Fig. 1.86. The final temperature dependence of the frictional coefficient of plastics is thus the combined effect of an increase in A due to a decrease in elastic modulus and a decrease in shear strength with a rise in temperature. However, the relationships between the elastic modulus and temperature, or between shear strength and temperature, are not simply directly proportional as shown in Fig. 1.86. There are often complex and discontinuous relationships in the region of glass transition and melting temperatures. Generally, the glass transition temperatures of plastics cover a wide range from -120°C to 150°C, and show different effects depending on the types of plastics. Because of these complications, it is not therefore possible to offer a simple and definitive equation for the dependence of friction on temperature.

TABLE 1.20
Values of pv_{max} ($kg/cm^2 \cdot cm/sec$), τ_{max}(°C) and μ_k

	Plastic Against Itself			Opposing Material Properties: Steel			Cu			Zn		
Material	pv_{max}	τ_{max}	μ_k	pv_{max}	τ_{max}	μ_k	pv_{max}	τ_{max}	μ_k	pv_{max}	τ_{max}	μ_k
PTFE, Polytetra-fluoroethylene	12-19	-	0.1 -0.2	23°C:10 -13.5	100 -150	0.2 -0.28	12 -16	120	0.12 -0.38	9.6 -13.5	110 -140	0.11 -0.34
PF, Phenolics (cloth filled)	220	90	0.1 -0.3	* 2°C:850 23°C:750 *50°C:310	140 140 140	0.16 -0.33	850	105	0.12 -0.28	630	110	0.13 -0.32
POM Polyoxymethylene	50	55	0.2 -0.28	* 2°C:1400 *16°C:820 23°C:560 *50°C:240	95 100 100 100	0.14 -0.4	1000 700	105 110	0.18 -0.3	380	105	0.1 -0.4
PC Polycarbonate	46	60	0.1 -0.6	23°C:460	100	0.17 -0.35	580	90	0.18 -0.3	300	75	0.14 -0.5
ABS Resin	35	45	0.22 -0.6	23°C:500	85	0.2 -0.4	550	90	0.23 -0.4	360	90	0.3 -0.45

*Water cooling

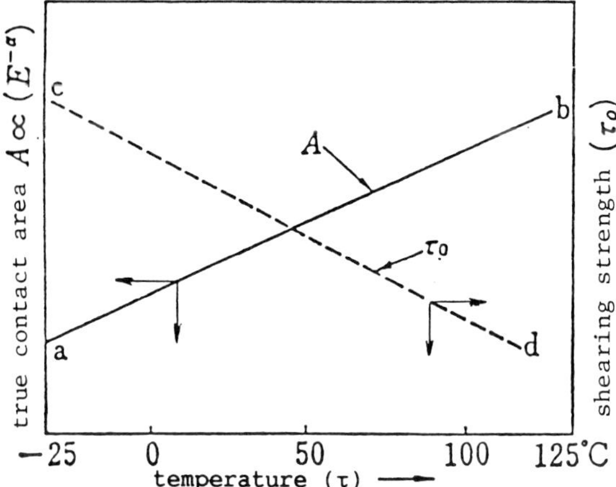

Fig. 1.86 Relationships between true contact area A or shear strength τ_0 and temperature τ

(ii) *Experimentation on temperature dependence*

Figure 1.87, as reported by Sieglaff [71], shows relationships between the frictional resistance F and load P for PVC at different temperatures. It can be seen that F becomes greater in the range above the glass transition temperature of 80°C. Figure 1.88, as reported by Watanabe [72], shows the relationship between μ_k and temperatures from 50°C to 210°C in Nylon 6, and shows a maximum value of μ_k at about 150°C, near the melting temperature of 216°C. Figure 1.89, as reported by Ludema [73], shows the relationship between temperature and μ_k for various plastics against steel, and that each plastic has a special characteristic relationship.

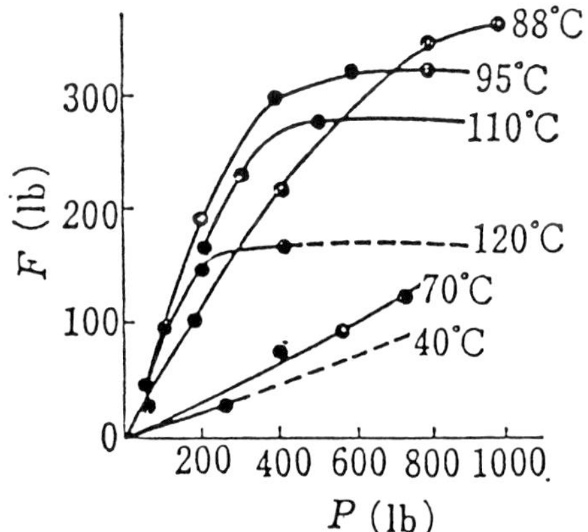

Fig. 1.87 Relationships between frictional resistance F and load P for PVC at different temperatures

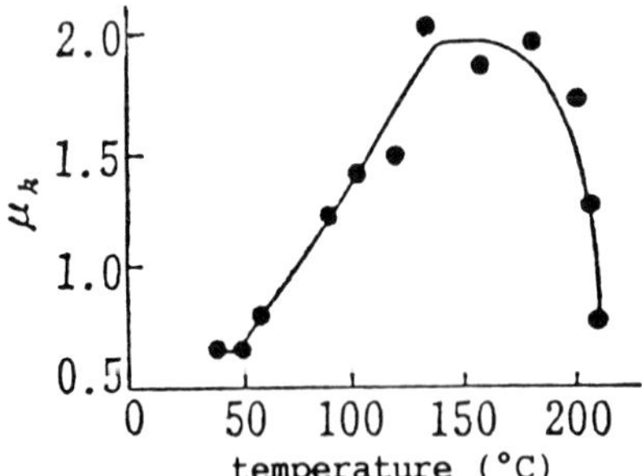

Fig. 1.88 Relationship between μ_k and temperature τ for Nylon

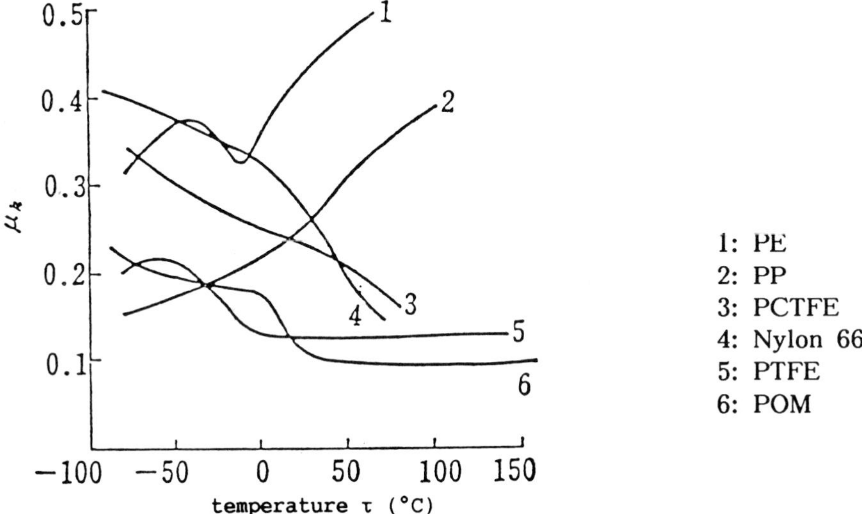

Fig. 1.89 Relationship between μ_k and temperature at low sliding speeds for various plastics against a steel ball

Table 1.21 shows the details and characteristic values of five types of plastics tested by the present author. Figures 1.90 to 1.94 show the relationships between the friction coefficient μ_k, or temperature rise τ, and the duration of sliding t at various sliding speeds. In these figures, the dotted line represents the relationship between τ and t and the solid line is that between μ_k and t.

TABLE 1.21
Specimens

Material	Code		Glass Transition Temp. t_g(°C)	Melting Temp. t_m(°C)	Tensile Strength (kgf/cm^2)
Nylon 6	PA	Ube 1022B	50 (37-57)	225	520
Polytetra-fluoroethylene	PTFE	Mitsui Frolo Chemical	30, 125 7-90, 20-30 126-130	326	150
Polycarbonate	PC	Teizin Panlite	150 (117-150)	260	550
Polystyrene	PS	Asahi Daw Stylon 666	86 (80-100)	-	360
Unsaturated polyester	UP	Showa Kobunshi Riglak	69 (69-87)	-	-

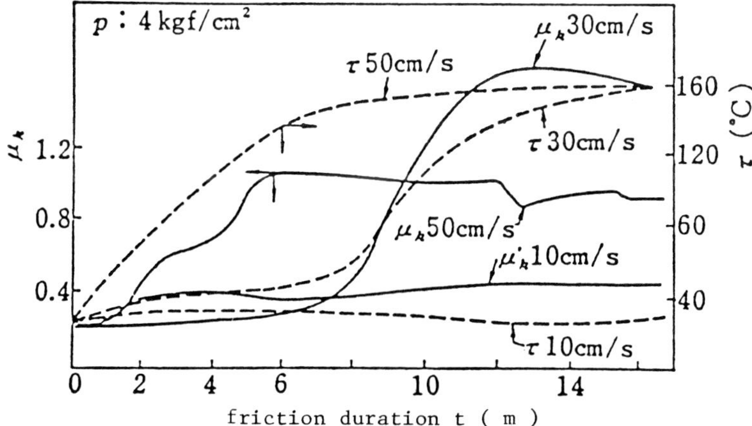

Fig. 1.90 Relationship between μ_k or τ and t for Nylon

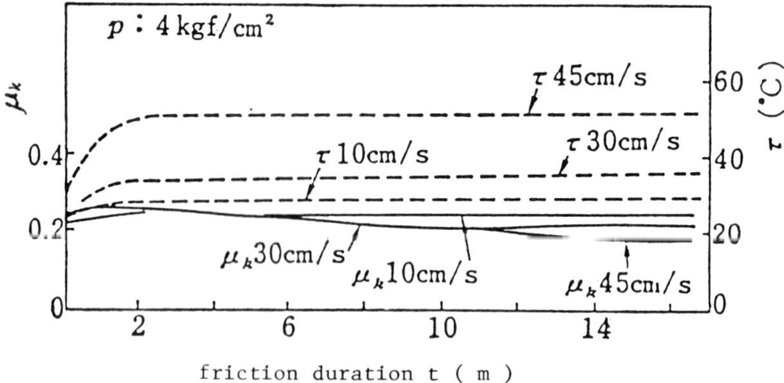

Fig. 1.91 Relationships between μ_k or τ and t for PTFE

It is clear from these figures that temperature and μ_k values are greatly increased with increasing sliding speed in nylon, but for PTFE, μ_k is stable or decreases slightly, while the temperature τ rises only very slightly. In polycarbonate, as Fig. 1.92 shows, both τ and μ_k increase stably with increasing speed, but τ rises greatly beyond 120°C and μ_k becomes maximum value at this temperature which is near the glass transition temperature of 150°C. In polystyrene (PS) as shown in Fig. 1.93, τ rises but μ_k is almost constant with increasing sliding speed. In unsaturated polyester (UP) as shown in Fig. 1.94, it appears that μ_k is low at a high speed of 35 cm/s, and increases with a rise in temperature. Figure 1.95 shows the relationship between μ_k and environmental temperature for various plastics; it can be seen that these relationships are not simple, with the exception of PTFE.

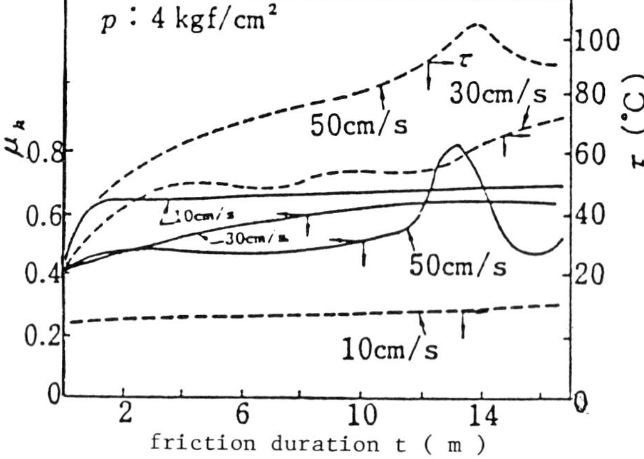

Fig. 1.92 Relationships between μ_k or τ and t for PC

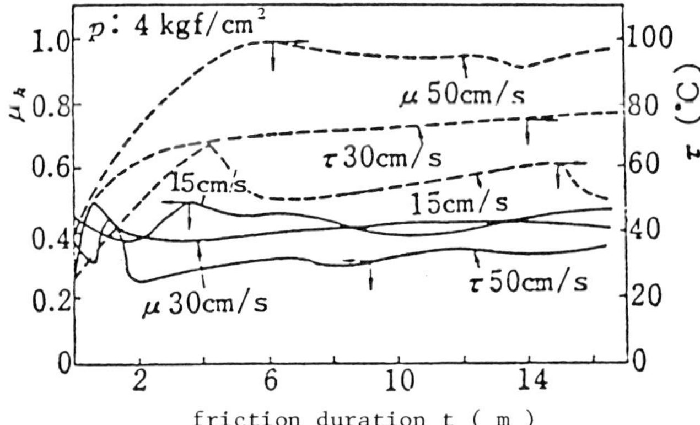

Fig. 1.93 Relationships between μ_k or τ and t for PS

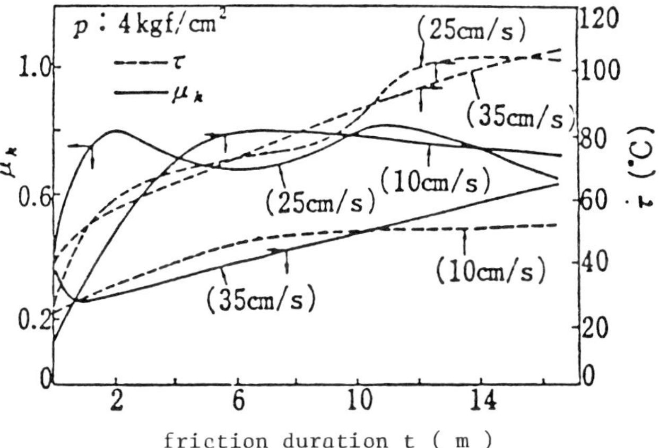

Fig. 1.94 Relationships between μ_k or τ and t for UP

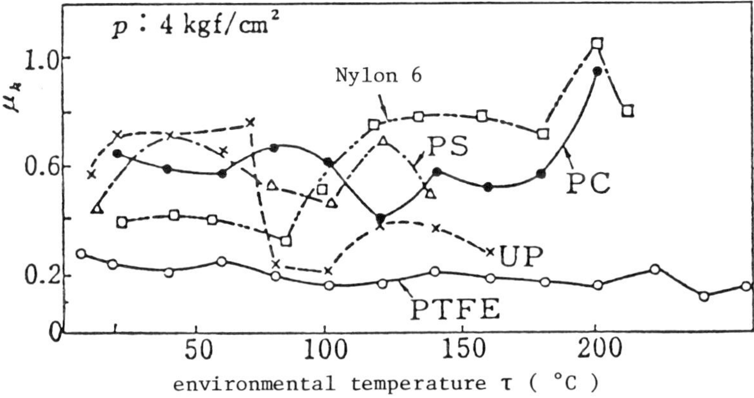

Fig. 1.95 Relationships between μ_k and the environmental temperature τ for various plastics

(iii) *Humidity dependence*

The theoretical dependence of μ_k on humidity is still unclear, but Figs. 1.96 and 1.97 give some experimental results. Figure 1.96, as reported by Watanabe [73], shows that μ_k becomes larger with increasing humidity for nylon, which absorbs moisture, but is unaffected for PE. Figure 1.97 [70] shows an example for PTFE, which does not absorb water, where the value of μ_k in dry conditions is greater than in humid conditions.

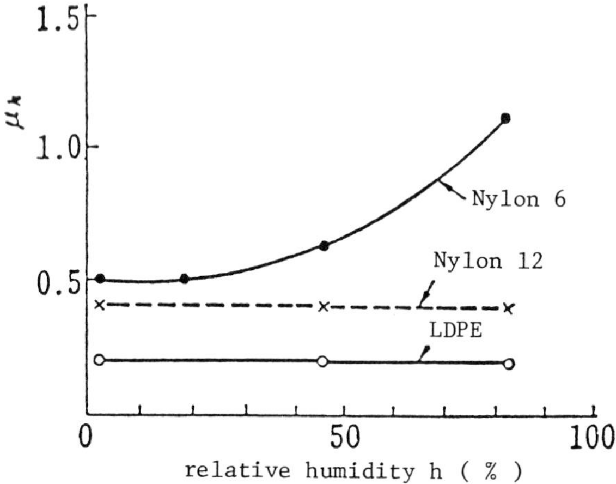

Fig. 1.96 Relationships between μ_k and relative humidity h (%) for various plastics

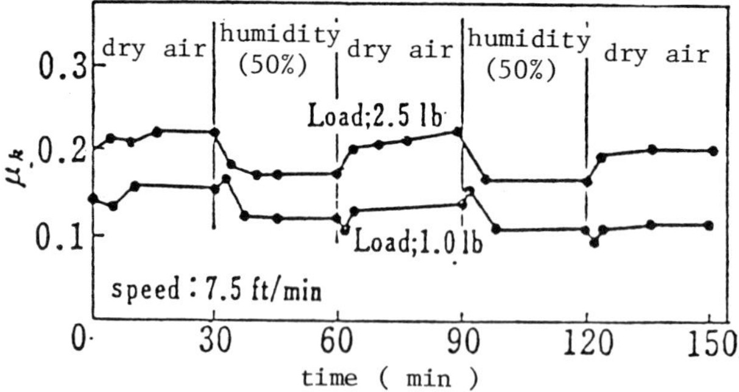

Fig. 1.97 Effect of temperature and humidity on μ_k of PTFE

REFERENCES

1 Bowden, F.P. and D. Tabor, The Friction and Lubrication of Solids, Part II, Oxford, Clarendon Press, 1964, p. 214-241.
2 Tabor, D., Advances in Polymer Friction and Wear, Vol. 5A. Plenum Press, 1974, p. 5-30.
3a Lee, L.H., Advances in Polymer Friction and Wear, Vol. 5A. Plenum Press, 1974, p. 31-68
3b Lee, L.H., Poly. Sci. Tech., 5A (1974) 33.
4 Hertz, H., Reine Angew. Math., 92 (1881) 156.
5 Meyer, E., VDI, 52 (1899) 649.
6 Pascoe, M.W. and D. Tabor, Proc. Phys. Soc., A, 235 (1956) 210.
7 Shooter, S.M. and D. Tabor, Proc. Phys. Soc., B, 65 (1952) 661.
8 Lincoln, B., Brit. J. Appl. Phys., 3 (1952) 260.
9 Yamaguchi, Y., Data of Katasa-Kenkyukai, 32 (1961) 216.
10 Yamaguchi, Y., Mechanical Properties of Plastic Materials, Nikkan Kogyo Press Co., 1961, p. 216.
11 Yamaguchi, Y., I. Sekiguchi, H. Tsuta and H. Suganuma, Preprint of conf. of JSLE, 1967.
12 Yamaguchi, Y., J. Japan Soc. Lub. Eng., 23 (1978) 802-807.
13 Reynolds, O., Phil. Trans. Roy. Soc., 166 (1876) 155.
14 An Example of Lubrication Handbook of JSLE, Yokendo Co., 1970, p. 673-683.
15 Special Edition: Rolling Friction, J. JSLE, 23, 11 (1978)
16 Bowden, F.P, and D. Tabor, The Friction and Lubrication of Solids, Part II, Oxford, Clarendon Press, 1964, p. 275-319.
17 Flom, D.G. and A.M. Bueche, J. Appl. Phys., 30 (11) (1959) 1725.
18 Flom, D.G., J. Appl. Phys., 31(2) (1960) 306.
19 Flom, D.G., J. Appl. Phys., 32(8) (1961) 1426.
20 May, W.D., E.L. Morris and D. Atack, J. Appl. Phys., 30(11) (1959) 1713.
21 Norman, R.H. Brit. J. Appl. Phys., 13 (1962) 358.
22 Takemura T. and K. Minato, Reports Eng. Fac. Kyushu Univ., 38(2) (1965) 165.
23 Minato, K., Y. Miyamaoto and T. Takemura, Reports Eng. Fac. Kyushu Univ., 39(2) (1965) 171.
24 Johnson, K.L., Wear, 9 (1966) 4.
25 Gremer, A., VDI, 97 (1955) 509-515.
26 Tanaka, K. Kobunshi-butzurigaku, Chizin-shokan Co., (1963) p. 289-293.
27 Matsubara, K. in: Y. Yamaguchi (ed.), Point of Selectioning Plastics, JSS, (1976), p. 172-183
28 Yamaguchi, Y. and I. Sekiguchi, J. JSLE, 11 (1966) 485.
29 Yamaguchi, Y. and I. Sekiguchi, Preprint of JSLE, (177).
30 Claus, F.J., Solid Lubricant and Self-Lubricating Solids, Academic Press, 1973, p. 45, 77.

31 Yamaguchi, Y. and I. Sekiguchi, J. Soc. Mat. Sci. Jap., 21 (1972) 228.
32 Yamaguchi, Y. and I. Sekiguchi, J. JSLE, 22 (1977) 58.
33 Yamaguchi, Y. and I. Sekiguchi, J. JSLE, 11 (1966) 486.
34 Materials and Testing, Soc. Mat. Sci. Japan, 1975, p. 23.
35 Kawai, R., Kobunshi, 6 (1957) 348.
36 Yamaguchi, Y., Mechanical Properties of Plastics, Nikkan Kogyo Press, 1967, p. 128.
37 Yamaguchi, Y., JSLE, 11 (1978) 89 (preprint).
38 Sekiguchi, I. and Y. Yamaguchi, JSLE, 5 (1980) (preprint).
39 Yamaguchi, Y. and I. Sekiguchi, J. JSLE, 11 (1979) (preprint).
40 Matsunaga, M. (ed.), Solid Lubricant Handbook, Saiwai Co., 1978, p. 394.
41 Tanaka, K. and Y. Uchiyama, J. JSLE, 19 (1974) 828.
42 Pratt, G.C. in: E.P. Braithwaite (ed.), Lubrication and Lubricant, Elsevier, Amsterdam, 1967, p. 377.
43 Lontz, J.F. and M. C. Kumick, ASLE, Trans 6 (1963) 276.
44 Bely, V.A., V.G. Savkin and A.I. Sviridyonok, Wear, 18 (1971) 11.
45 Tanaka, K. and Y. Uchiyama, Wear, 23 (1973) 153.
46 Shooter, K.V. and P.H. Thomas, Research, 2 (1949) 533.
47 Shooter, K.V. and D. Tabor, Proc. Soc., B65 (1952) 661.
48 Shooter, K.V. and Rabinowicz, Proc. Phys. Soc., B65 (1952) 671.
49 Matsubara, K. Reports of Kikaishikensho, 52 (1963).
50 Sada, T. and Mizuno, M. Report of Kaken, 33 (1957) 54.
51a Tanaka, K., J. JSLE, 12 (1967) 31.
51b Tanaka, K., J. JSLE, 14 (1969) 33.
52 Flom, D.G. and N.T. Porile, J. Appl. Phys., 26 (1955) 1088.
53 Sielgoff, G.L. and M.E. Kucsma, J. Appl. Phys., 34 (1963) 342.
54 Yamaguchi, Y., Y. Oyanagi and I. Sekiguchi, Kagakuin Univ. Report, 22 (1967) 38.
55 Shooter, K.V., Proc. Roy. Soc., A212 (1952).
56 Bower, R.C., W.C. Clinton and W.A. Zisman, Mod. Plastics, Feb. (1954) 131.
57 Matsubara, K., Int. Macro Mol. Chem. Symp., VIII (1966) 115 (preprint).
58 Matsubara, K., Point of Selectioning of Plastics, JSS (1976) 121.
59 Grosh, K.A., Proc. Roy. Soc., A274 (1963) 21.
60 Ludema, K.C. and D. Tabor, Wear 9 (1966) 329.
61 Yamaguchi, Y. and I. Sekiguchi, J. JSLE, 11 (1966) 495.
62 Shooter, K.V., Proc. Roy. Soc., A212 (1952) 488.
63 Lomax, J.Y. and J.T. O'Rourke, Machine Design, 23 (1966) 158.
64 Awaya, J. and R.Kimura, Trans. JSME, 34(260) (1968) 569.
65 Takano, K., T. Kawanishi and T. Okada, Plastics 20(3) (1969) 37.
66 Lewis, R.B., J. Amer. Soc. Lub. Eng., 356 (1969).
67 Yamaguchi, Y. and I. Sekiguchi, Goseijushi, 17(9) (1971) 49.
68 Miyake, S. and Y. Katayama, J. JSLE, 18 (1973) 404.
69a Yamaguchi, Y. and K. Kashiwagi, Poly. Eng. Sci., 22 (1982) 248.

69b Yamaguchi, Y. and K. Kashiwagi, J. JSLE, 19 (1974) 833.
70 Handbook of Mech. Eng., JSME, 1968, p. 11-21.
71 Sieglaff, C.L. and M.E. Kucsma, J. Appl. Phys., 34 (1963) 342.
72 Watanabe, M., Wear, 12 (1968) 185.
73 Ludema, K.C. and D. Tabor, Wear, 9 (1966) 329.
74 Sekiguchi, I. and Y. Yamaguchi, JSLE (1979) (preprint).
75 Watanabe, M. Okayama JSLE (1979) (preprint).
76 Reichen, G.S., Lub. Eng., 20 (1964) 409.

CHAPTER 2

WEAR

2.1 SIGNIFICANCE AND CLASSIFICATION OF WEAR

The OECD (Organization for Economic Cooperation and Development) defines wear as "the progressive loss of substance from the operating surface of a body occurring as a result of relative motion at the surface" [1]. The following phenomena are the causes or results of wear [2]

(1) adhesion and transfer
(2) abrasion or cutting
(3) plastic deformation
(4) fatigue
(5) surface fracture
(6) tearing
(7) melting

In practice, wear is the result of complicated combinations of these phenomena and different types and amounts of wear will occur depending on the various conditions. Accordingly, it is necessary to define the ways in which various types of wear are affected by the contact conditions. A classification may be made as follows:

2.1.1 *Combination of shape and surface conditions*

When the actual contact surfaces are plane, cylindrical or cubic, there are six surface combinations which can be made: plane-plane, plane-cylindrical, plane-cubic, cylindrical-cylindrical, cylindrical-cubic and cubic-cubic.

2.1.2 *Presence of abrasives*

(i) *Adhesive wear or sliding wear*

Wear during relative motion between surfaces without any abrasives in the contact zone is defined as adhesive or sliding wear. It may be broken down into three types, depending on whether there is continuous or discontinuous motion over the surface:

(1) continuous adhesive wear between two surfaces (Fig. 2.1(a))
(2) discontinuous adhesive wear at the contact caused by the opposing surface (Fig. 2.1(b))
(3) discontinuous adhesive wear between the surfaces of both specimens (Fig. 2.1(c))

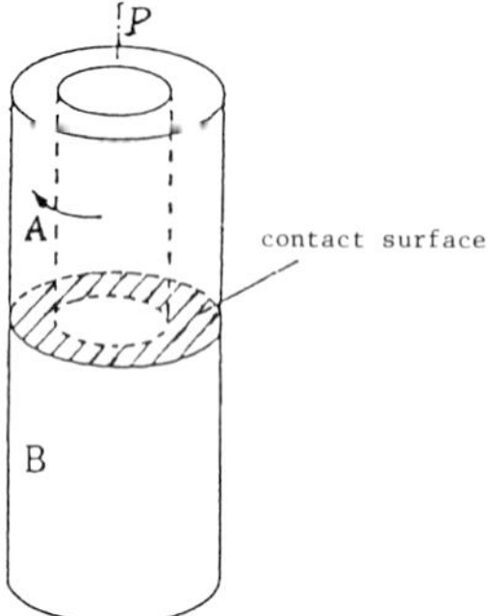

(a) continuous adhesive wear between cylinder end surfaces (conforming contact)

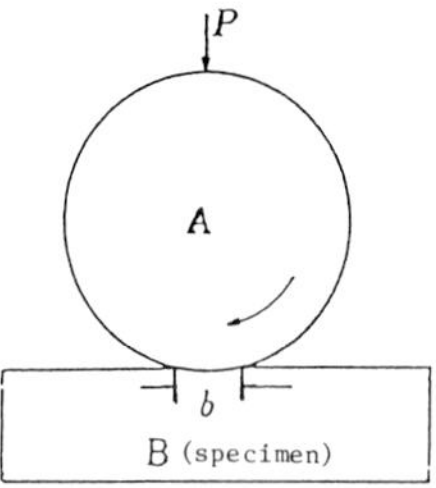

(b) discontinuous wear on a cylindrical counterface (non-conforming contact)

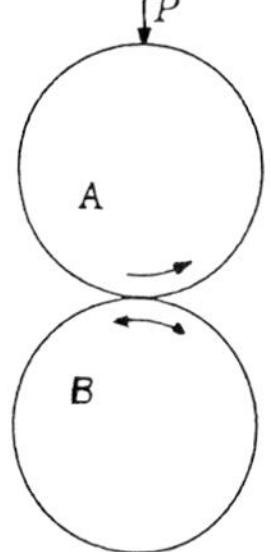

(c) discontinuous adhesive wear between two cylindrical surfaces

Fig. 2.1 Adhesive wear-test configurations

(ii) *Abrasive wear*

Wear during relative motion between contacting surfaces with abrasives is defined as abrasive wear, and it may be classified into the following three types depending on the condition of the abrasives:

(1) loose abrasive wear
(2) bonded abrasive wear
(3) erosive/impact wear

2.1.3 *Relative frictional motion*

The following are the two types of relative motion which cause wear of materials:

(1) unidirectional motion
(2) reciprocating motion

2.1.4 *Presence of a lubricant*

Wear also occurs in materials depending on the lubrication conditions:

(1) wear under unlubricated or dry conditions
(2) wear under lubricated conditions

2.1.5 *Contrasting materials*

Wear of a material can also be classified in terms of the types of opposing surface materials such as metals, plastics, concrete or sand.

In these classifications, adhesive and abrasive wear have different characteristics from those discussed in item 2.1.2 above. Wear is therefore now divided into two broad categories, adhesive wear and abrasive wear, which are discussed in the following section.

2.2 ADHESIVE WEAR CHARACTERISTICS

2.2.1 *Experiments on adhesive wear*

There are many kinds of testing methods for the various types of wear. The two main ones for adhesive wear currently in use depend on the type of friction surface. One method involves adhesive wear between plane surfaces and the other wear between cylindrical surfaces.

(i) *Wear between plane surfaces*

(1) *Adhesive wear between cylinder-end surfaces.* This method is used for studying continuous wear as well as friction, as shown in Fig. 2.1(a).

For metals, the test is known as the Suzuki type whilst for cast irons it is the Ito type [4].

(2) *The pin on disc arrangement.* This method examines adhesive wear caused by frictional motion between the surface of a large circular disc and the end of a pin, such as a small cylinder or a cone, as shown in Fig. 2.2. In general, the specimen of interest is used as the pin. The method is similar to that reported in ASTM-D-2716 and to those used by Tanaka et al. [3], Mizuno [4], an international cooperative programme [5], Matsunaga and Tsuya [6] and Amsler [7].

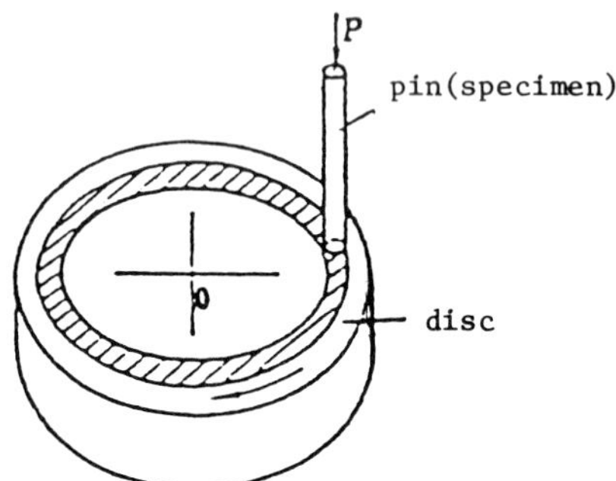

Fig. 2.2 Adhesive wear of pin on disc

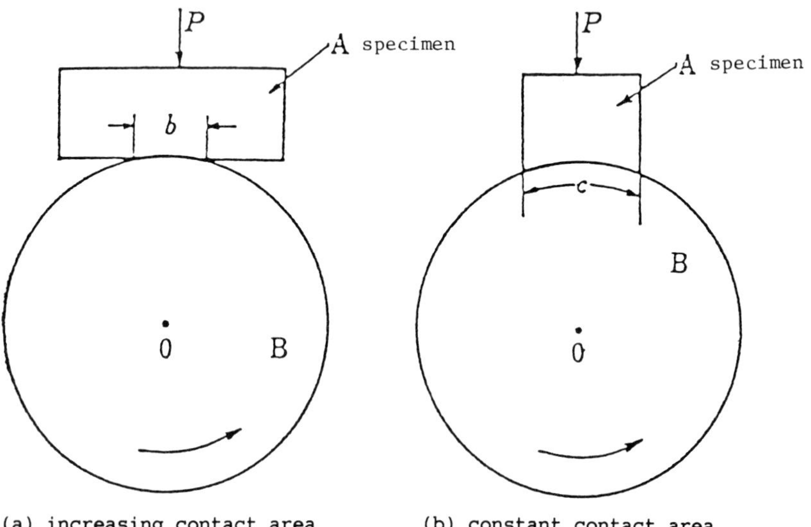

Fig. 2.3 Adhesive wear against a single cylindrical surface

(ii) *Wear with cylindrical surfaces*

Adhesive wear testing methods of this type are divided into two types; one with a single cylindrical element and the other with two cylindrical elements.

(1) *Wear on a single cylinder.* This test is carried out between a block specimen and an opposing cylindrical surface, as shown in Fig. 2.3, and the method is generally similar to that in Fig. 2.1(b). The surface of the block is plane and the contact is initially a line with the cylinder. However, as wear progresses, the contact surface on the block becomes curved and increases in area. This can therefore be referred to as an "increasing contact area" type of wear. The method is used with an "Ogoshi type, rapid wear-testing machine" [4], in which the contact load P is increased as wear progresses in order to maintain a constant contact pressure p. Figure 2.3(b) shows an adhesive wear test method between a block specimen A with a concave surface which conforms to that of the opposing cylinder B. In this case, the contact surface area, and pressure under constant load, remain constant irrespective of the progress of wear. It is important, however, that the two cylindrical surfaces should conform very closely to each other. This type of testing is generally carried out with an Amsler apparatus [7].

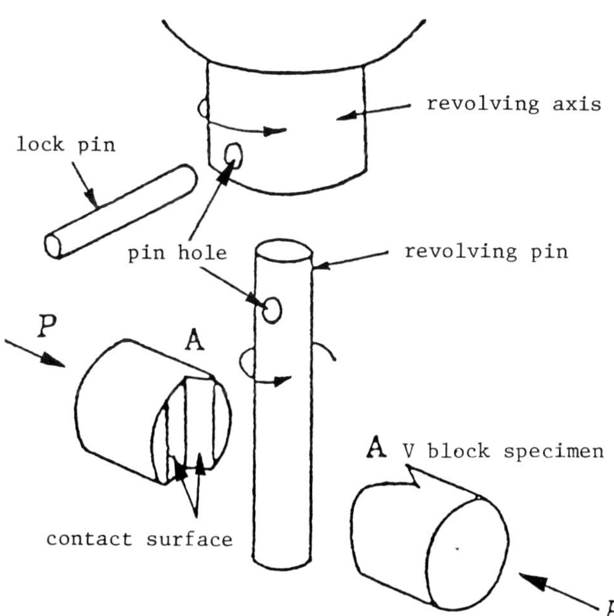

Fig. 2.4 Adhesive wear using a double V-block arrangement ("Falex")

(2) *Wear with two cylindrical surfaces.* As Fig. 2.4 shows, this type of wear test is carried out using block specimens V on both sides of a revolving pin under a pressure at right-angles to the pin axes. This method is known as a "Falex-type" or "Vee Block" test.

2.2.2 *Characteristics of adhesive wear*

(i) *Theoretical considerations*

Since adhesive wear results from a combination of all the various phenomena described above, theoretical considerations become complex, and it is therefore difficult to derive one simple equation which is applicable to all the various conditions. Some theoretical equations previously proposed are as follows:

(1) *Holm and Rhee.* Holm and Lewis [9] have proposed the following equation for adhesive wear between a metal surface and a plastic

$$V = KPvt \tag{2.1}$$

where V is the wear volume, P is the load, v is the sliding velocity, t is the duration of sliding, and K is known as a wear factor.

Rhee [10] suggested a more rigorous equation than that of (2.1)

$$\Delta W = KP^a v^b \cdot t^c \tag{2.2}$$

where ΔW is the weight loss by wear, a, b and c are parameters and K is a wear factor similar to that of equation (2.1). In one experimental result for the wear of plastics composites comprising 60% asbestos, 10% of inorganic material and 30% resin sliding against Cr cast iron, the values of a, b and c were all found to be equal to 1.6 and the value of K was 3.9×10^{-11}. The tests were made at 100 psi, 300 rpm and μ_k was equal to 0.35.

(2) *Peterson's adhesive wear theory.* Peterson et al. [2] have deduced an equation for adhesive wear based on an adhesion-shear-transfer theory as follows: If V is the total adhesive wear volume, a_r is the radius of contact area of an asperity, h is the thickness of the material being sheared, ℓ is the sliding distance to produce a transferred fragment, P is the normal load, H_i is the indentation hardness, n is the number of contact points, L is the total sliding distance, and v is the wear volume per asperity encounter, then

$$v = \pi a_r^2 h$$

therefore $\dfrac{v}{\ell} = \pi\, a_r^2\, \dfrac{h}{\ell}$

$$\text{and } \frac{V}{L} = n\,\pi\,a_r^2\,\frac{h}{\ell}$$

$$\text{but } H_i = \frac{P}{A_r} = \frac{P}{n\pi a_r^2} \text{ or } n\,\pi\,a_r^2 = \frac{P}{H_i}$$

$$\text{therefore } \frac{V}{L} = \frac{P}{H_i} \cdot \frac{h}{\ell} \tag{2.4}$$

and h/ℓ is 1/3 to 1/6. Therefore, if K is the wear factor and h/ℓ is 1/3, V is given by

$$V = K\,\frac{PL}{3H_i} \tag{2.5}$$

(3) *Jain's theory.* Jain and Bahadur [11] have deduced the following equation for adhesive wear when also considering fatigue aspects:

$$V = K_2 \cdot \frac{PL}{S_o} \tag{2.6}$$

where V is the wear volume, P is the normal load, L is the sliding distance, S_o is the fatigue stress related to the failure stress during a single stress cycle, and K_2 is an adhesive wear factor which is a complex function of Poisson ratio, the coefficient of friction, the elastic modulus, the shape of the asperity and the fatigue characteristics. Since most of these parameters are affected by temperature, the value of K_2 must be obtained from actual wear experiments.

(ii) *Discussion based upon experimentation*

Experimental research in the field of adhesive wear of plastic materials has been conducted for the past thirty years as follows: Kawamoto et al. [12] on phenolics; Solda et al. [13], Matsubara [14,15], Awaya et al. [16], Takeuchi et al. [17] and Clauss [18] on nylon; Matsubara [14], Takeuchi et al. [19] and Clauss [18] on polytetrafluoroethylene; Matsubara [14], Awaya et al. [16] on polyethylene; Bongiovanni [20] on polypropylene; Yamaguchi et al. on PVC [21], polyphenylenesulfide (PPS) [22], U polymer (polyarylate), polysulfone [23], polyacetal (POM) [24], Matsubara et al. [25] and Yamaguchi et al. [26] on polyimide; and Yamaguchi et al. for epoxy resin [27], unsaturated polyester [27] and diallylphthalate resin [28].

These reports have presented the results of experiments carried out individually on each plastic. It is difficult to determine the general adhesive wear properties of plastics from these data because the testing methods, conditions and scope of each experiment were different. The author has, therefore, carried out adhesive wear tests on three different plastics using three different, geometrical arrangements, as listed below.

(1) Continuous contact between two plane surface ("thrust-washers")

(2) Discontinuous contact between a large rotating cylinder and a smaller one ("pin-ring") with a constant contact area

(3) As (2) but where the contact area increases as a consequence of wear

The following sections compare the results of these tests with the theoretical considerations mentioned above.

(1) *Wear in a "thrust-washer" arrangement.* Fig. 2.5 shows the apparatus used for this type of wear testing. The test specimen was a hollow cylinder, as shown in Fig. 2.6, and adhesive wear was generated on the plane end surface of this cylinder during unidirectional rotation under an axial normal load P. Experiments were made with three types of plastics: cloth-filled phenolic (PF), Nylon 6 and polycarbonate, as described in Table 2.1. In these experiments, the specimen on top (the counterface) was a hollow steel cylinder (S50C) which rotated against the static, plastic cylinder. Each plastic was tested under normal loads and speeds.

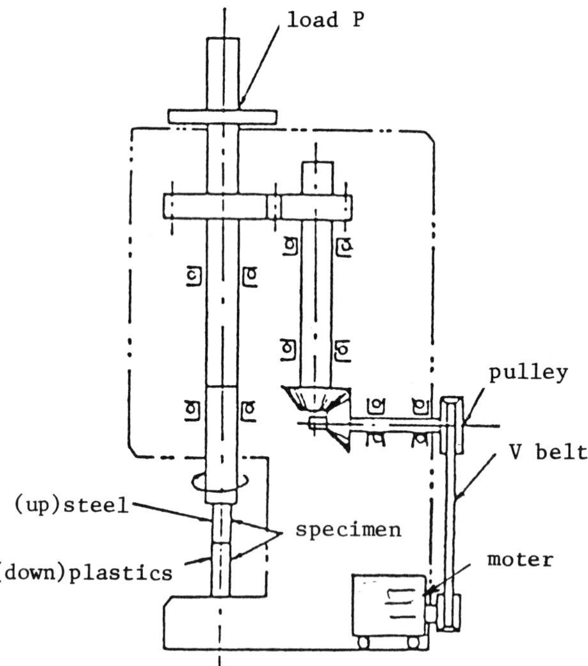

Fig. 2.5 Test apparatus for continuous adhesive wear between cylinder end surfaces

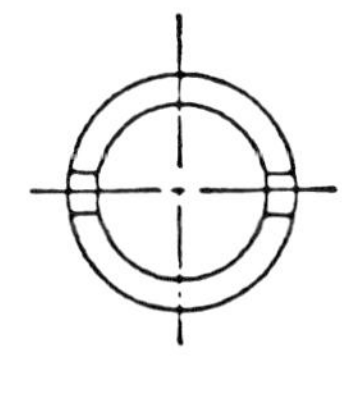

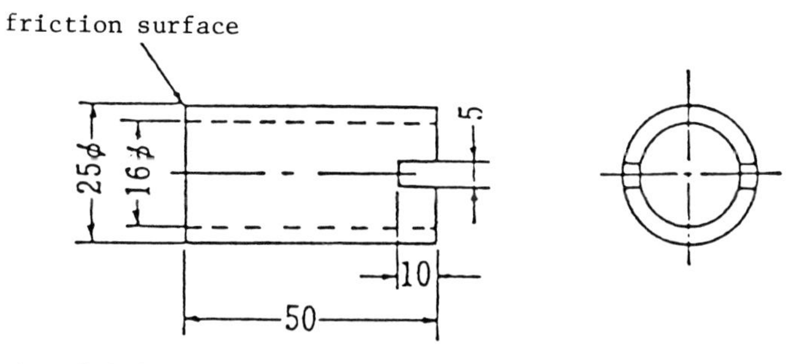

Fig. 2.6 Shape and dimensions of a specimen for continuous adhesive wear testing

TABLE 2.1
Specimen description

Materials	Producer and Trade Name	
Cotton-filled phenolic	Sumitomo - Bakelite	Laminated
Nylon 6	Toray - CM 1031	Extruded
Polycarbonate	Teijin - Panlite	Extruded

Counterface Material: Steel (S50C)

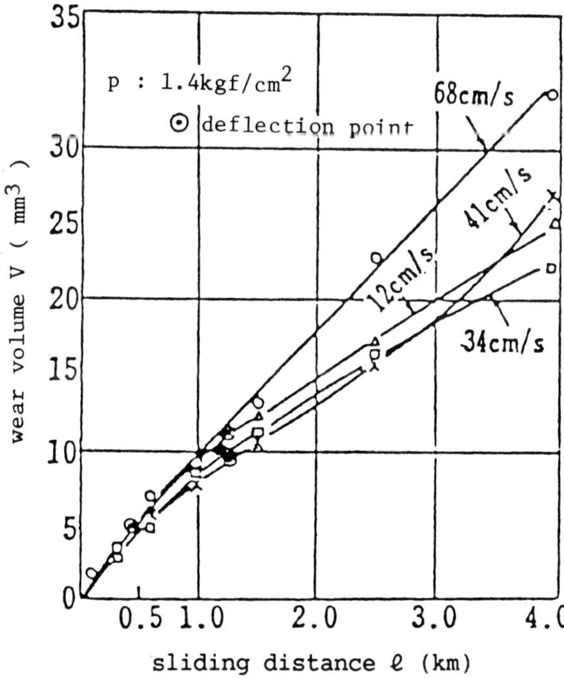

Fig. 2.7 Relationships between adhesive wear volume and sliding distance for cotton cloth-filled phenolics (continuous wear)

The wear volume per unit sliding distance (V/ℓ) may be referred to as the wear rate. From this relationship, the values of the initial wear rate W_1 corresponding to k_1 in the initial wear region, the steady-state wear rate W_2 corresponding to K_1' in the steady-state region and the ratio W_1/W_2 at a sliding distance of 4 km were obtained at various velocities and are given in Table 2.2. The values of W_1/W_2 are generally greater than 1. Figure 2.11 shows the relationships between the distance of sliding at the deflection point and velocity for each plastic.

TABLE 2.2
Adhesive wear rate (mm^3/km) and deflection point (sliding distance, km)

Sliding velocity (cm/s)		Nylon 6	Poly-carbonate	Phenolics
12	initial wear rate, W_1	5.0	11.3	13.7
	steady-state wear rate, W_2	2.9	2.7	2.9
	W_1/W_2	1.7	4.2	4.7
	sliding distance at deflection point	0.6	1.4	1.3
34	initial wear rate, W_1	3.7	9.0	9.7
	steady-state wear rate, W_2	1.8	2.7	4.3
	W_1/W_2	2.0	3.3	2.3
	sliding distance at deflection point	2.5	1.6	1.2
41	initial wear rate, W_1	4.3	7.0	11.1
	steady-state wear rate, W_2	1.2	3.0	7.5
	W_1/W_2	3.6	2.3	1.5
	sliding distance at deflection point	1.8	1.5	1.3
68	initial wear rate, W_1	9.3	11.0	11.1
	steady-state wear rate, W_2	0.73	2.5	7.4
	W_1/W_2	12.7	4.2	1.5
	sliding distance at deflection point	0.8	0.6	0.4

surface contact pressure: 1.4 kgf/cm^2; total sliding distance: 4 km

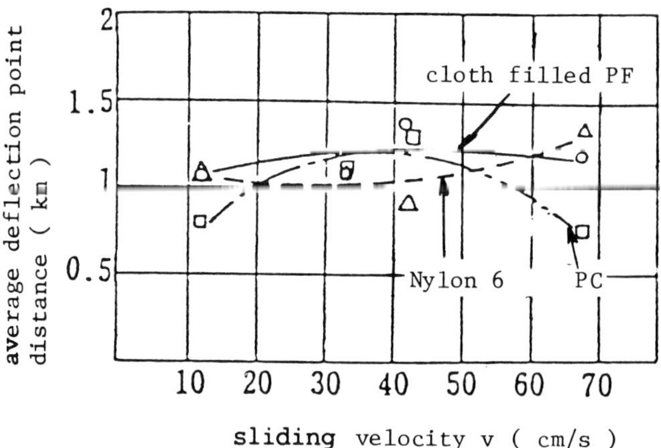

Fig. 2.11 Relationship between average distance to the deflection point and velocity for various plastics

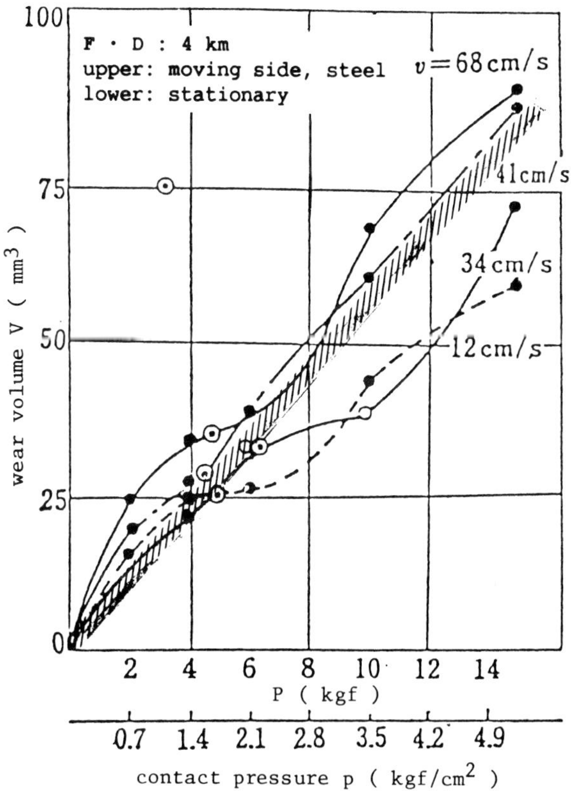

Fig. 2.12 Relationships between V and p for cloth-filled phenolics at different speeds

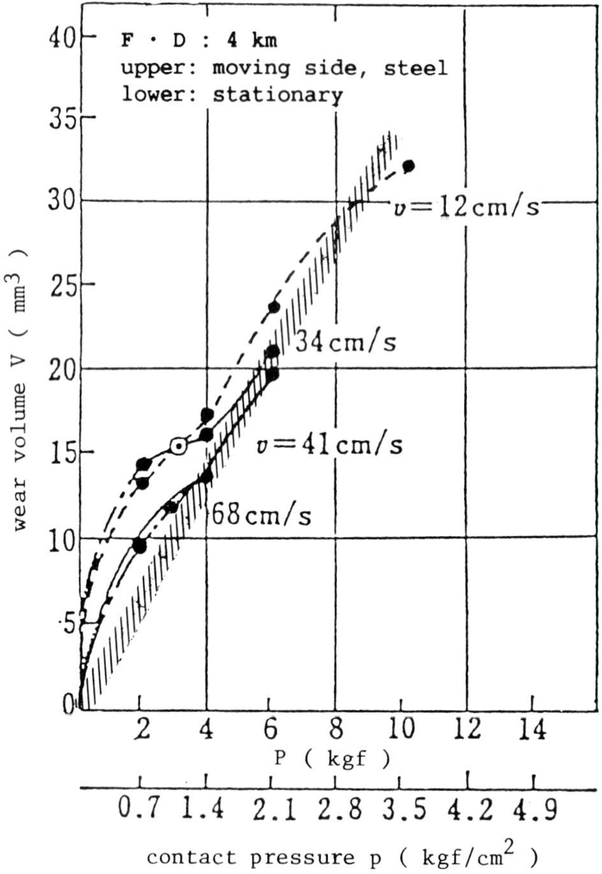

Fig. 2.13 Relationships between V and p for PC at different speeds

(b) Relationship between V and p

Figures 2.12, 2.13 and 2.14 show the relationships between the adhesive wear volume V and surface contact pressure p at a sliding distance of 4 km and at velocities from 12 to 68 cm/s for each of three plastics. It is clear from the curves in these figures that V is almost directly proportional to p over a wide range, although there is a complex curvilinear relationship when p is less than about 1.4 kgf/cm². In general, however, the following equation is applicable

$$V = k_2 \cdot p \qquad (2.8)$$

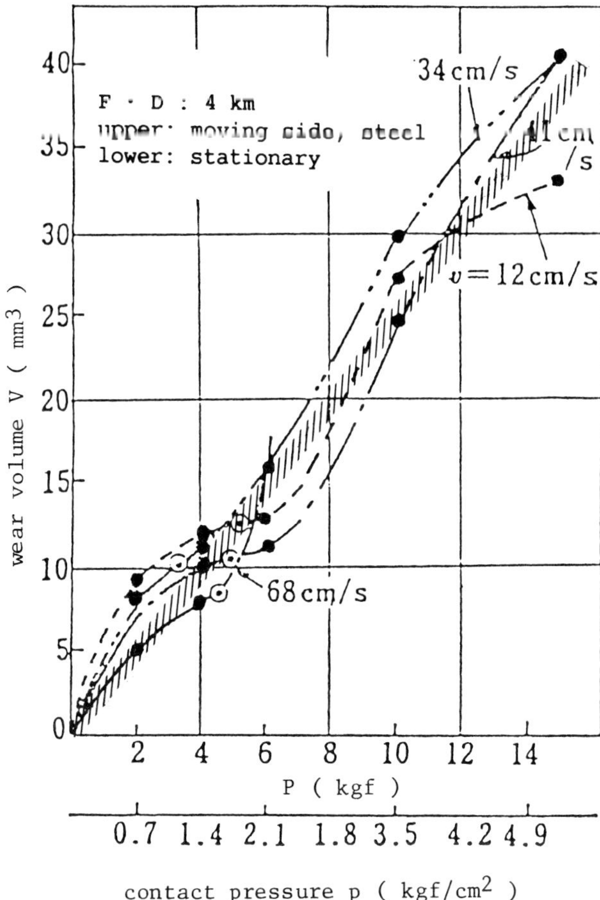

Fig. 2.14 Relationships between adhesive wear volume and contact pressure p for Nylon 6 at various speeds

(c) Relationship between V and v

Figures 2.15, 2.16 and 2.17 show the relationship between wear volume V, or wear rate W (mm^3/km), and velocity v at sliding distances up to 4 km for each of three plastics. It can be seen from the curves that V, or W, either increases or decreases with an increase in v depending on the type of plastic and the magnitude of the contact pressure. That is, the value of the index b in equation (2.2) becomes positive or negative according to the testing conditions for a particular plastic. Accordingly, it is not possible to conclude simply that V is proportional to sliding velocity, as shown by equation (2.2).

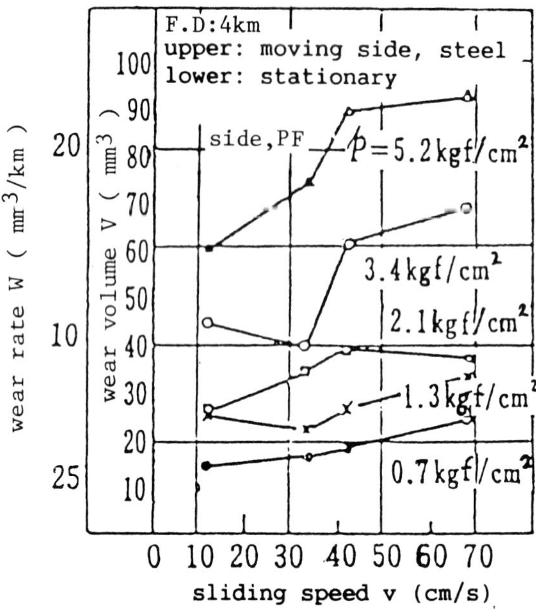

Fig. 2.15 Relationships between adhesive wear rate W or volume V and speed v for cotton-filled phenolics

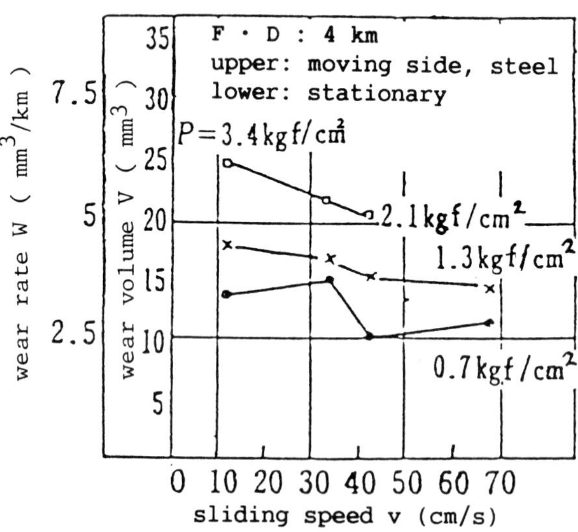

Fig. 2.16 Relationships between adhesive wear volume V or wear rate W and speed v for PC

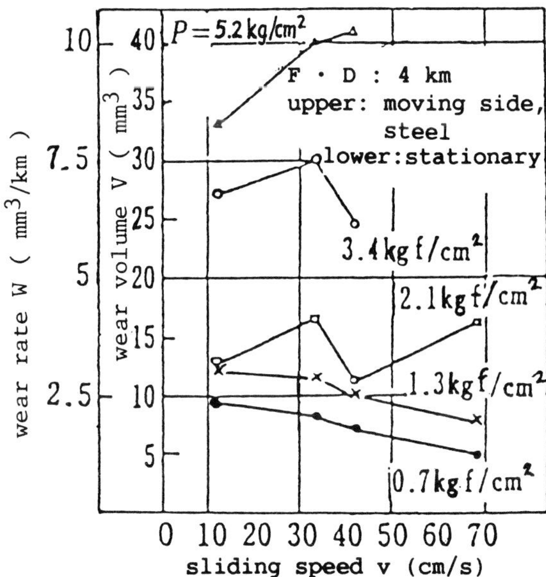

Fig. 2.17 Relationships between adhesive wear volume V or wear rate W and speed v for Nylon 6

(d) Synthesis

From the data in sections (a), (b) and (c) above, which conform to equations (2.7) and (2.8), we can replace the constants K_1 and K_2 in these equations by a general constant, K, and obtain the following equation

$$V = K \cdot p \cdot \ell \qquad (2.9)$$

Since the adhesive wear volume V is directly proportional to the contact pressure p, the contact surface area A and the sliding distance ℓ, the adhesive wear volume per unit distance km, unit contact pressure kgf/cm^2 and surface area cm^2 may be defined as a "specific adhesive wear rate" V_s. As pA = P, then

$$V_s = \frac{V}{p \cdot A \cdot \ell} = \frac{V}{P \cdot \ell} \qquad (2.10)$$

The units of V_s thus become mm^3/kgf·km,

$$V_s = K/A \qquad (2.11)$$

therefore $V = V_s P \ell$ (2.12)

The previous data can now be presented in terms of the specific adhesive wear rate V_s. Individual values of the initial wear rate V_{s1} and steady-state wear rate V_{s2} for each of three plastics located in the bottom position against steel in the top position at various velocities are shown in Table 2.3.

TABLE 2.3
Initial and steady-state specific adhesive wear rate V_s (mm^3/kgf, km)

Velocity	a		b		c		d		e		f	
(cm/sec)	V_{s1}	V_{s2}	$V_{s1'}$	$V_{s2'}$	V_{s1}	V_{s2}	$V_{s1'}$	$V_{s2'}$	V_{s1}	V_{s2}	$V_{s1'}$	$V_{s2'}$
12	0.9	0.5	2.4	1.4	6.0	3.7	2.2	1.1	2.2	0.4	1.5	0.6
34	0.8	0.7	2.1	1.6	12.0	5.7	1.6	1.1	1.2	0.3	0.8	0.7
41	1.3	1.0	2.0	1.0	5.1	2.5	1.9	0.9	1.3	1.0	1.7	0.6
68	2.0	0.6	3.0	1.7	2.0	1.8	-	-	1.1	0.6	2.1	0.5

Top/Bottom: a - phenolics/steel; b - steel/phenolics; c - PC/steel; d - steel/PC; e - Nylon 6/steel; f - steel/Nylon 6

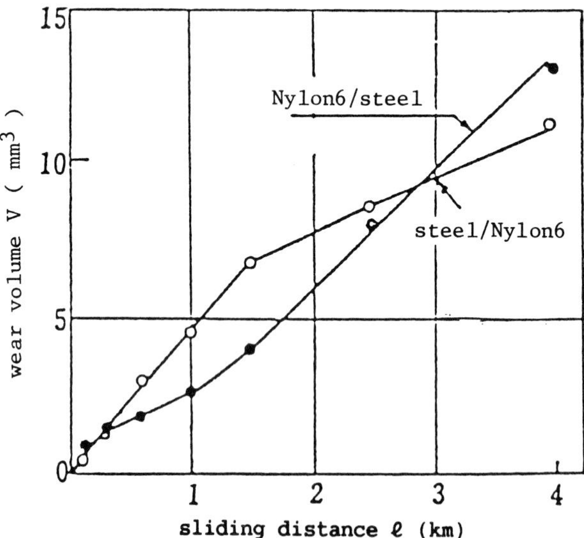

Fig. 2.18 Relationships between adhesive wear volume and sliding distance for Nylon 6 against steel (v: 4.1 cm/s; p: 1.3 kgf/cm^2)[31] (continuous wear)

(e) Specimens located in the inverse position

Figure 2.18 shows a comparison of the relationships between wear volume V and sliding distance ℓ for specimens with the rotating plastic in the top position and the stationary steel counterface in the bottom position, and vice versa. These relationships are similar to those in Figs. 2.7, 2.8 and 2.9. However, in the relationships between V and v for phenolics (top) and steel (bottom), Fig. 2.19 shows that V reaches a maximum value at a velocity of about 45 cm/s.

The values of the initial specific adhesive wear rate V_{s1}, and the steady-state wear rate V_{s2} at various velocities for the inverse arrangement (plastic - top, steel - bottom) are also shown in Table 2.3.

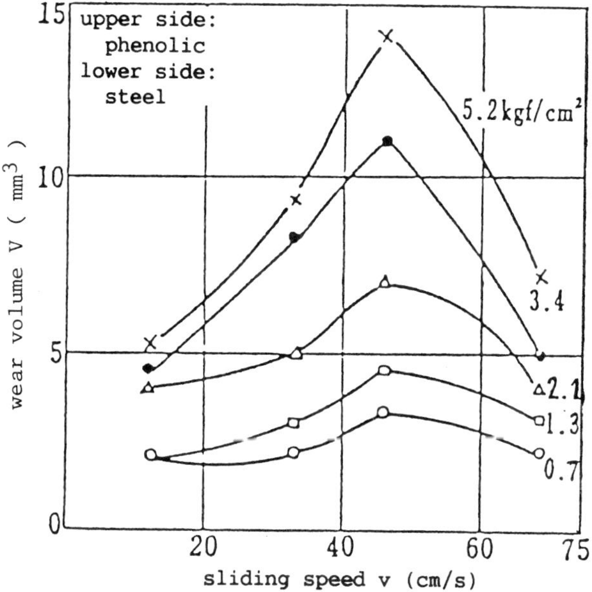

Fig. 2.19 Relationships between adhesive wear volume and speed for phenolics against steel (L: 1.0 km)[31] (continuous wear)

(2) *Wear in a pin-ring arrangement - constant contact area.* As Fig. 2.20 shows, when sliding occurs between a plastic specimen cylinder conforming to the curved surface of a rotating steel disc, the disc is in discontinuous contact with the plastic and the contact area and pressure may be held constant. The adhesive wear characteristic of this arrangement was examined experimentally as follows. Cotton cloth filled phenolic (PF), polycarbonate (PC) and polyacetal (POM) were tested against a counterface disc of steel (S50C). Table 2.4 presents a summary of the materials.

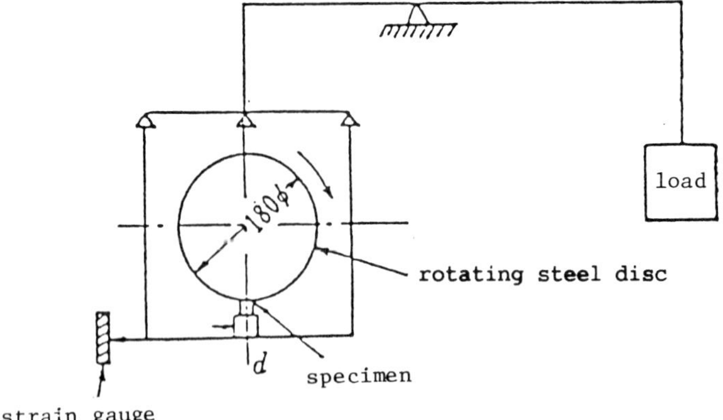

Fig. 2.20 Apparatus for discontinuous adhesive wear testing/non-conforming contact with a constant area of the cylinder end surface

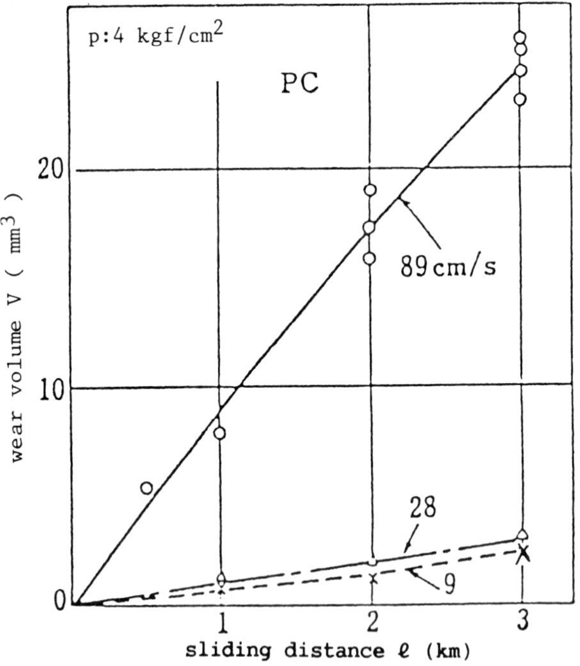

Fig. 2.21 Relationships between wear volume and sliding distance at various speeds for PC (discontinuous adhesive wear, pin/ring)

TABLE 2.4
Specimens

Materials	Manufacturer/Trade Name	Moulding Method
Thermosetting Plastics		
Cotton-filled phenolic	Sumitomo Bakelite	Extruded
Thermoplastics; Semicrystalline		
Polycarbonate	Teijin; Panlite	Extruded
Polyacetal	Dupont; Delrin	Extruded

(a) Relationship between V and ℓ

An example of the relationship between adhesive wear volume V and sliding distance ℓ is shown in Fig. 2.21 for polycarbonate (PC) at 4 kgf/cm² contact pressure at sliding velocities of 9, 18 and 89 cm/s. It appears from the curves that V is almost directly proportional to ℓ, $V = k_1\ell$, and it is difficult to detect any deflection point.

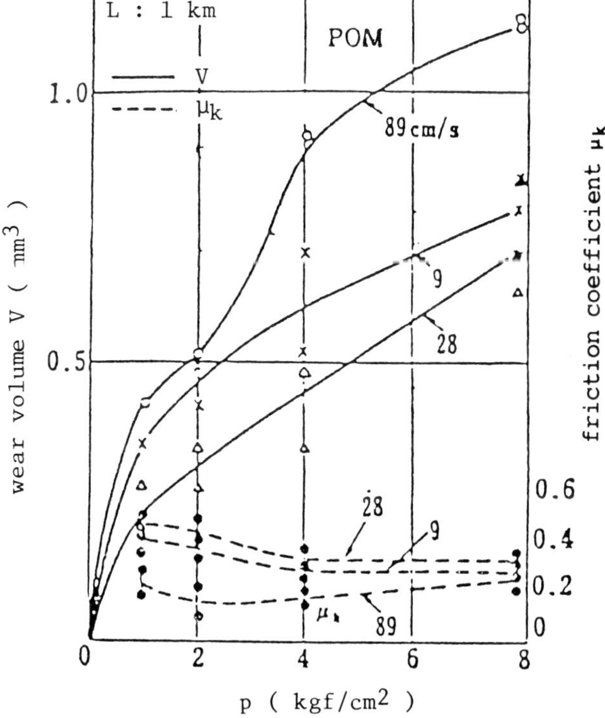

Fig. 2.22 Relationships between wear volume or friction coefficient and contact pressure for polyacetal (POM) at various speeds (discontinuous adhesive wear, pin/ring)

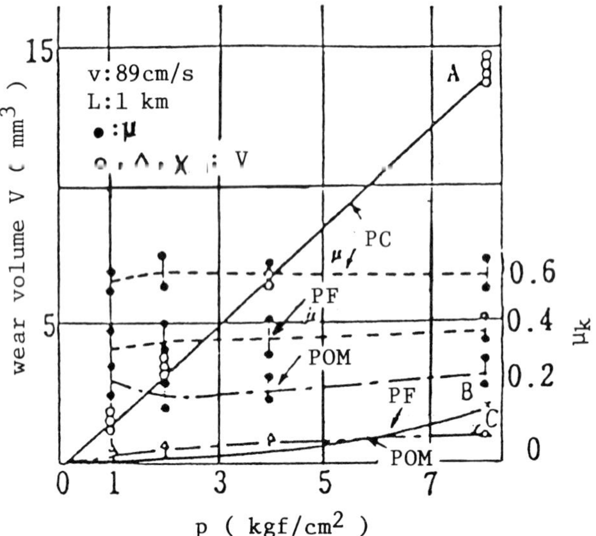

Fig. 2.23 Relationships between wear volume or friction coefficient and contact pressure p for three types of plastics (discontinuous adhesive wear, pin/ring)

(b) Relationship between V and p

Figure 2.22 shows an example of the relationship between adhesive wear volume V and contact pressure p, for POM at up to 1 km sliding distance at various velocities. It can be seen that V is almost directly proportional to p, $V = k_2P$, similar to the relationship found during continuous adhesive wear (thrust-washer arrangement) but with some scatter. In the figure, the dotted lines show the way in which the friction coefficients μ_k vary with p. Figure 2.23 shows the V-p relationships at 85 cm/s velocity for each of three plastics (curves A, B and C), and these results also show an approximate direct proportionality.

(c) Relationship between V and v

Figures 2.24 and 2.25 show the relationships between the adhesive wear volume V at up to 1 km sliding distance and velocity v for POM and PC at contact pressures of 1, 2, 4 and 8 kgf/cm². It appears that there is no correlation between these relationships and those found during continuous adhesive wear (thrust-washer).

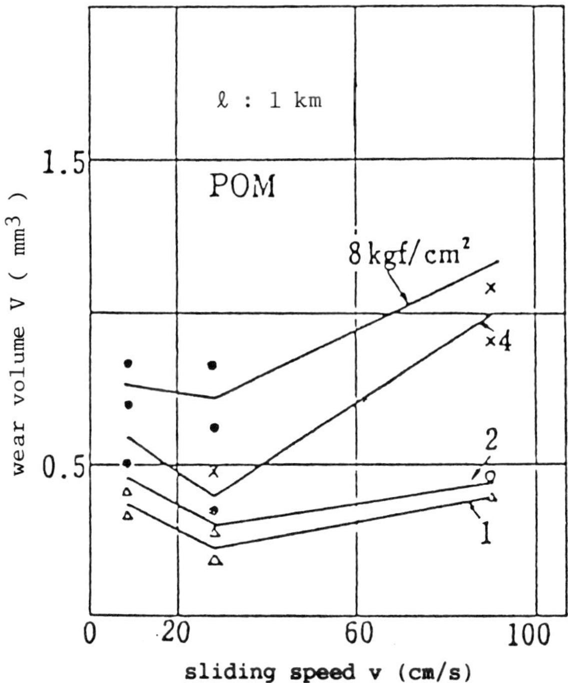

Fig. 2.24 Relationships between adhesive wear volume and speed at various contact pressures p for POM (discontinuous adhesive wear, pin/ring)

(d) Comparison with continuous adhesive wear

The solid lines in Fig. 2.26 show the relationships between the discontinuous specific adhesive wear rate V_s' and velocity v for each of the three plastics, whilst the dotted lines show those between the initial, continuous, specific adhesive wear rate V_{s1} and v for PF and PC. It is clear from this figure that V_{s1} is much larger than V_s', and the ratio of V_{s1} to V_s' is about 1.8 for PF and about 5 for PC.

(3) *Wear in a plate-ring arrangement - contact area increasing.* This section discusses discontinuous adhesive wear between cylindrical surfaces where the contact area increases. Experiments were made by Yamaguchi et al. [33] using a rapid wear-testing machine with a plastic plate in contact with a rotating steel disc, shown in Fig. 2.3(a). Four plastics were used, as listed in Table 2.5.

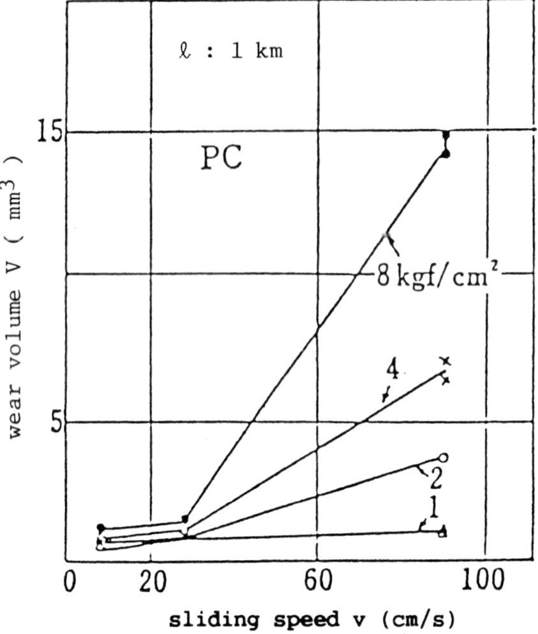

Fig. 2.25 Relationships between adhesive wear and speed at various contact pressures for PC (discontinuous adhesive wear, pin/ring)

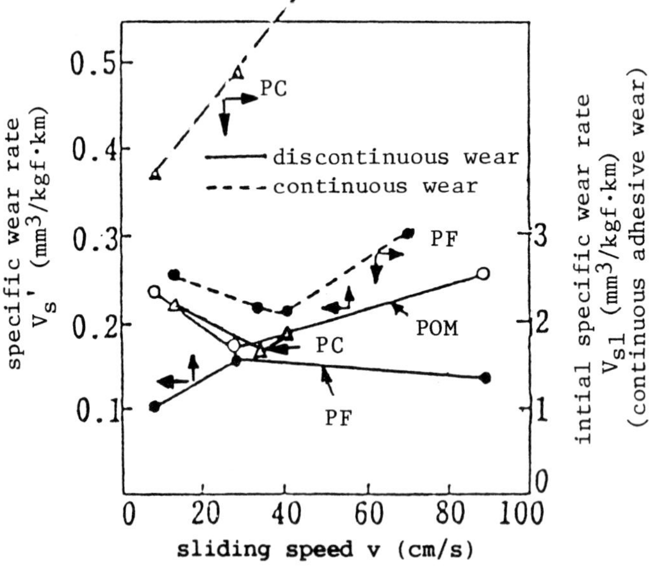

Fig. 2.26 Variation with speed of the specific adhesive wear rates for various plastics against steel in continuous (conforming) contact V_{s1} or discontinuous contact (plate/ring), V_s'

TABLE 2.5
Types of specimens

Material	Manufacturer/ Trade Name	Specimen Preparation
Cotton-filled phenolic (PF)	Sumitomo; Sumikon PM	Machined
Polycarbonate (PC)	Teijin; Panlite	from
ABS resin	Toray; Toyo Luck 600	extruded
Polyacetal	Dupont; Delrin	circular bars

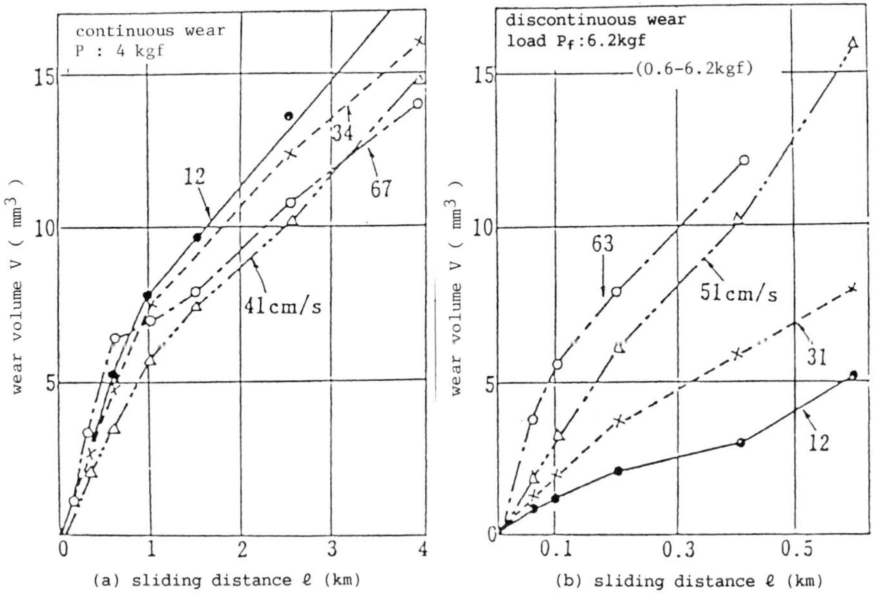

Fig. 2.27 Wear volume-distance relationships for PC at various speeds (a) continuous wear (conforming contact), (b) discontinuous wear (pin/ring)

(a) Relationship between V and ℓ

Figures 2.27 and 2.28 show the relationships for PC and POM between adhesive wear volume V and sliding distance ℓ for (a) continuous wear (thrust washers), and (b) discontinuous wear between a steel disc and the plastic plates at constant contact pressure and velocity. It appears from these figures that each curve has a deflection point, a, which divides the initial adhesive wear region from the steady-state region, as already shown by curve A in Fig. 2.10. In the steady-state region, the results conform to the equation $V = k_1\ell$.

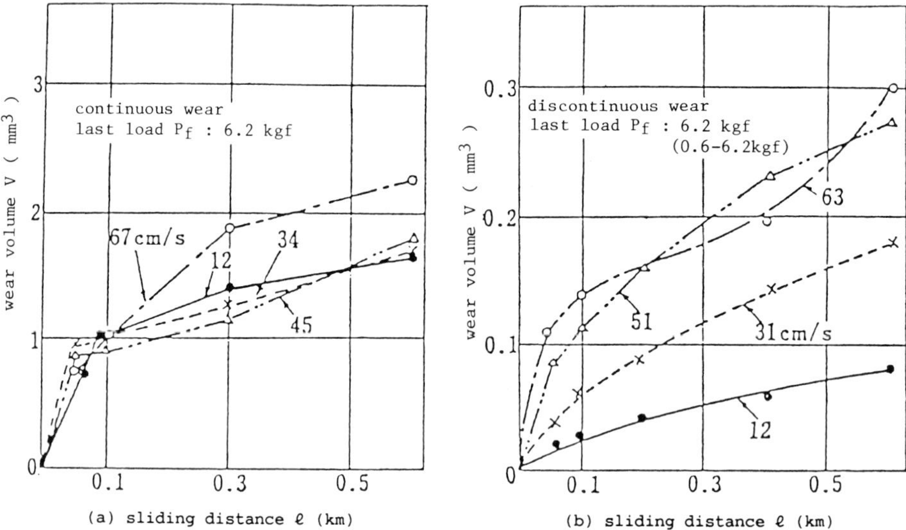

Fig. 2.28 Wear volume-distance relationships for POM at various speeds (a) continuous wear (conforming contact), (b) discontinuous wear (plate/ring)

(b) Relationship between V and p

Using PC and ABS resin as examples, Figs. 2.29 and 2.30 show the relationship between the adhesive wear volume V at up to 0.3 km sliding distance, and the final load P_f at various velocities. It is clear from the curves that V is almost directly proportional to P_f, $V = k_2p_f$, similar to the result found during continuous adhesive wear.

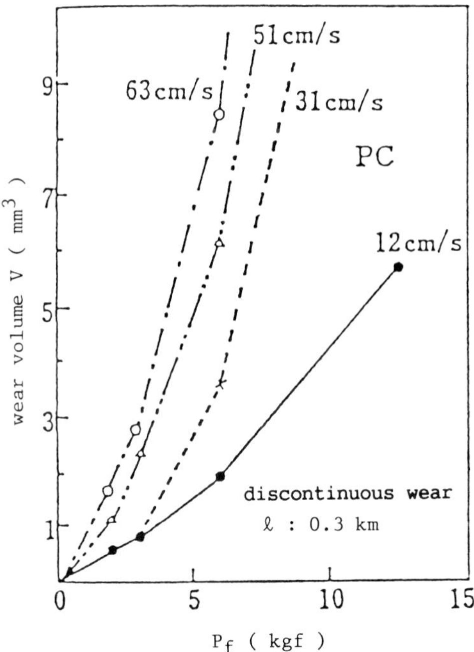

Fig. 2.29 Relationships between the adhesive wear volume and load P_f for PC at various speeds (discontinuous wear, plate/ring)

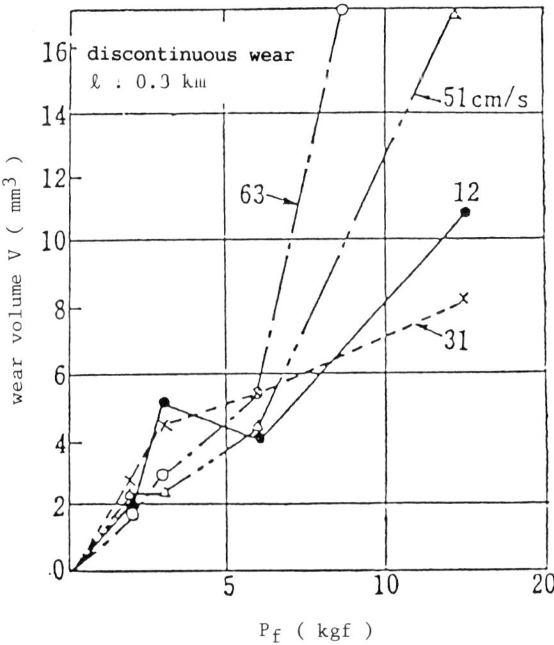

Fig. 2.30 Relationships between the discontinuous adhesive wear volume and last load P_f at various speeds for ABS resin

(c) Relationship between V and v

For ABS resin, Fig. 2.31(b) shows the variations of adhesive wear volume V with velocity v at a sliding distance up to 0.2 km at different final loads, Pf. Figure 2.31(a) shows the corresponding variations for the case of continuous sliding between cylinder end surfaces (thrust-washers). There seem to be no well-defined relationships between V and v.

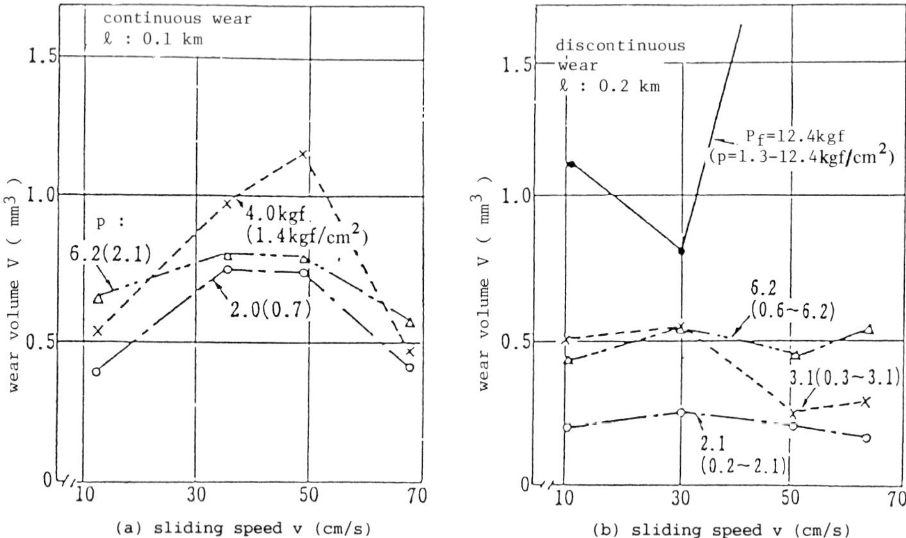

Fig. 2.31 Wear volume-speed relationships for ABS resin at various pressures (a) continuous wear (conforming contact); (b) discontinuous wear (plate/ring)

(d) Comparisons of specific adhesive wear rates

Figure 2.32 shows the relationship between specific adhesive wear rate, V'_s, during discontinuous sliding (plate-ring) and that for continuous sliding, V_s, between cylinder-end surfaces for four kinds of plastic. It is clear from the figure that V_s is almost directly proportional to V'_s, and the value of V_s is almost two to seven times, or an average of 4.5 times, that of V'_s.

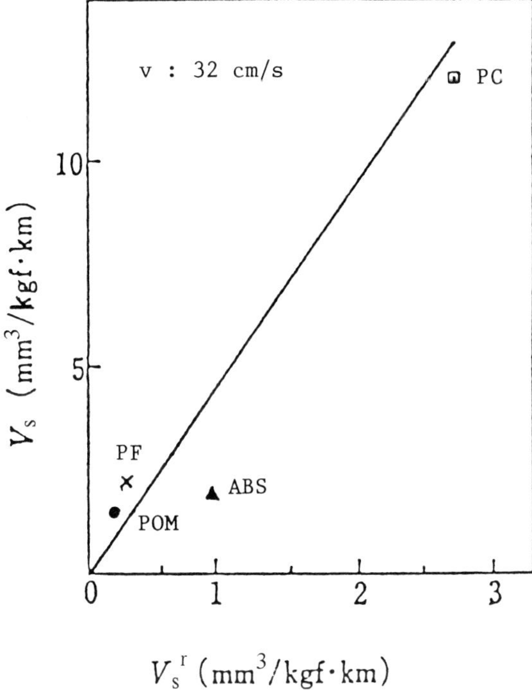

Fig. 2.32 Relationship between V_s (continuous specific adhesive wear rate) and V_s'

2.3 ABRASIVE WEAR CHARACTERISTICS

2.3.1 *Significance and classification*

According to the OECD (Organization for Economic Cooperation and Development), abrasion or abrasive wear is defined as "wear by displacement of material caused by hard particles or hard protuberances". Abrasive wear may be classified into the following three types, depending upon the form of abrasives or protuberances.

(i) *Abrasion by loose abrasive.* In this type of wear, loose abrasives displace material from the surface of the specimen. Examples of relevant testing methods for this type of wear are outlined in ASTM-D-1242A and JIS-K-7205.

(ii) *Abrasion by bonded abrasives.* Wear by bonded abrasives, such as a grindstone or emery paper, can be characterised by tests using the methods outlined in ASTM-D-1242 (B-type) and JIS-K-7204 (Taber abrader), for example.

(iii) *Erosion.* In this type of wear, material is displaced from the surface by the impact of loose abrasives against the surface and can be studied using the testing methods of ASTM-D-673 or an Elmendorf wear testing machine.

2.3.2 *Theories*

There are two analytical approaches to abrasive wear. The first is a mechanical one and the second a morphological one for scratched grooves.

As an example of the first method, Lancaster [35] has described original Russian work by Ratner in which an equation for abrasive wear was deduced on the basis that the work done to detach material must be directly proportional to the break strength σ_o of the material and the strain ε at the breaking point, taking into account the hardness of the material H and the friction coefficient μ_k as follows

$$V_{sa} = C \cdot \frac{\mu_k}{H\sigma_o\varepsilon} \tag{2.13}$$

where V_{sa} is the specific abrasive wear rate and C is a constant depending on the type of abrasion and the material. He attempted to verify this equation using the experimental results shown in Fig. 2.33 which show that V_{sa} is directly proportional to $1/\sigma_o\varepsilon$ for some plastics.

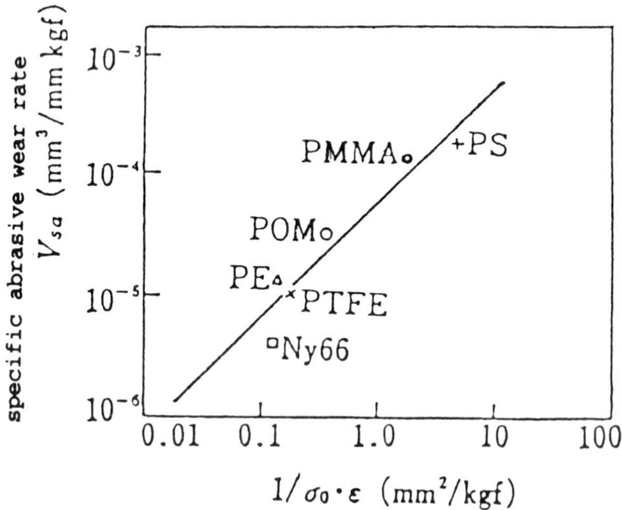

Fig. 2.33 Relationship between the specific abrasive wear rate V_{sa} and $1/\sigma_o\varepsilon$[35]

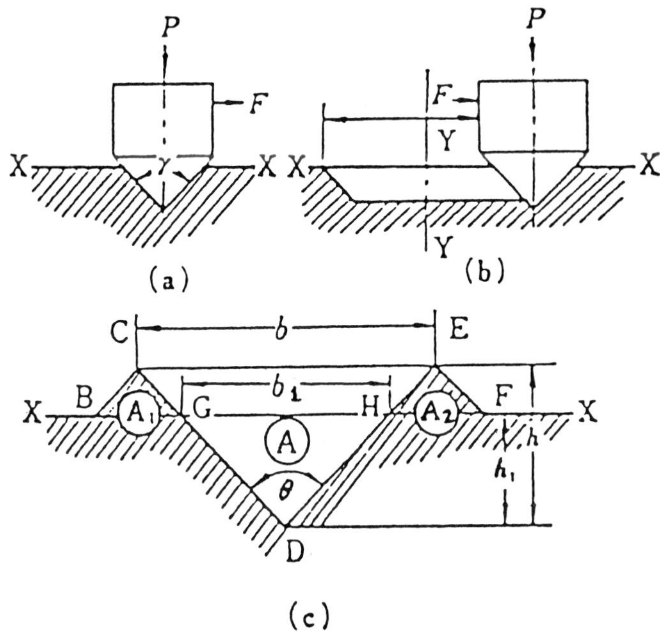

Fig. 2.34 Simple model of abrasive wear

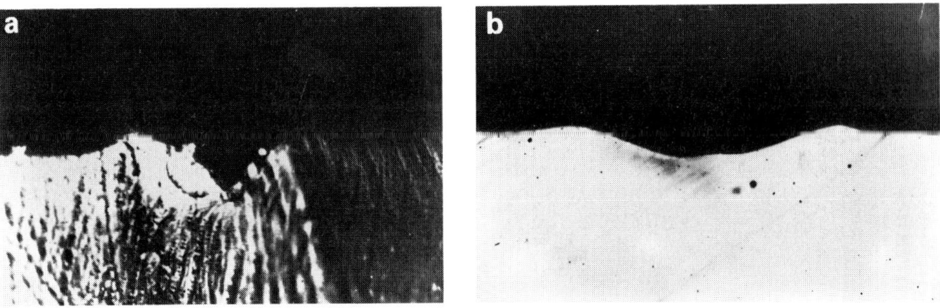

Fig. 2.35 Photographs of scratched grooves in plastic sections produced by Marten's scratch-hardness tester (x 560)

As an example of the second method, the research results of Yamaguchi et al. [36,37] will now be discussed. Figure 2.34 shows a simple model of the morphology of an abrasion groove caused by relative motion of a diamond cone with apex angle γ (such as that used for scratch hardness tests) over a material surface under a normal load P. The volume removed from the surface by the cone is the abrasive volume. If the cross-sectional area of the groove is A_o and the distance moved is ℓ, then the abrasive volume V_o is

$$V_o = A_o \ell \qquad (2.14)$$

The shape of a groove section in a plastic material is shown in Fig. 2.34(c). There is a pile-up of material on each side of the groove, shown as A_1 and A_2 in the figure, and the base angle θ becomes greater than the cone angle γ in plastic materials which show significant viscoelastic behaviour. In general, the raised areas A_1 and A_2 and the difference between θ and γ are comparatively small for thermosetting or amorphous plastics, as illustrated in the photograph in Fig. 2.35(a), which shows a groove section for polymethyl methacrylate. However, the rise of point D, the difference of θ from γ, and the areas of A_1 and A_2 become much greater for semicrystalline polymers in a rubber-like condition at room temperature. This is illustrated in the photograph in Fig. 2.35(b) which shows a groove section for Nylon 6. In the groove section shown in Fig. 2.34(c), the relationship between the true cross-sectional area A_o, which is removed, the raised sectional areas A_1 ($\triangle$BCG) and A_2 ($\triangle$HEF), and the groove area A ($\triangle$GDH) is as follows

$$A_o = A - (A_1 + A_2) \qquad (2.15)$$

and

$$V_o = [A - (A_1 + A_2)]\ell \qquad (2.16)$$

Put $\alpha = A_o/A$

and then $V_o = \alpha A \ell$ (2.17)

From the figure, the sectional area A becomes

$$A = \frac{b_1}{2} \cdot h_1 = \frac{b_1^2}{4} \cot \frac{\theta}{2}$$

Let $\frac{b_1}{b} = \frac{h_1}{h} = n$

$$b_1^2 = b^2 n^2$$

The scratch hardness $H_c = P/b^2$ (kgf/mm²) [33], where P is the normal load. Therefore,

$$b_1^2 = \left(\frac{P}{H_c}\right) n^2$$

and

$$A = \frac{P}{4H_c} \cdot \cot\frac{\theta}{2} \cdot n^2 \qquad (2.18)$$

From equations (2.17) and (2.18)

$$V_o = \frac{\alpha}{4H_c} \cdot \cot\frac{\theta}{2} \cdot n^2 \cdot P \cdot \ell \qquad (2.19)$$

In actual abrasion, abrasives and protuberances do not have edges as sharp as a diamond cone and contacts occur at many points simultaneously. Thus, each abrasive does not scratch as effectively as shown in the simple model. One can define a "scratch efficiency" η as the ratio of the actual abrasive wear volume V_a to that of the ideal wear volume V_o given by equation (2.19). Thus $V_a = V_o \eta$, and

$$V_a = \frac{\alpha}{4H_c} \cdot \cot\frac{\theta}{2} \cdot n^2 \cdot P \cdot \ell \cdot \eta \qquad (2.20)$$

If we denote $\alpha \cdot \cot \theta/2 \cdot n^2/4H_c$, as a wear factor, β, then

$$V_a = \beta \cdot P \cdot \ell \cdot \eta \qquad (2.21)$$

The specific abrasive wear rate V_{sa} is defined as the wear volume (mm^3) per unit contact pressure (kgf/cm^2), per unit sliding distance (km) and per unit contact area (cm^2). Thus, from equation (2.21)

$$V_{sa} = \beta \cdot \eta \times 10^6 \ (mm^3/km,kg) \qquad (2.22)$$

$$\text{Therefore } V = V_{sa} \cdot P \cdot \ell \qquad (2.23)$$

2.3.3 *Testing methods*

There are many kinds of abrasive wear-testing methods depending on the type of abrasives or protuberances. The three main methods involve loose abrasives, bonded abrasives and impacting abrasives (erosion).

(i) *Loose abrasives*

A representative device for this type of testing is shown in Fig. 2.36. A rotating specimen contacts with a rotating steel surface under a load of 4.5 kgf. Alumina powders with a grain size of about No. 80 are dropped onto the surface, and the specimen moves up and down by a distance of about

1.6 mm. Standards include "Testing method for abrasive resistance of plastics by abrasives" in JIS-K-7205 and ASTM-D-1242 (A type) and the method is also used with Olsen and Amsler machines.

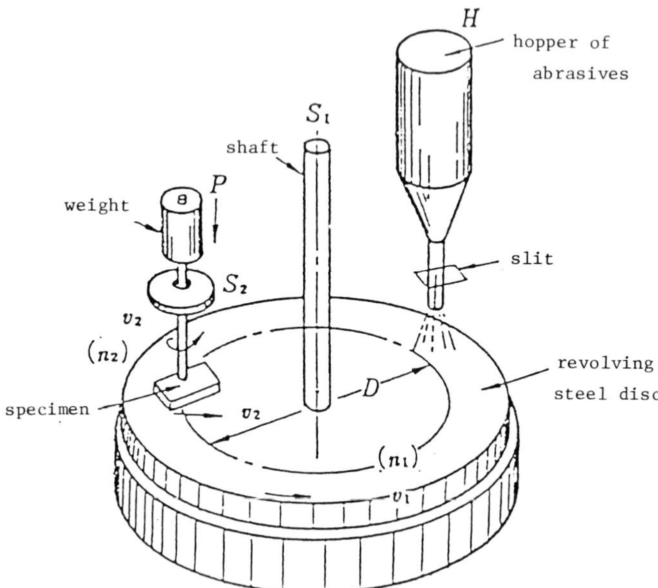

Fig. 2.36 Abrasion test machine using loose abrasives (ASTM-A type or JIS K7205)

(ii) *Bonded abrasives*

(1) *Abrasion testing using a Taber abrader.* The Taber abrader, developed by the Taber Co., and shown in Fig. 2.37, is widely used for abrasion testing between a plane surface material and two cylindrical grindstones of bonded abrasives. With this apparatus, a large disc specimen is placed on a motor-driven rotating disc and a pair of abrasive wheels (grindstones), which turn freely on their axes, are applied to the test specimen at a specified position under a specified load. As shown in Fig. 2.37(b), the abrasive wheels revolve with speed component V_r of circular speed V of specimen and abrade the specimen with speed component V_s. Abrasive powder is removed by a vacuum cleaner. Figure 2.37(a) shows the apparatus for testing decorative sheet in JIS-K-6902, in which rubber discs covered

with emery paper are used in place of the abrasive wheels. This type of abrader, as shown by the schematic in Fig. 2.37(b), has been standardized in JIS-K-7204 as "Testing method for abrasion resistance of plastics by an abrasive wheel", and is specified as a Taber abrader for the testing of the transparent resistance of plastics in ASTM-D-1044.

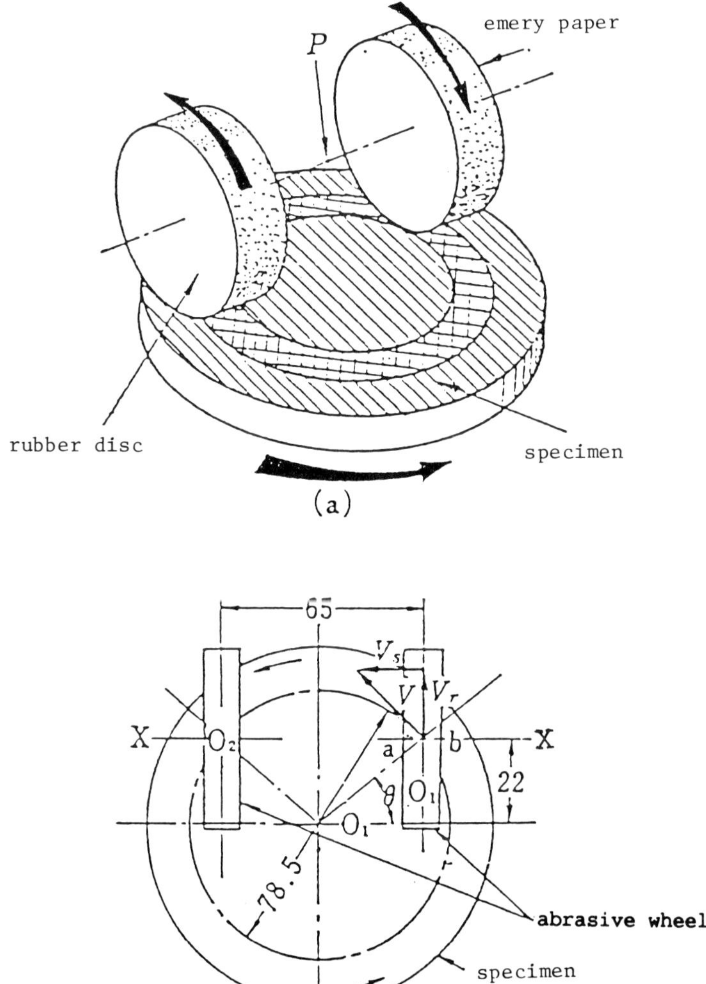

Fig. 2.37 Taber type (or JIS-K-7204) abrasion test machine using bonded abrasives (abrasive wheel)

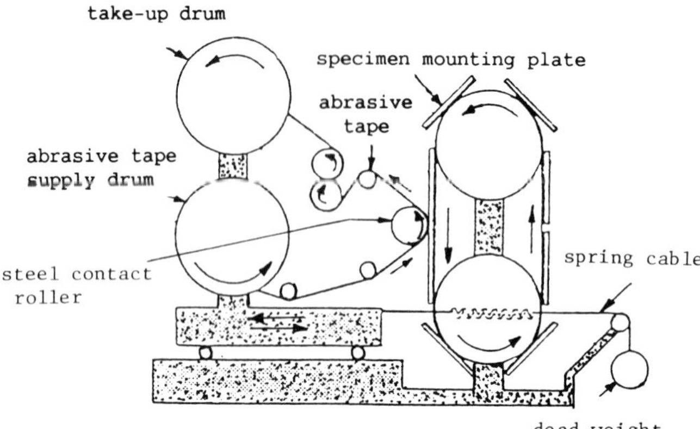

Fig. 2.38 Armstrong-type, bonded-abrasive, abrading machine (ASTM-D-1242)

(2) *Armstrong abrader.* Figure 2.38 shows an Armstrong abrader, which consists of a sliding abrasive tape and eight plate specimens which are mounted on a moving band. Pressure on the system is exerted by a spring cable and the abrasive tape is wound around a take-up drum following contact with the plates. This method is an example of a bonded abrasive test machine as specified in ASTM-D-1242.

(3) *Other bonded abrasive abraders.* Two other types of equipment using bonded abrasives are the William's abrader and the ISO film abrader. With the William's abrader, a pair of specimens contact the side plane of a rotating disc and it is used primarily for abrasion testing of metallic materials. With the ISO film abrader, abrasion takes place between a reciprocating plane specimen and a cylindrical grindstone revolving in a stepped fashion.

(iii) *Erosion*

One of the methods used to measure abrasion generated by the collision between loose abrasives and a material surface is standardized in ASTM-D-673, "Test of mar resistance". In this, carborundum powder is dropped onto a specimen surface inclined at 45° to the horizontal. Another example is Elmendorf's machine, shown in Fig. 2.39. With this apparatus, abrasion is generated by the impact of four rotating plate specimens against loose abrasives dispersed in an octagonal test chamber.

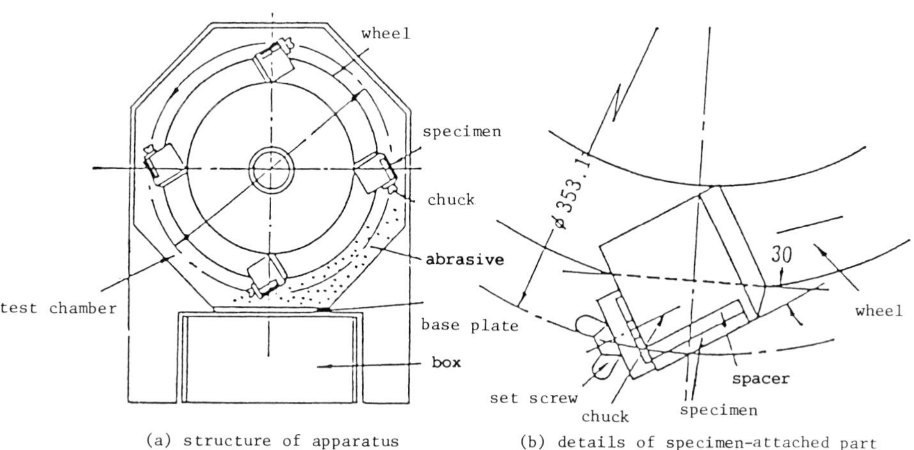

Fig. 2.39 Elmendorf-type abrasion test machine using impacting abrasives

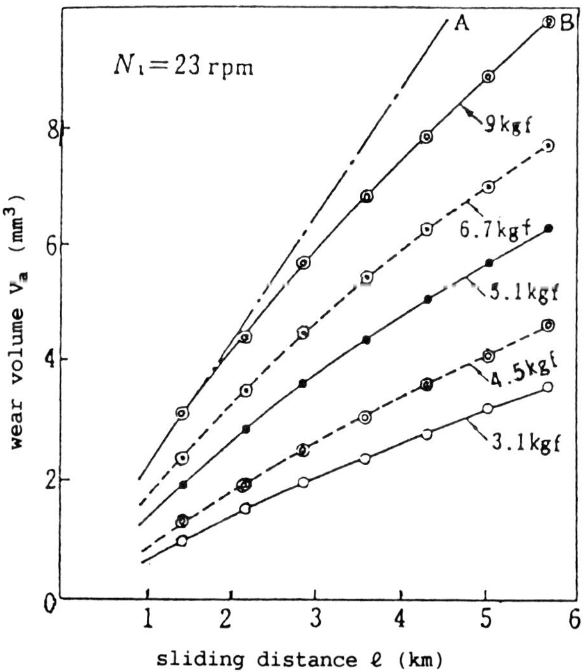

Fig. 2.40 Relationships between abrasive wear volume V_a and sliding distance ℓ for polystyrene using loose abrasives under various normal loads

2.3.4 *Characteristics of abrasive wear*

In this section, abrasive wear characteristics will be discussed in relation to sliding distance, contact pressure, velocity and the condition of the abrasives. In addition, equations (2.20) to (2.23), deduced from a simple model, are examined experimentally.

(i) *Effect of sliding distance*

Figure 2.40 shows the relationship between abrasive wear volume V_a (cm^3) resulting from the presence of loose abrasives (as in ASTM-D-1242 or JIS-K-7205) and sliding distance ℓ at various contact loads. The example given is for polystyrene. Figure 2.41 shows similar relationships for various plastics using a Taber abrader under a constant normal load. It is clear from both of these figures that V_a is almost directly proportional to ℓ, i.e. $V_a = k_1 \ell$. Figure 2.42 shows the relationship between abrasive wear volume V_a and the total number of revolutions of specimen N using an Elmendorf abrasion test machine, and approximate proportionality is again obtained. It should be noted that the value of V_a is initially negative due to embedding of the abrasives into the specimen.

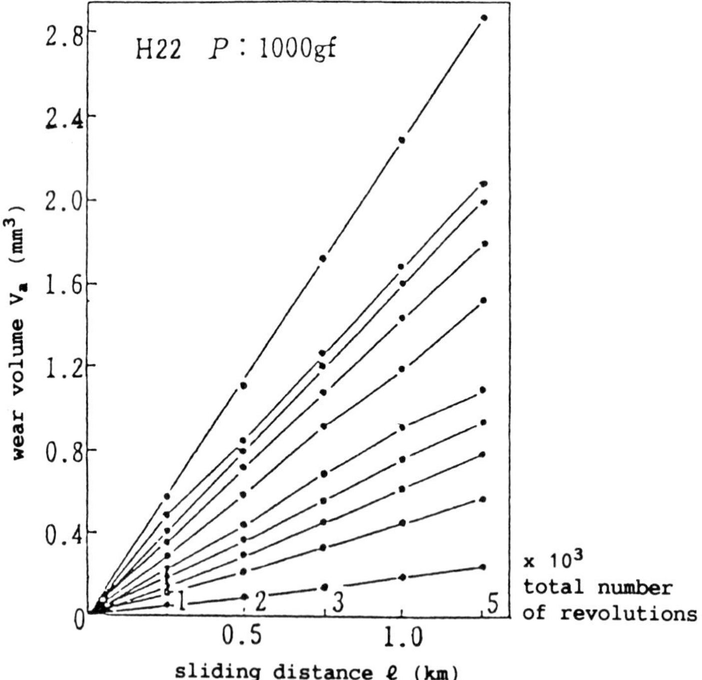

Fig. 2.41 Relationships between abrasive wear volume V_a and sliding distance ℓ for various plastics using bonded abrasives under a normal load of 1000 gf

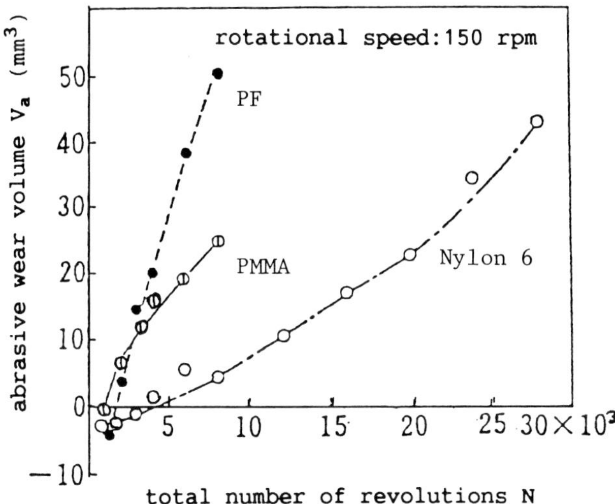

Fig. 2.42 Relationship between abrasive wear volume V_a and the total number of revolutions N for various plastics, using an Elmendorf abrasion test machine

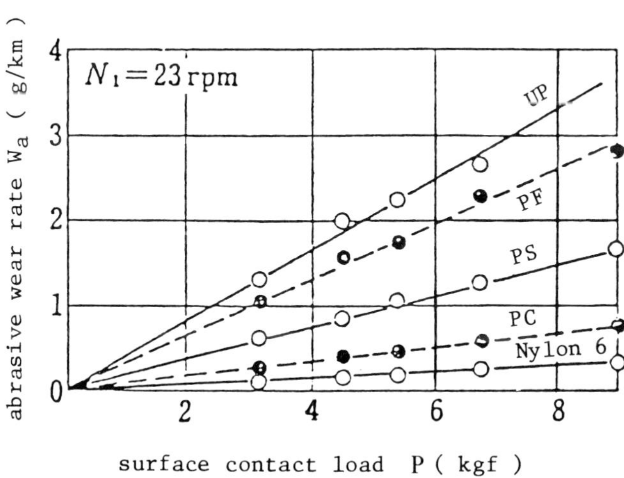

Fig. 2.43 Relationship between the abrasive wear rate Wa of various plastics and the suface contact load P using loose abrasives

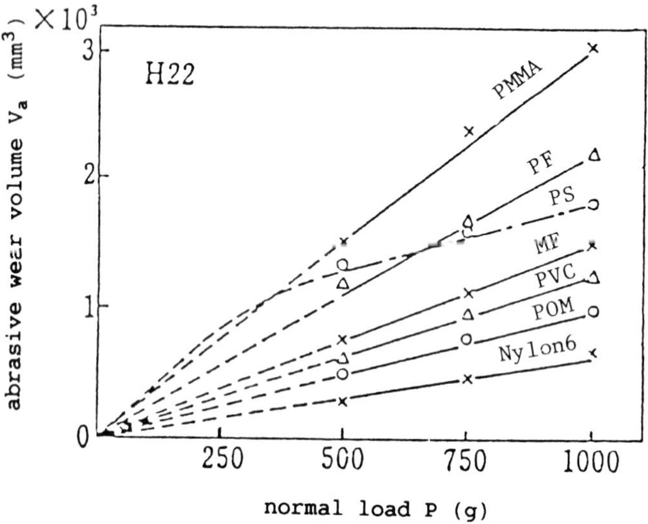

Fig. 2.44 Relationships between abrasive wear volume V_a (up to 5,000 revolutions) and normal load P for various plastics using bonded abrasives

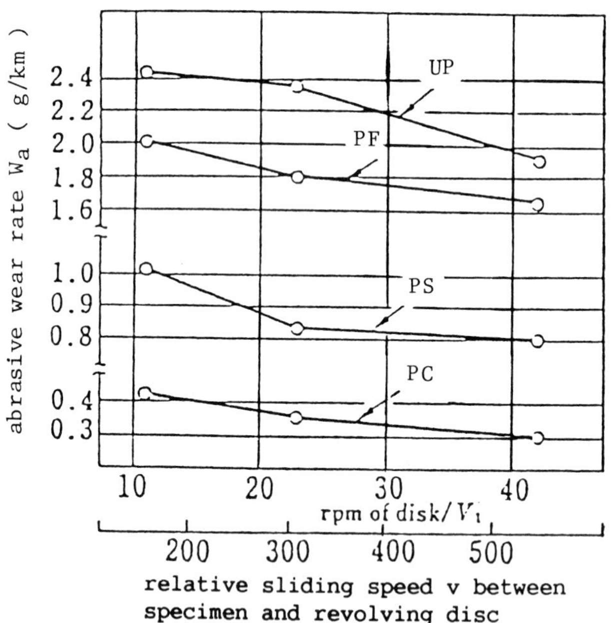

Fig. 2.45 Relationship between the abrasive wear rate Wa and relative sliding speed v for various plastics using loose abrasives

(ii) *Effect of contact load*

Figure 2.43 shows the relationships between the abrasive wear rate Wa by loose abrasives (ASTM-D-1242-A) and the contact load P for five different plastics. Figure 2.44 shows similar relationships for each of seven plastics when using bonded abrasives with a Taber abrader. From this figure it can be seen that the abrasive wear volume V is almost directly proportional to the contact load P. Polystyrene is an exception, however, because of filling-up of the depressions in the abrasive wheel with debris. Thus

$$V_a = k_2 P \qquad (2.25)$$

where K_2 is constant depending on the type of material.

(iii) *Effect of speed*

Figure 2.45 shows the relationships between abrasive wear rate Wa by loose abrasives (ASTM-D-1242-A) and the mean relative speed v of the specimen with respect to the steel disc surface for each of four plastics. It is clear from these curves that Wa decreases slightly with an increase in v.

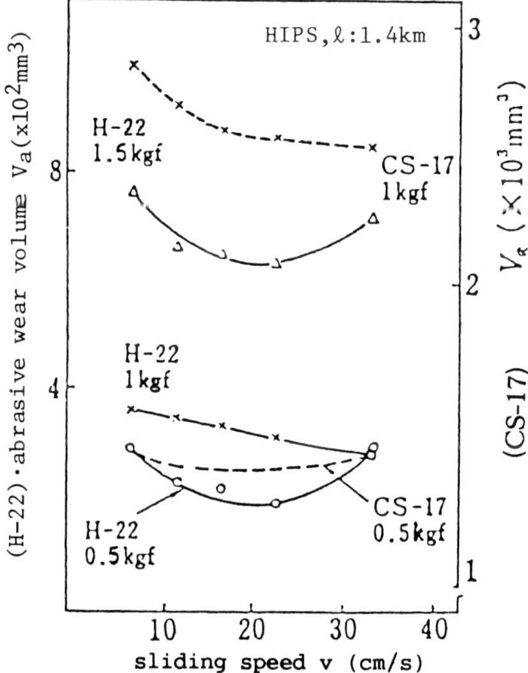

Fig. 2.46 Relationships between abrasive wear volume V_a and relative sliding speed v for HIPS using a Taber abrader

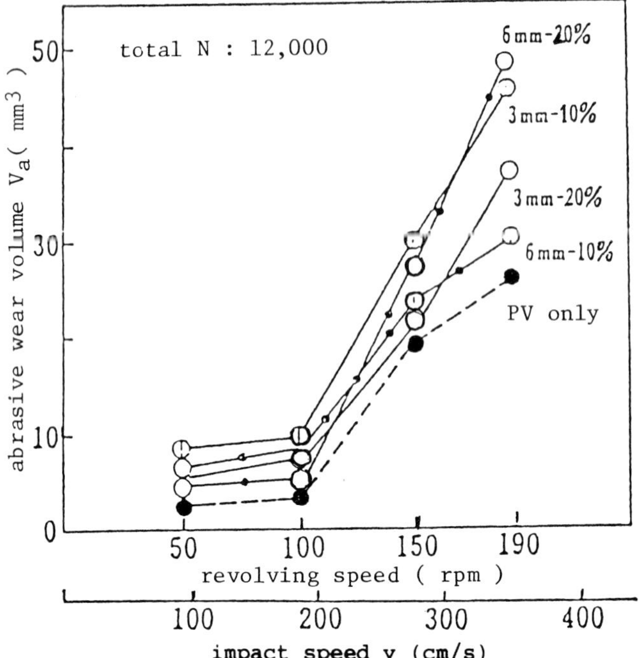

Fig. 2.47 Relationship between abrasive wear volume V_a and rotational speed of an Elmendorf abrader for various FR-PVC composites (GF: length, 3mm, 6 mm, 10-20%)

Figure 2.46 shows the variation with speed v of the abrasive wear volume V_a caused by two different abrasive wheels (H-22 or CS-17) on a Taber abrader under different loads for high impact polystyrene (HIPS). V_a appears to reach a minimum at about 20 cm/s.

Figure 2.47 shows the relationship between wear volume V_a during erosion using an Elmendorf machine and the impact speed v for various FR-PVC. It appears from these curves that V_a is nearly proportional to the square of v, since the impact energy given to the specimen from the abrasive particles is directly proportional to the square of v. From all of these results, it is not possible to deduce any simple relationship between V_a and v which is of general validity.

(iv) *Experimental confirmation*

In this section, experiments will be described to confirm equations (2.14) to (2.23). Figure 2.35 has already shown examples of photomicrographs of groove sections for two out of six types of plastics which were scratched by a 90° diamond cone in a Marten's hardness tester. These plastics were phenolics (PF), unsaturated polyester (UP), polymethylmethacrylate (PMMA), polystyrene (PS), polycarbonate (PC) and Nylon 6. The scratch hardness Hc and the areas of A, A_1 and A_2 in Fig. 2.34 were measured from these photographs, and then the values of α {= A_o/A = [A - (A_1 + A_2)]}, groove base-angle θ, cot $\theta/2$ and $n^2(=b_1^2/b^2)$ were obtained. From this data, values of the abrasive wear factor β (= $\alpha \cdot \cot \theta/2 \cdot n^2/4Hc$) were then calculated. All the results are shown in Table 2.6.

TABLE 2.6
Constants of scratched groove sections and the specific abrasive wear rate V_{sa} for various plastics

Materials	a	Hc[1]	$\theta°$	cot $\theta/2$	n^2	β[ref.2]	V_{sa}[3]
Phenolic (PF)	0.82	33.7	111	0.69	0.42	$1.77x10^{-3}$	396
Unsaturated polyester (UP)	0.66	16.2	95	0.92	0.64	$2.54x10^{-3}$	480
Polymethylmethacrylate (PMMA)	0.21	17.3	98	0.87	0.49	$1.50x10^{-3}$	277
Polystyrene (PS)	0.11	9.4	87	1.03	0.41	$1.2\ x10^{-3}$	238
Polycarbonate (PC)	0.12	9.2	128	0.49	0.32	$0.41x10^{-3}$	90
Nylon 6 (PA)	0.25	11.9	149	0.27	0.24	$0.32x10^{-3}$	32

1 Kgf/mm^2
2 mm^2/kgf; [(α/4Hc)(cot $\theta/2 \cdot n^2$)]
3 mm^3/km · kgf

The specific abrasive wear rates V_{sa} of six types of plastics were measured by a loose abrasives abrasion tester (ASTM-D-1242-A) under 4.5 kgf contact load for 1,000 disc revolutions, or a sliding distance of 5.75 km. These results are also shown in Table 2.6. The relationship between β and V_{sa} is shown in Fig. 2.48, and the direct proportionality observed is consistent with the abrasive wear theory presented in equations (2.20) to (2.23). In this case, the value of the abrasive wear efficiency calculated from the equation (η = V_{sa}/β x 10^6) is about 30%.

From these experimental results, the theoretical equation (2.21) seems reasonable and equations (2.24) and (2.25) can be combined to give equation (2.23), $V_a = V_{sa} P\ell$. Accordingly, it is justifiable to use a specific abrasive wear rate V_{sa} (mm^3/km · kgf) (volume per unit contact load (kgf) and per unit sliding distance (km)) to characterise abrasive wear behaviour.

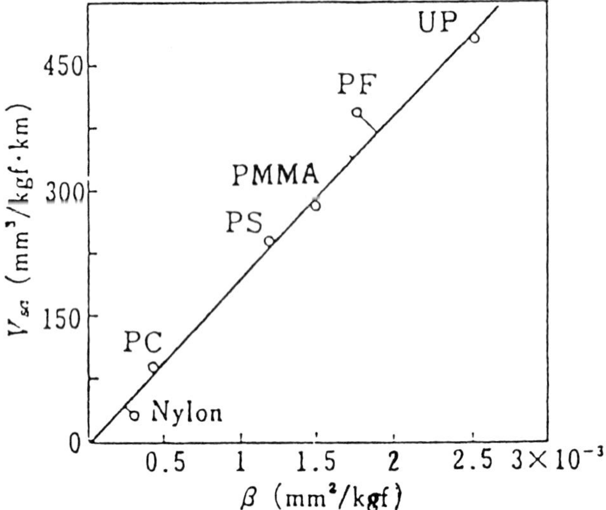

Fig. 2.48 Relationship between specific abrasive wear rate V_{sa} and abrasive wear factor β for various plastics

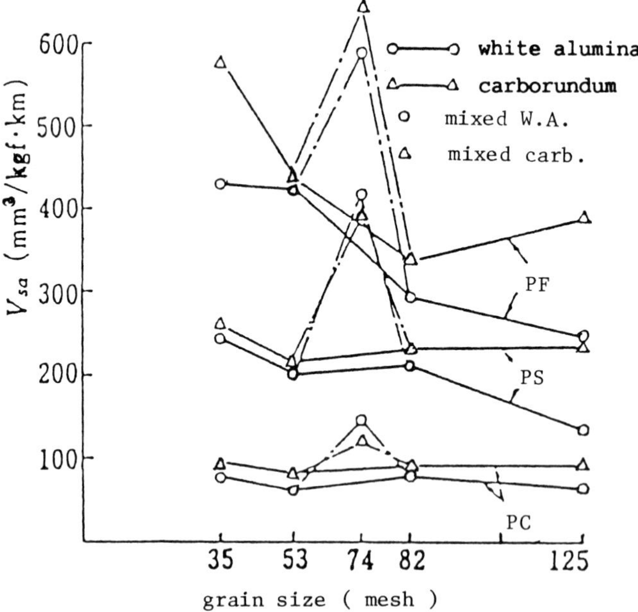

Fig. 2.49 Effect of grain size of white alumina and carborundum on the specific abrasive wear rate V_{sa} of various plastics

(v) *Effect of abrasives*

It is to be expected that abrasive wear will be influenced by the type, shape and condition of abrasives. Figure 2.49 [42] shows experimental relationships for three different plastics between the specific abrasive wear rate V_{sa} and grain size for two kinds of abrasives, white alumina and carborundum. It is clear from this figure that the abrasive efficiency η of carborundum is slightly greater than that of white alumina, that of V_{sa} increases with increasing grain size and that wear by combined grain sizes is larger than that by a certain grain size.

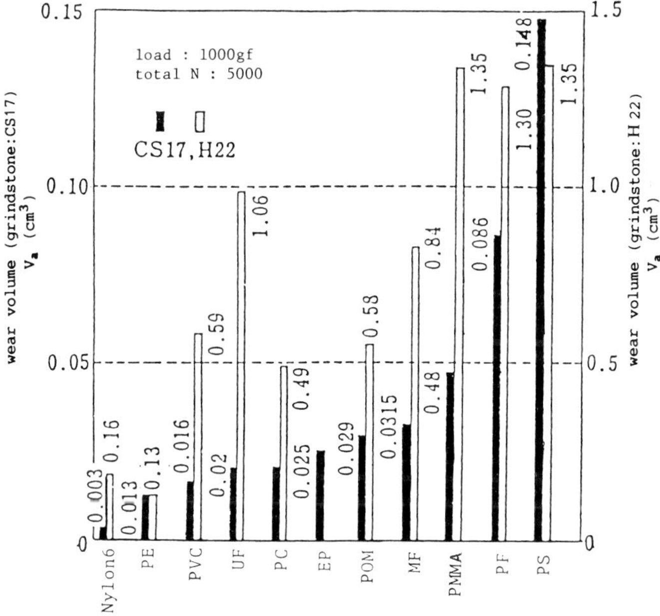

Fig. 2.50 Abrasive wear ranking of various plastics using a Taber abrader

Figure 2.50 [43] shows the abrasive wear volume for each of eleven plastics produced by two kinds of bonded-abrasive wheels, H-22 and CS-17, in a Taber abrader. It appears from this figure that the wear volume due to the H-22 wheel is generally much greater than that due to the CS-17 wheel. However, the ranking of the different plastics does not remain the same on the two abrasive wheels. In a loose abrasive test, wear efficiency decreases with an increase in use because the abrasive edges become worn.

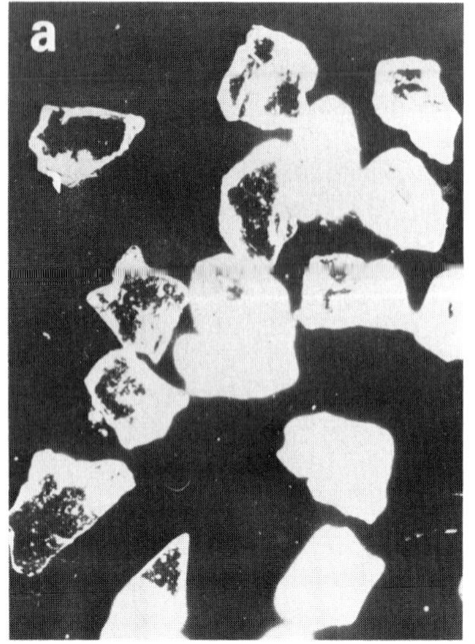

(a) fresh almina powder

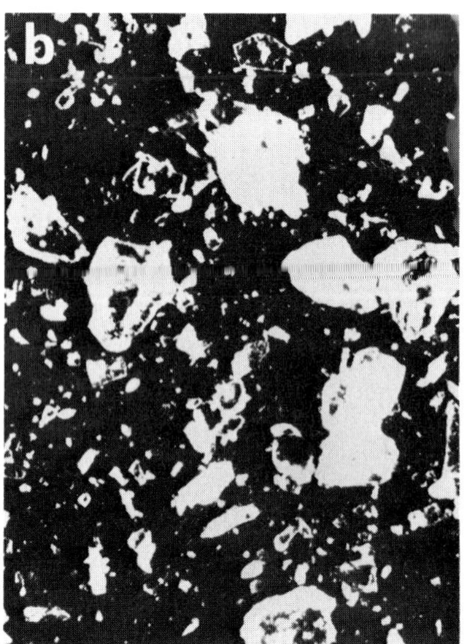

(b) almina powder after abrasion of 5 km

Fig. 2.51 Photomicrographs of alumina abrasives (x 100)

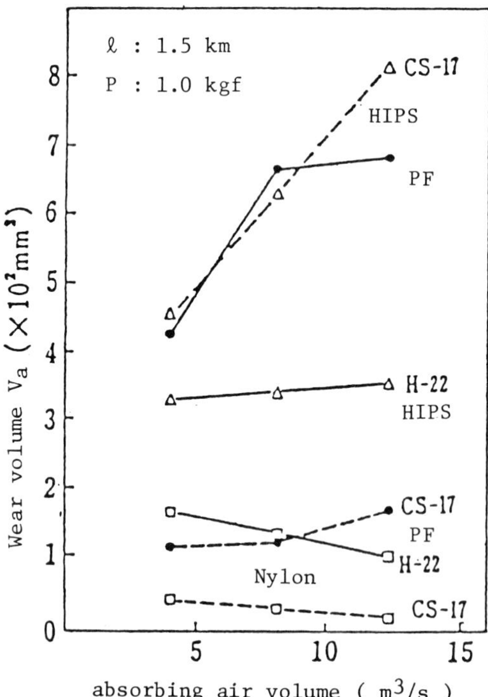

Fig. 2.52 Relationships between abrasive wear volume V_a and absorbing air volume for various plastics using a Taber abrader with grindstones H-22 or CS-17

Figure 2.51 shows the shape of alumina grains before and after an abrasive wear test. This phenomena explains the shape of the curves in Fig. 2.40 which indicate a decrease in the amount of abrasion occurring with increasing sliding distance. Figure 2.52 shows the effect of the amount of air containing abrasive dust absorbed into a vacuum cleaner on the abrasive wear volume using a Taber abrader. It should be noted that the abrasive wear volume varies widely even under the same experimental conditions. Figure 2.53 shows the considerable variety of experimental abrasive wear results obtained by various laboratories which tested each of the materials under the same conditions.

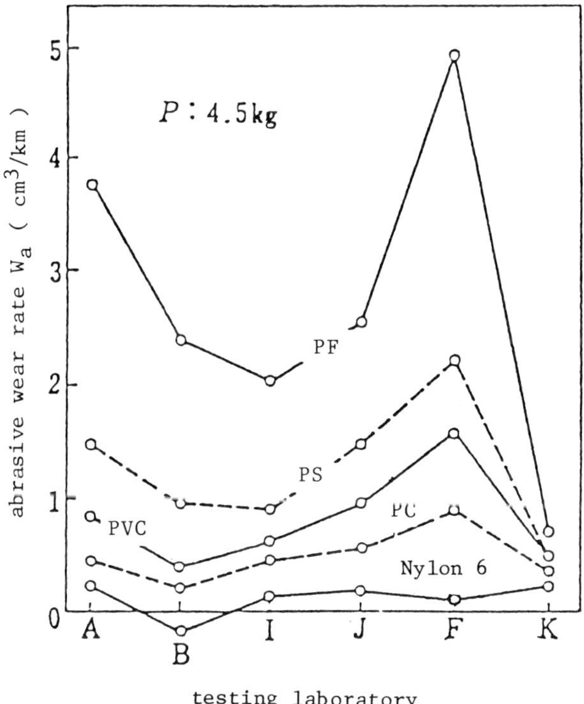

Fig. 2.53 Results of tests by different laboratories on the abrasive wear rates W_a of various plastics, using a loose-abrasive wear-testing apparatus

(vi) *Comparison of abrasive wear tests*

Figure 2.54 [44] shows the relationships for five plastics between the specific abrasive wear rate V_{sa} with two abrasive wheels, H-22 or CS-17, on a Taber abrader and V_{sa} by loose abrasives on an ASTM-D-1242-A type abrader. The plastics were Nylon 6, polycarbonate (PC), polyvinylchloride (PVC), epoxy resin (EP) and phenolic (PF). It is clear from the figure that each relationship is almost linear.

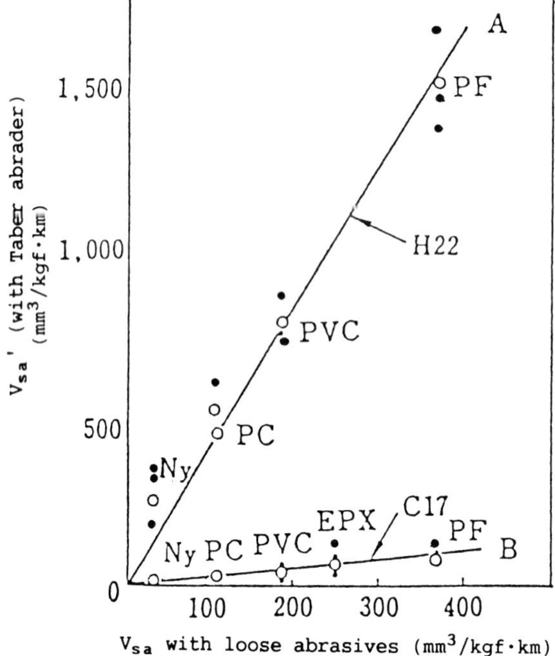

Fig. 2.54 Relationships between V_{sa} (using a Taber abrader with grindstones H-22 or CS-17) and V_{sa} (using a loose-abrasive abrader (ASTM-D-1242A)) for various plastics

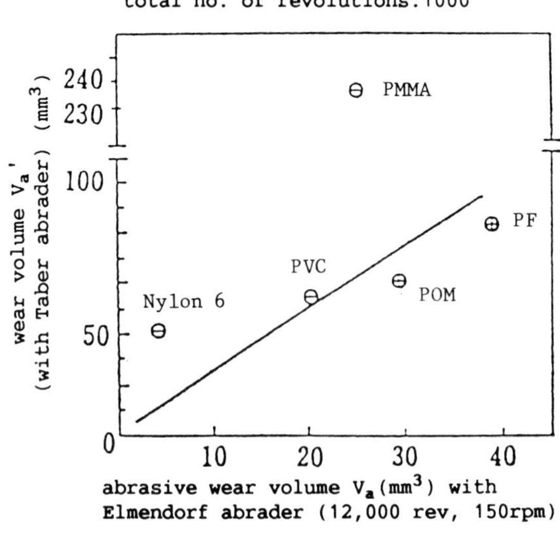

Fig. 2.55 Relationship between abrasive wear Volume V_a' using a Taber abrader and V_a using an Elmendorf abrader for various plastics

Figure 2.55 shows the relationship for five plastics between abrasive wear volume V_a using bonded abrasives on a Taber abrader and that for erosion in an Elmendorf abrader. It is clear that for PVC, POM and PF the relationship is almost linear, but the values of V_a for Nylon 6 and PMMA are anomalous.

REFERENCES

1 Glossary of Terms and Definitions in the Field of Friction and Lubrication, Part I, OECD, 1969, p. 64.
2 Peterson, M.B., M.K. Gabel and M.J. Devine, ASTM Standardization News, 9-12 Sept. 1974.
3 Tanaka, K., Y. Uchiyama et al., J. JSLE, 19 (1974) 823
4 Mizuno, M., J. JSLE, 22 (1967) 152.
5 International Study Committee for Wear Test of Flooring Materials, Wear, 4 (1961) 479.
6 Matsunaga, M., and Y. Tsuya, Handbook of Solid Lubrication, Saiwai Shobo Co., 1978, p. 368.
7 Amsler, A.J. VDI, 66 (1922) 15.
8 Friction and Wear Test Machine, Faville-Le Valley Corp.
9 Lewis, R.B., Mech. Eng., 86 (1964) 32.
10 Rhee, S.K., Wear, 16 (1970) 431.
11 Jain, V.K. and S. Bahadur, Wear of Materials, ASME, 1979, p. 556.
12 Kawamoto et al., Trans. JSME, 18 (1952) 48.
13 Sada, T. et al., Rep. Kaken, 33 (1957) 45.
14 Matsubara, K., Rep. Kikaishikensho, 52 (1964).
15 Matsubara, K., J. JSLE, 2 (1957) 533.
16 Awaya, J. et al., Trans. JSME, 27 (1961) 113.
17 Takeuchi, E. et al., Appl. Phys., 28 (1959) 233.
18 Clauss, F.J., Solid Lubricant and Self-Lubricating Solids, Academic Press, New York, 1972, pp. 139-287.
19 Tanaka, K. et al., J. JSLE, 12 (1967) 15.
20 Bongiovanni, G., Mod. Plast., May 44 (1967)
21 Yamaguchi, Y., I. Sekiguchi et al., Kogakuin Univ. Rep., 35 (1973) 30.
22 Yamaguchi, Y., Plastics, 26 (1975) 5-74.
23 Sekiguchi, I., Y. Yamaguchi et al., Kogakuin Univ. Rep., 44 (1978) 34.
24 Yamaguchi, Y., I. Sekiguchi et al., J. Materials, 21 (1971) 826.
25 Matsubara, K., M. Watanabe and M. Karasawa, J. JSLE, 14 (1969) 43.
26 Yamaguchi, Y., I. Sekiguchi et al., J. JSLE, 22 (1977) 58.
27 Yamaguchi, Y., I. Sekiguchi et al., Kogakuin Univ. Rep., 30 (1971) 59.
28 Yamaguchi, Y., I. Sekiguchi and K. Tsuru, Kogakuin Univ. Rep., 29 (1971) 58.

29 Yamaguchi, Y. and I. Sekiguchi, J. JSLE, 14 (1969) 154.
30 Matsunaga, M. and Y. Tsuga, Handbook of Solid Lubrication, Saiwai Shobo Co., 1978, p. 160.
31 Yamaguchi, Y., I. Sekiguchi et al., Kogakuin Univ. Rep., 24 (1968) 53.
32 Yamaguchi, Y., I. Sekiguchi et al., Kogakuin Univ. Rep., 27 (1970) 51.
33 Yamaguchi, Y., I. Sekiguchi et al., Kogakuin Univ. Rep., 28 (1970) 21.
34 Ogoshi, S., T. Sata and M. Mizuno, Trans. JSME, 21 (1955) 555.
35 Lancaster, J.K., in: A.D. Jenkins (Ed.) Friction and Wear, Chapter 14, Polymer Science. North-Holland Pub. Co., 1972.
36 Yamaguchi, Y. and Y. Oyanagi et al., J. Materials, 14 (1965) 218.
37 Yamaguchi, Y. and Y. Oyanagi et al., J. Materials, 14 (1965) 212.
38 Catalogue of Tokyo-Shikenki-Production, sheet no. 30403.
39 Catalogue of Gas-Shikenki Co., Model NUS-ISO-I type tester.
40 Yamaguchi, Y., Tosei News, 37 (1974) 1.
41 Yamaguchi, Y. et al., JSMS (1966) (preprint).
42 Yamaguchi, Y. and I. Sekiguchi, J. JSLE, 11 (1966) 396.
43 Yamaguchi, Y., Plastics, 20 (3) (1969) 25.
44 Yamaguchi, Y., Polymer Friends, 11 (1967) 713.

CHAPTER 3

IMPROVEMENT OF LUBRICITY

Plastics possessing self-lubricating properties are useful for many sliding components and improved sliding properties, such as low friction, high wear resistance and a high pv_{max} value are often required. There are three methods used to improve the sliding properties of plastic materials; through the molecular and superstructure of the polymer itself, by the blending of polymers and by producing composites of plastics with various fillers, such as metals, inorganic and organic materials. In this chapter, attention is concentrated on the last two methods: polymer blending and composites.

3.1 POLYMER BLENDING

3.1.1 *Polystyrene, polyacrylonitrile and polybutadiene [1,2]*

(i) *Experiment*

An example of the improved self-lubricating characteristics which can be achieved through the blending of high polymeric materials, such as polystyrene (PS), polyacrilonitrile (PAN) and polybutadiene (PB), will now be described. Table 3.1 shows the contents of each homopolymer component, PS, PAN and PB, and the conditions for blending and moulding various mixtures, PAN/PS, PB/PS, AS/PB and PS/NBR (acrylonitrile butadiene copolymer), which were then subjected to friction and wear tests. All the blended specimens shown in Table 3.1 were mixed at 165°C to 175°C for 10 to 30 minutes by means of a test roller and then compression moulded at 150°C to 230°C.

In the friction and wear tests, carbon steel (S45C) was used as a counterface material with a surface roughness of 2 to 3 μm. The static coefficient of friction, μ_s, was measured under standard conditions of 23°C and 50% R.H. by means of the increasing inclination method shown in Fig. 1.59. For the measurement of the kinetic coefficient of friction, μ_k, and in the wear tests, a cylinder-end wear test device was employed, as shown previously in Fig. 2.1(a). The characteristic value of μ_s, is given by the mean of ten tests, and that of μ_k, by the mean value during a sliding

TABLE 3.1
Composition and moulding conditions for mixtures of PS, PAN, and PB

Plastics		Code	Composition (wt%)				Blending Conditions		Moulding Conditions	
			PS	PAN	PB	AS	Temp.(°C)	Time(min)	Temp.(°C)	Time(min)
Base polymers	PS	PS	100						150	5
	PAN	PAN		100					230	5
	PB	PB			100					
	*AS (PAN/PS)	AS	75	25		100			200	5
	AS **PAN+PS	AS1	90	10			165-175	15	200	5
		AS2	80	20			165-175	15	205	5
Blended polymers	BS **PB+PS	BS1	90		10		165-170	20	200	5
		BS2	80		20		165-170	20	200	5
		BS3	70		30		175	20	200	5
	ABS AS+PB	ABS1			10	90	165	10	200	5
		ABS2			20	80	165-168	10	200	5
		ABS3			30	70	165-170	15	200	5
	*NBR+PS (NBR:AN/B)	ABS4	50	17.5	32.5		165	30	200	5
		ABS5	60	14.0	26.0		165	30	200	5
		ABS6	70	10.5	19.5		165	10	200	5

* AS (PAN/PS) and NBR (AN/B) are commercially-available materials
** PAN+PS, etc., are the plastics blended in this work

distance of 3000 metres. For comparison purposes, another type of wear test was also conducted, using the plate-ring arrangement already shown in Fig. 2.1(b).

Wear characteristics are expressed using the specific wear rate, V_s or V'_s (mm^3/kgf·km) for the two geometrical arrangements in Figs. 2.1(a) and 2.1(b) respectively. All experiments were made for 25 minutes at a pressure of 3.9 kgf/cm^2, a velocity of v=30 cm/s and thus a pv = 117 kgf/cm^2·cm/s.

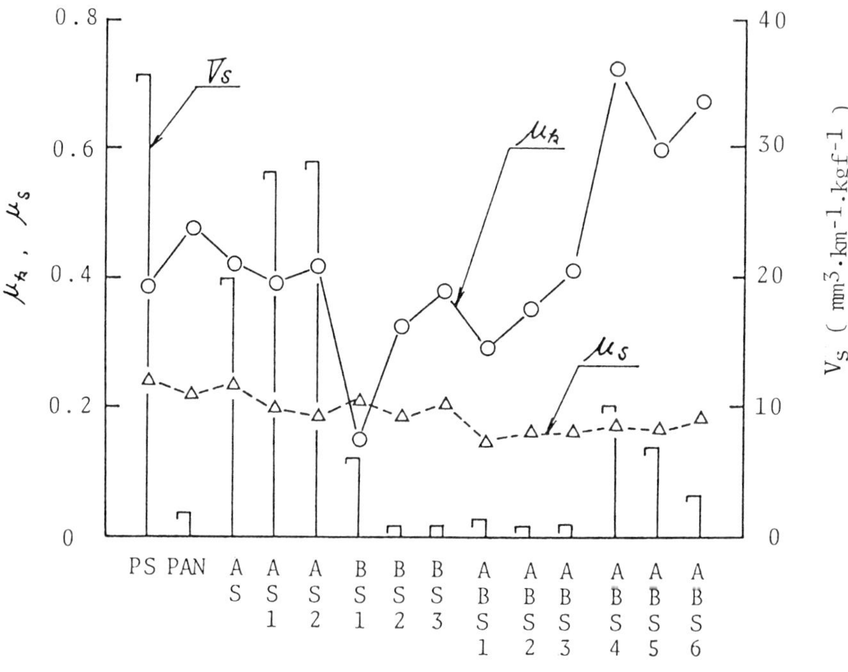

Fig. 3.1 Values of μ_k, μ_s and V_s for homopolymers and blended plastics

(ii) *Results*

(1) *General results.* Figure 3.1 summarizes the average values of the static coefficient of friction, μ_s, the kinetic coefficient of friction, μ_k, and the continuous specific sliding wear rate, V_s, of the various plastics against carbon steel. The relationships between these values and the various polymer mixtures are shown in Figs. 3.2, 3.3 and 3.4.

(2) *Effect of PS blended with PAN.* From friction and wear tests with 100% PS, PS blended with 10% and 20% PAN and 100% PAN, the relationships between μ_s, μ_k, V_s and the PAN content were determined, as shown by the three curves in Fig. 3.2. Each value of μ_s, μ_k and V_s was an intermediate value of PS or PAN, and was considered to form a continuous relative curve with no special characteristics having been observed.

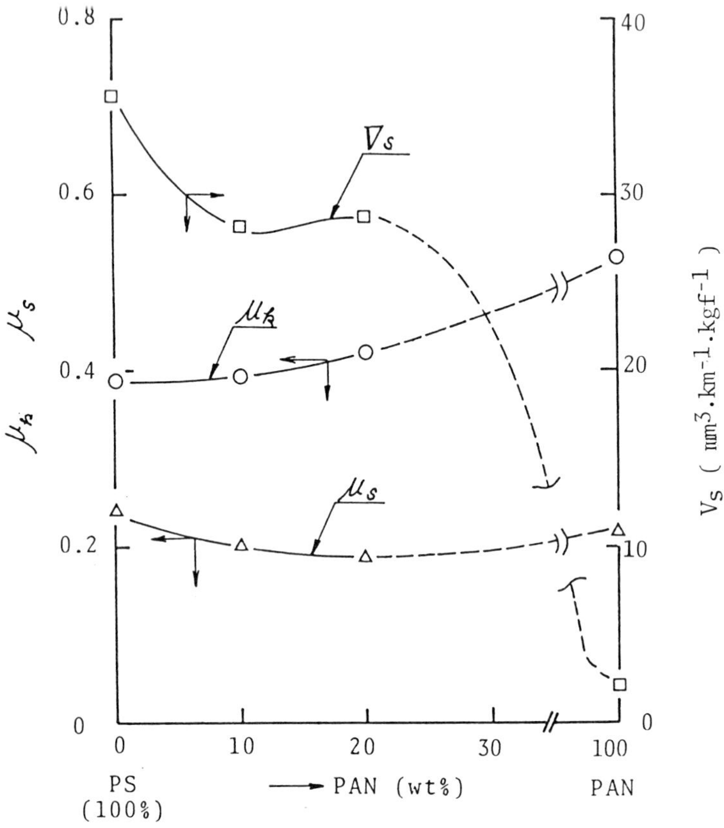

Fig. 3.2 Relationships between μ_k, μ_s and V_s and the blending percentage of PS and PAN

(3) *Effect of PS or AS blended with PB.* Figure 3.3 shows the relationships between μ_s, μ_k and composition for PS or AS blended with PB at 10, 20 and 30% concentrations. It can be seen that μ_s decreases slightly with an increase in the PB content, whilst the μ_k value decreases sharply to a minimum value at approximately 10% PB; V_s decreases sharply when the PB content is 10% or more, and the wear resistance thus greatly increases.

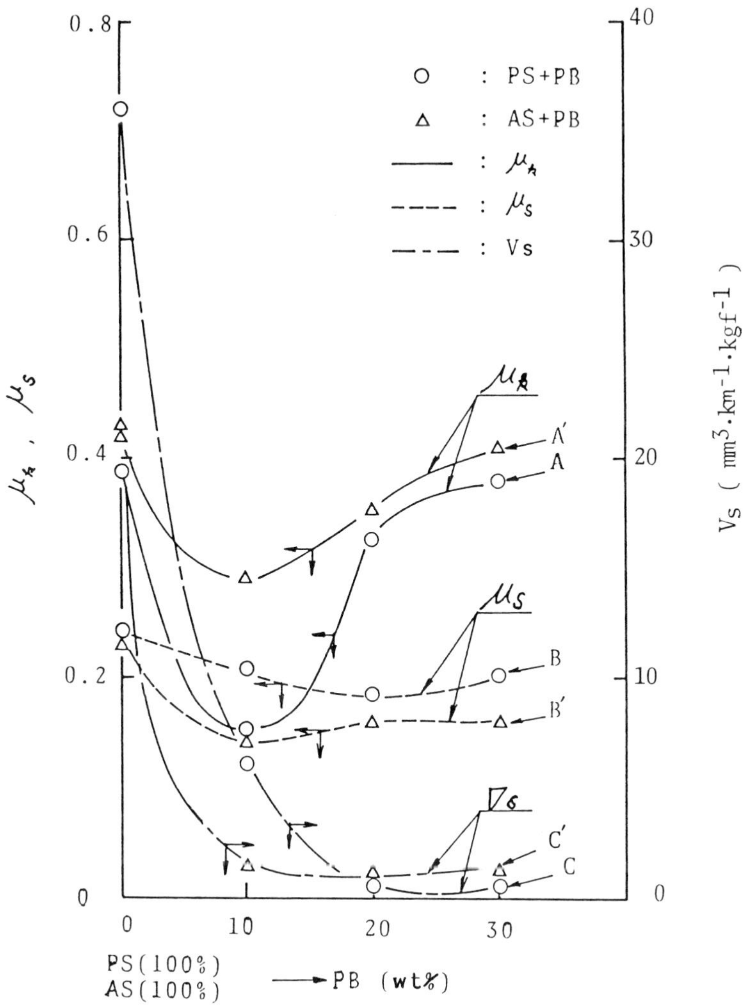

Fig. 3.3 Relationships between μ_k, μ_s and V_s and the blending percentage of PB in PS or AS

(4) *Effect of PS blended with NBR (AB).* The relationships between μ_k, μ_s, V_s and composition for PS blended with NMR (AB) are shown by the three curves in Fig. 3.4. PS blended with NMR (AB) forms a product equivalent to ABS resin. The effect of such blending on μ_s is minimal, but μ_k decreases at about 10% NMR. The decrease in V_s is significant at 10 to 30% NBR, but this decrease is no smaller than that for PS blended with PB only. In the ABS resin described above, when only the lubricating characteristics are considered, it is clear that μ_k and V_s are greatly reduced by blending

with 10 to 20% PB, and the friction characteristics and wear resistance are thus improved.

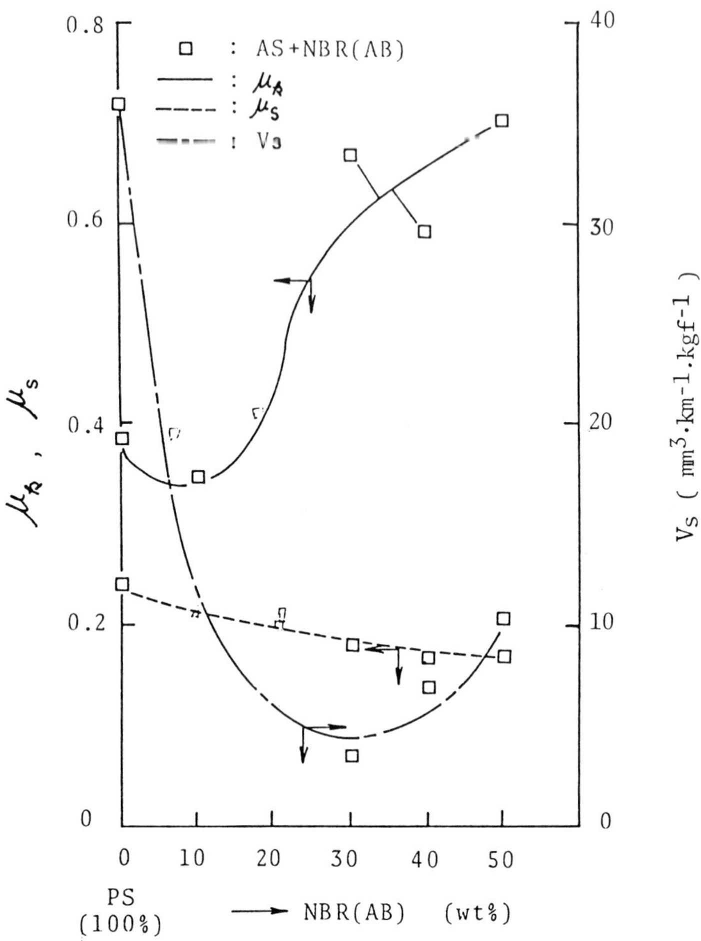

Fig. 3.4 Relationships between μ_k, μ_s and V_s and the blending percentage of NBR (AB) in PS

3.1.2 *Polyamide and polyacetal with polybutadiene and polytetrafluoroethylene [2,3]*

(i) *Experiment*

Table 3.2 shows the content of each component and the conditions of blending and moulding for polyacetal (POM) and polyamide (PA) blended with polybutadiene (PB) or polytetrafluoroethylene (PTFE). Experiments to obtain the characteristic values, μ_k, μ_s and V_s of these materials were conducted as described in the previous section. The conditions of the continuous sliding test were: contact pressure, 5 to 9.2 kgf/cm^2; velocity, 80 cm/s; and sliding distance, 3000 to 5000 metres. The limiting pv values (pv_{max}) at which unstable friction recurred were obtained by increasing the contact pressure from 1.2 to 12.2 kgf/cm^2 using the device shown in Fig. 2.1(a).

TABLE 3.2
Composition and moulding conditions of blending for POM and PA

Base Polymer	Blending Polymer Type	Blending Polymer Percentage(%)	Blending Conditions Temp (°C)	Blending Conditions Time (min)	Moulding Conditions Temp (°C)	Moulding Conditions Time (min)
Polyacetal	PB	0,5,10,15,20,25	195	15	190	5
POM	PTFE	0,10,20	195	15	190	5
Polyamide	PB	0,5,10,15,20	240	15	240	5
PA (Nylon 6)	PTFE	0,10,20	240	15	240	5

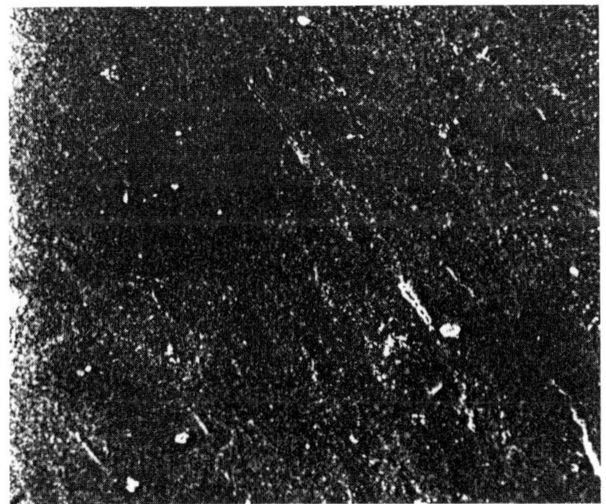

Fig. 3.5 Surface of PA (100%) by SEM before wear test (x 100)

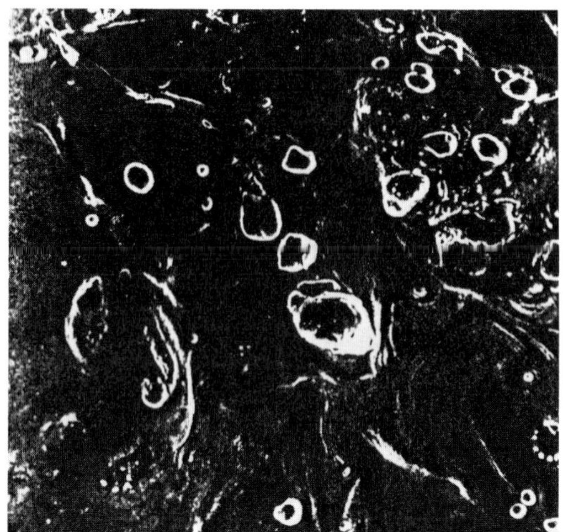

Fig. 3.6 Surface of polyblend PA with 15% PB, by SEM before wear test (x 100)

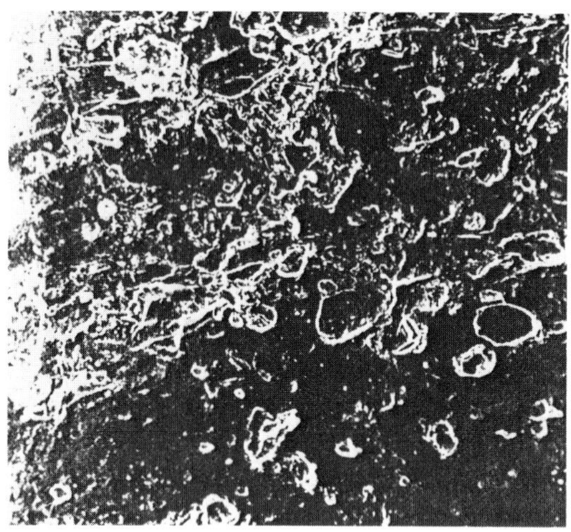

Fig. 3.7 Surface of polyblend PA with 20% PTFE, by SEM before wear test (x 100)

(ii) *Results*

(1) *Morphological review.* Scanning electron micrographs (SEMs) of polyamide (PA) and its blended surfaces are shown in Figs. 3.5 to 3.9. Figure 3.5 shows a SEM of PA (100%) surface before the wear test and Figs. 3.6 and 3.7 show those of PA blended with 15% PB and 20% PTFE respectively, also before the wear test. Figures 3.8 and 3.9 show PA blended with 15% PB

and 10% PTFE after the wear test. In the SEMs for blended PA before wear, the added polymer, PB or PTFE, is distinguishable as separate spherical particles. However, as shown in Figs. 3.8 and 3.9, these polymers can no longer be clearly distinguished after wear.

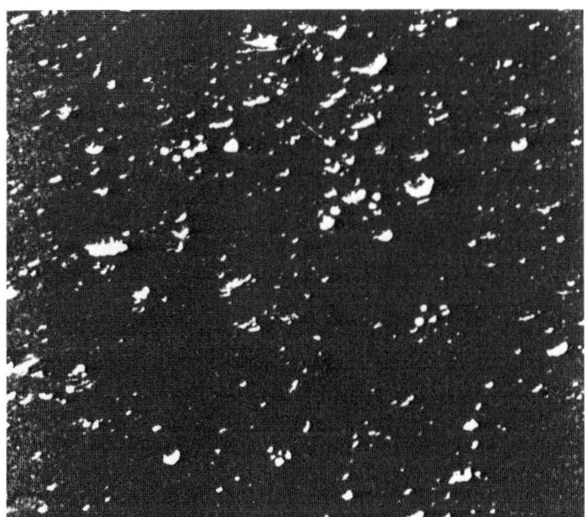

Fig. 3.8 Surface of polyblend PA with 15% PB, by SEM after wear test (x 100)

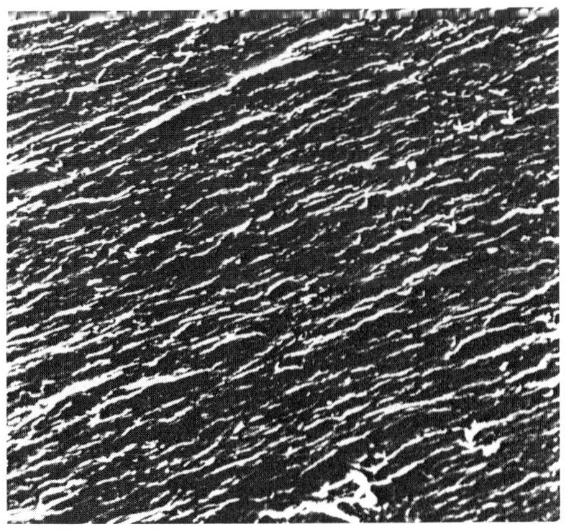

Fig. 3.9 Surface of polyblend PA with 10% PTFE, by SEM after wear text (x 100)

(2) *POM and PA blended with PB.* Figure 3.10 shows the relationships between μ_k, V_s composition for POM and PA blended with 5% to 20% PB. In both curves, μ_k is a minimum when PB is approximately 5% and V_s is considerably reduced from 5% to 15% PB. When POM or PA was blended with 5% to 25% PB, μ_k and V_s decreased significantly, as in the case of PS. However, the blended composition of PB for the minimum value of μ_k and V_s is somewhat different to that of POM or PA.

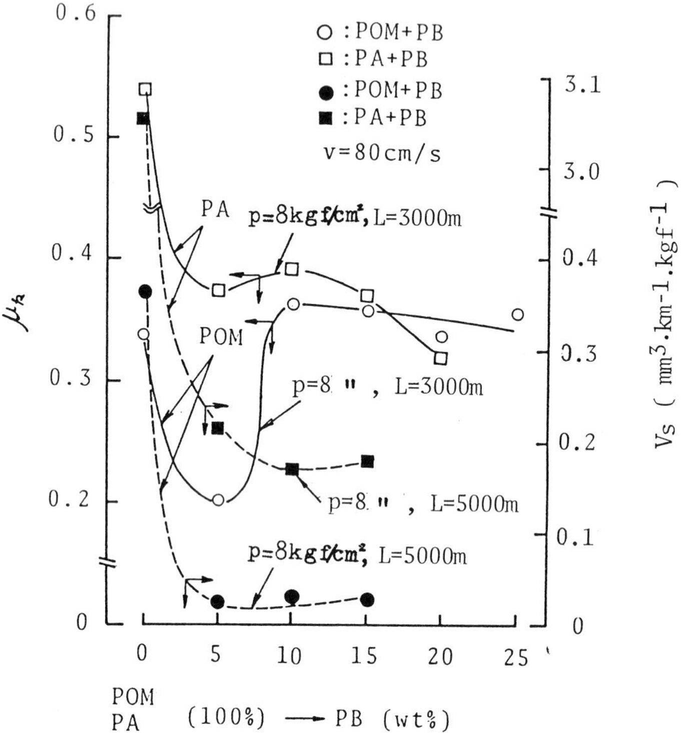

Fig. 3.10 Relationships between μ_k, V_s and the blending percentage of PB in POM or PA

(3) *POM and PA blended with PTFE.* Figure 3.11 shows the relationship between μ_k, V_s and composition for POM and PA blended with PTFE from 0% to 20%. The main effects of the PTFE additions to both polymers are that beyond about 10% μ_k decreases slightly and V_s decreases considerably.

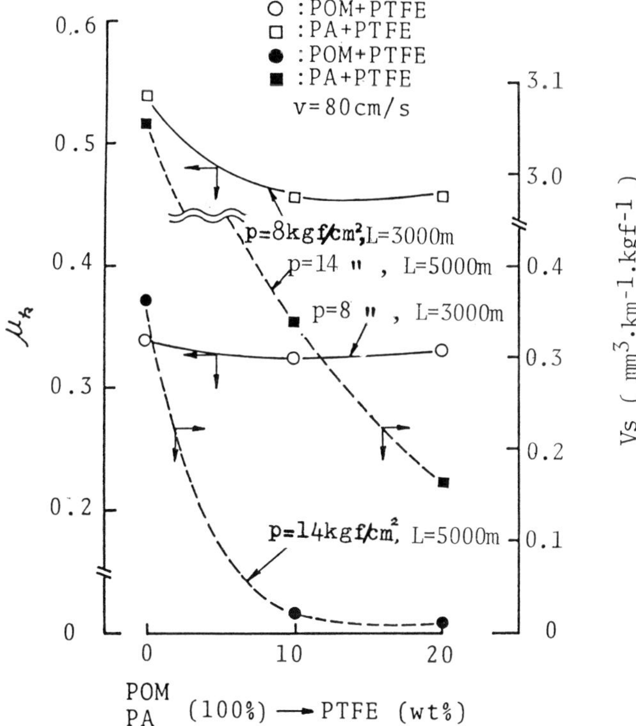

Fig. 3.11 Relationships between μ_k, V_s and the blending percentage of PTFE in POM or PA

(4) *Effect of POM and PA blended with PB or PTFE on V_s'.* Figure 3.12 shows the relationships between the specific adhesive wear rate V_s' and composition for POM blended with PB and PTFE (lower curves) and PA blended with PB and PTFE (upper curves). The wear tests were made with the discontinuous sliding (pin-ring) arrangement shown in Fig. 2.1(b) at 80 cm/s, a load of 6.2 kgf and for a sliding distance of 600 metres. For POM, V_s' decreases when the PB or PTFE content is less than approximately 15%, and tends to increase above this value. The same trend can be seen for PA.

(5) *Effect of PB and PTFE on hardness.* Figure 3.13 shows the relationship between the Rockwell Hardness. H_R, of the moulded surface of POM and PA blended with PB or PTFE, and the blending rate. The two upper curves are for POM with PB or PTFE, and the two lower curves for PA. In both cases, the hardness is greatly reduced by blending. The reduction with PB is substantially larger than that caused by PTFE.

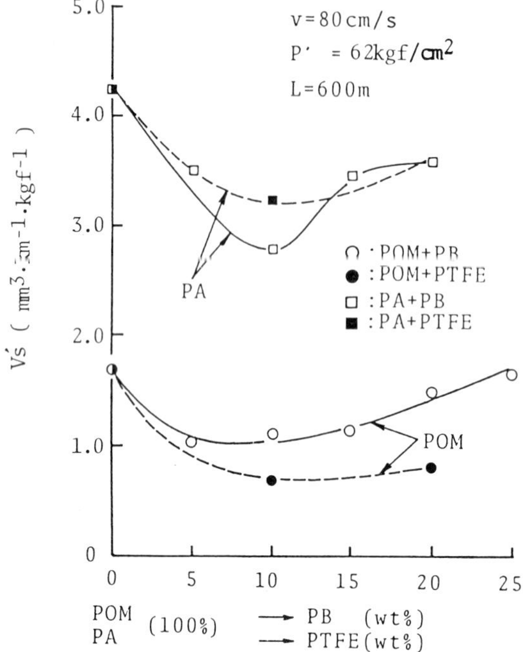

Fig. 3.12 Relationship between V_s' and the blending percentage of PB or PTFE in POM or PA

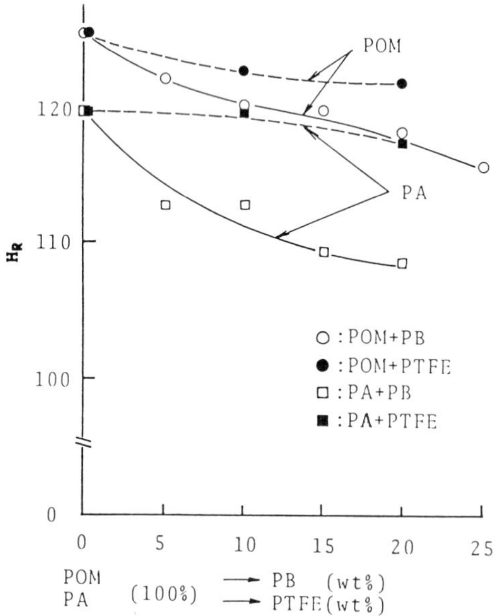

Fig. 3.13 Relationship between the Rockwell hardness, H_R and the blending percentage of PB or PTFE in POM or PA

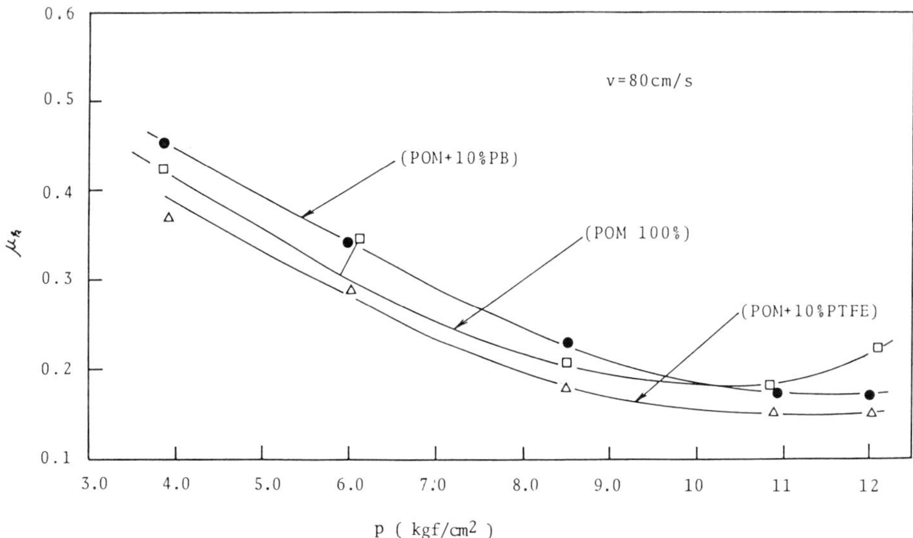

Fig. 3.14 Relationship between μ_k and pressure, p, for POM and POM blended with PB or PTFE

(6) *Effect of blending on the limiting pv value.* When the pv value during sliding is too high the rise in temperature caused by frictional heating may exceed the limiting service temperature. This results in an inability to continue with normal, stable friction operation. The pv value in this condition is called the "limiting pv value", as described in Chapter 1.6.1. In this chapter, however, it is defined as the limiting pressure at the point, and the limiting pv value is the value of the limiting pressure multiplied by the friction velocity at the same point. Figure 3.14 shows μ_k-p curves at a velocity of 80 cm/s for 100% POM, and POM blended with 10% PB, or 10% PTFE. In the curve for 100% POM, the limiting pressure is 10.9 kgf/cm^2; thus, the limiting pv value is approximately 872 kgf/cm^2·cm/s. However, for POM blended with 10% PB or 10% PTFE, the limiting pv value exceeds 1020 kgf/cm^2·cm/s.

Figure 3.15 shows the μ_k-p curves for 100% PA and PA blended with 10% PB or 20% PTFE. Again it can be seen that the limiting pv value improves with blending, because of an increase in the limiting pressure.

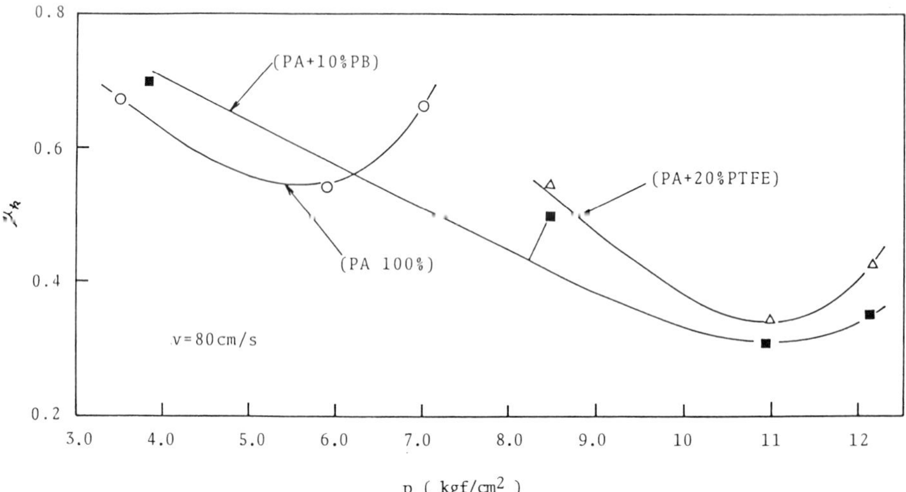

Fig. 3.15 Relationship between μ_k and pressure, p, for PA and PA blended with PB or PTFE

3.2 COMPOSITES

A very wide range of different types of plastics and fillers are potentially useful for tribological applications and the selection of the most appropriate combination for any particular application is far from simple. There are few theoretical guidelines and it is necessary to rely mainly on experimental information derived from friction and wear testing.

Examples of relevant experimental work included that by Lomax [4] and Play [5] on composites of some plastics with polytetrafluoroethylene (PTFE) and by Giltrow [6], Lancaster [7,8], Miyake [9], Tanaka [10] and Briscoe [11] on composites of plastics with carbon or carbon fibre. There is also work on plastics-mica composites by Eiss [12], plastics with boron fillers by Plumb [13], plastics and glass fibers by Tsuzuki [14], plastics and asbestos by Pogosian [15], plastics and potassium titanate fibres by Halberstadt [16], Nylon 6 or PTFE with various fillers by Clauss [17,28], and plastics and oil by Dzhanakmedov [19].

In this book, some experimental results are described for nine combinations of five thermosetting and six thermoplastic materials with various fillers, together with the effects of plastic coating of a metal surface, as shown in Table 3.3.

TABLE 3.3(a)
Combinations of plastic composites for sliding use

No.	Plastics	Fillers	Test Methods
1 [ref.20]	Diallyl Phthalate Resin (DAP)	Asbestos Fibres (1-%-20%) Carbon Fibres (1-5%) MoS_2 4F (PTFE) Glass Fibres (20%)	Tension, flexural Ogoshi's wear tester JIS-K 7205
2 [ref.21]	Epoxy Resin (EP) Unsaturated Polyester (UP)	Silica Powder (10-60%)	Bending Continuous sliding Friction and wear tests Ogoshi's wear tester JIS-K 7205
3 [ref.22]	Phenolics (PF) and (DAP)	Asbestos Glass Fibres	Flexural Non-conforming, friction and wear tests (pin-ring)
4 [ref.23]	Polyimide (PI)	Graphite (25-40%) Asbestos (10%) Glass Fibres (10%) MoS_2 (25%) PTFE (20%)	Flexural, compression, hardness Continuous sliding friction, wear Ogoshi's wear tester
5 [ref.28]	Polyacetal (POM)	Carbon Fibres (5-25%) PTFE (5-25%)	Continuous sliding friction, wear

TABLE 3.5
Specific wear rate of diallyl phthalate resin (DAP) (mm^3/kgf·km) [8]

Filler	Amt.%	Specific adhesive wear rate V'_s (plate-ring)	Specific abrasive wear rate V_{sa} (JIS-K 7205)
Diallyl Phthalate (DAP):			
Glass fibres	0	7.12	378
	20	0.15	571
Asbestos	1	0.10	407
	3	0.12	300
	5	0.15	263
Carbon fibres	1	1.77	374
	3	4.17	544
	5	5.08	578
MoS_2	1	2.82	382
	3	1.20	340
	6	3.04	298
PTFE	5	0.65	539
	10	0.27	380
	15	0.38	404
Asbestos filled phenolic		1.20	-
Asbestos, glass fibre filled DAP		0.10	-
Cloth filled phenolic		-	343
Polycarbonate		1.56	-
ABS resin		0.57	-
Polyacetal		0.09	-

3.2.2 *Composites of epoxy resin and unsaturated polyester [21]*

Epoxy resin (EP) and unsaturated polyester (UP) thermosetting plastics are suitable matrix materials for composites because, unlike phenolic resin, condensation products are not produced during curing and few bubbles remain in the final moulded product. As epoxy resin composites containing silica powder are too difficult to cut with a high-speed steel cutter, the effects of silica filler upon the friction and wear properties of EP and UP composites are discussed experimentally.

Table 3.6 shows details of the various materials used (matrix and filler), the filler content and the moulding conditions of specimens. The mean diameter of the silica powder was approximately 180 μm.

TABLE 3.6
Types and moulding conditions of specimen

Matrix	Filler	Amt.%	Details of Matrix	Moulding Condition
Epoxy Resin	Silica	0	Araldite B	Curing temperature:
(EP)	powder	10	(CT-200)	180°C
		20	Hardner HT 901	Curing time: 3 hrs.
		40	(100:30)	
		60		
Unsaturated	Silica	0	Ester G43	Curing temperature:
Polyester	powder	10	BP0 paste	60°C
(UP)		20	(100:15)	Curing time: 3 hrs.
		40		
		60		

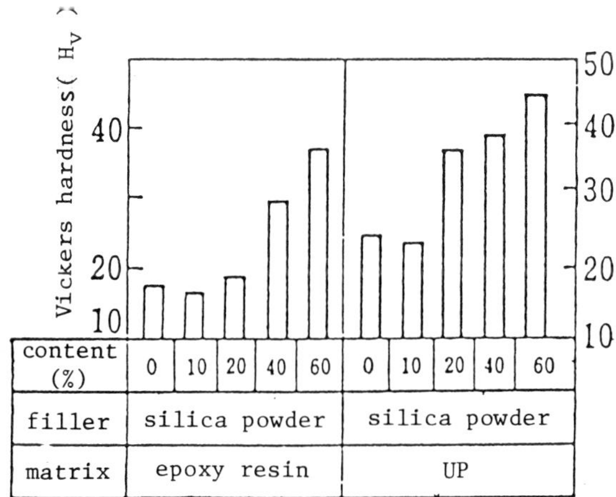

Fig. 3.16 Relationship between hardness and filler content of epoxy resin and UP

TABLE 3.7
Specific wear rate of some plastics (mm^3/kgf·km) [12]

Matrix	Filler	Amt.%	Specific adhesive wear rate V_s' (pin-ring)	Specific abrasive wear rate V_{sa} (JIS-K 7265)
Epoxy resin	Silica powder	0	2.59	85
		10	1.60	84
		20	0.76	73
		40	0.03	65
		60	0.05	255
Unsaturated polyester (UP)	Silica powder	0	5.62	198
		10	3.16	187
		20	2.61	321
		40	0.18	361
		60	0.12	502
Diallyl-phthalate resin (DAP)		0	7.12	378
	Glass	25	0.15	571
	PTFE	10	0.27	380
	Graphite	1	1.77	374
	Asbestos	3	0.12	300
	MoS_2	3	0.19	340
Cloth-filled phenolics			0.06	300
POM			0.09	
ABS resin			0.57	
PC			1.56	

Details of the asbestos component are shown in Table 3.4. Table 3.8 shows the details and some mechanical properties of each matrix (phenolic (PF) and diallyl phthalate (DAP) resins) and three types of these composites containing 60% asbestos in the former and 10% or 20% in the latter. The mechanical properties of the DAP composites containing glass fibres are generally higher than those of PF composites. Figures 3.20 and 3.21 show the relationships between μ_k and the velocity, v, at different contact pressures for a PF composite containing 60% asbestos and a DAP composite containing both 20% glass plus 20% asbestos fibres. The μ_k values were measured against a steel disc at various velocities, v, and contact pressures, p, using the friction and wear tester shown in Fig. 3.19.

TABLE 3.8
Specimen

Types	Details of Matrix	Hardness H_v	Flexural Strength ($kgf{\cdot}mm^{-2}$)	Charpy Impact Strength ($kgf{\cdot}cm{\cdot}cm^{-2}$)
60% Asbestos filled phenolic	Sumitomo Bakelite	46.6	5-6	3
10% Asbestos, 20% glass fibre filled DAP resin	Sumitomo Bakelite AM200,SF3	60.5	8-12	5
20% Asbestos 20% glass fibre filled DAP resin		48.8	8-12	15

Note - Moulding: compression pressure, 150 $kg{\cdot}cm^{-2}$; temperature, 150°C; time, 10 minutes

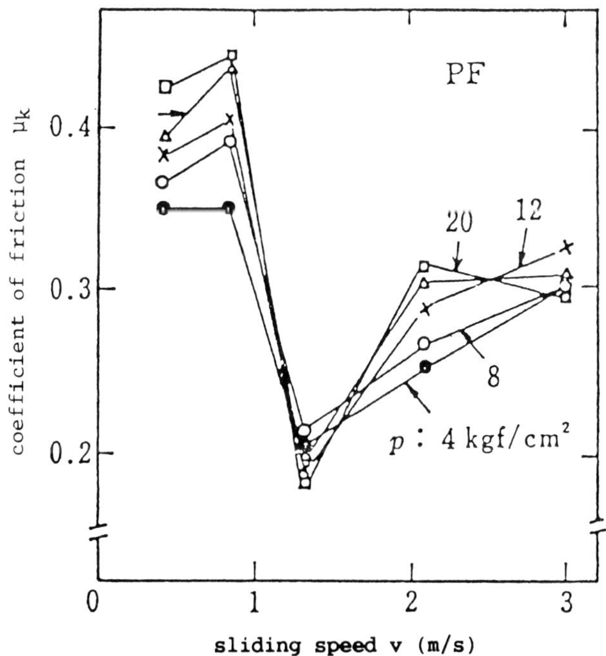

Fig. 3.20 Relationships between μ_k and sliding speed, v, at various pressures for a 60% asbestos-filled phenolic

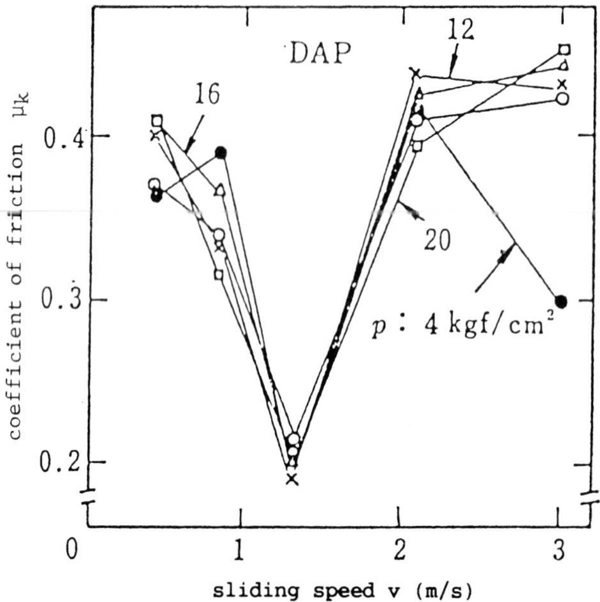

Fig. 3.21 Relationships between μ_k and sliding speed, v, at various pressures for DAP filled with 20% glass fibre and 20% asbestos

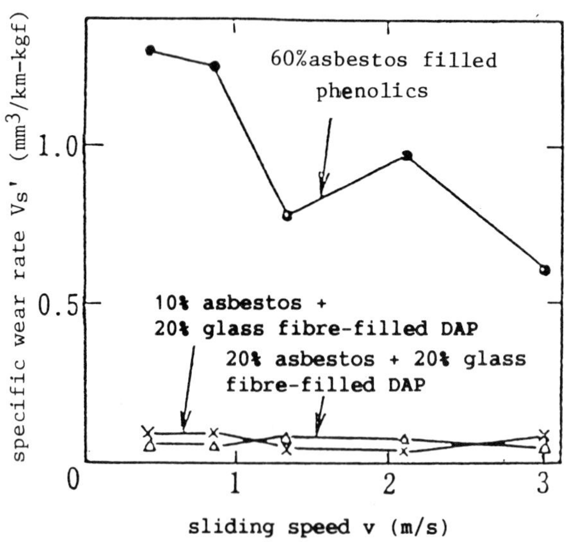

Fig. 3.22 Relationships between V_s and sliding speed, v, for three types of plastic composites

It is clear from these figures that the μ_k of both composites falls to a minimum at a velocity of 1.3 m/s, and is approximately 0.4 at a velocity of 0.2 m/s. Above 2 m/s, however, μ_k is approximately 0.28 for PF and 0.42 for DAP. Figure 3.22 shows the relationships between the adhesive wear rate, V_s' in non conforming contact (pin ring) and velocity, v. The value of V_s' is from 0.6 to 1.3 mm^3/kgf·km for PF composites, and 0.08 to 0.11 for DAP composites, the latter being about 1/7 to 1/10 of the former. It is concluded from these figures that asbestos fibres are generally effective as a filler to improve both the sliding frictional resistance and the adhesive wear resistance. The value of V_s' exceeds that of unfilled PF resin only when the asbestos content is over 60%. Both the mechanical strength and the adhesive wear resistance are improved by using asbestos fibres together with approximately 20% glass fibres.

3.2.4 *Composites of polyimide [23]*

Polyimide, a thermosetting plastic composed of linear polymers and first developed by the du Pont Company in 1959, possesses high heat resistance and excellent sliding properties. Currently, many types of polyimides are being produced in various countries for special uses, despite the high costs. Some studies of the sliding properties of polyimides have been made by Lewis [24], Matsubara [25,26] and Alvay [27].

The details of polyimide (addition copolymer polyaminobismaleimide, PABM, Kerimid) composites containing fillers of graphite, MoS_2, PTFE, asbestos and glass fibres, are shown in Tables 3.9 and 3.10. The specimens were made by the compression moulding process shown in Fig. 3.23. The specific adhesive wear rate, V_s or V_s', of the composites against steel, using both conforming (thrust-washers) and non-conforming (plate-ring) testers, and their mechanical properties, i.e. Rockwell hardness, flexural and compressive strengths, were measured experimentally.

TABLE 3.9
Details of fillers in polyimide - Kerimid 604-1000 (before Kinel 1000), Rhodia Co.

Filler	Details
Graphite	Hidachifunmatsu Yakin Co., GP 100
MoS_2	Hidachifunmatsu Yakin Co., MD 108
PTFE	Unon P 310
Asbestos	Nippon Asbestos Co. (crysotile)
Glass Fibre	Asahi Glass Fibre Co., CS0 3HB, 3mm

TABLE 3.10
Composition of Specimens

No. of Specimen	Polyimide (wt%)	Filler Type	Filler Amt.(wt%)
1	100	-	-
2	75	Graphite	25
3-1	60	Graphite	40
3-2	60	Graphite	30
		Asbestos	10
3-3	60	Graphite	30
		Glass Fibre	10
4	65	Graphite	10
		MoS_2	25
5	80	PTFE	20

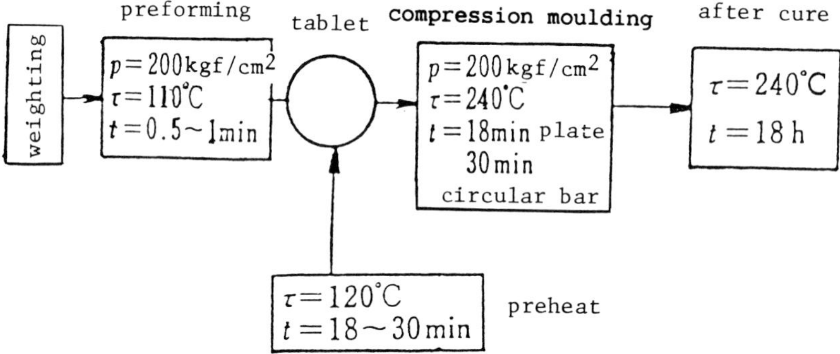

Fig. 3.23 Diagram of moulding process

Figure 3.24 shows the Rockwell hardness, H_RM, the flexural and compressive strengths, σ_b and σ_c, of each polyimide composite, and indicates that values decrease with the addition of fillers. Figure 3.25 shows an example of the relationship between temperature or μ_k, and the contact pressure, p, for composites No. 3.3. It can be seen that the value of μ_k decreases to as low as 0.05 and remains approximately constant even when the surface temperature rises to a comparatively high value of 150°C at high contact pressures.

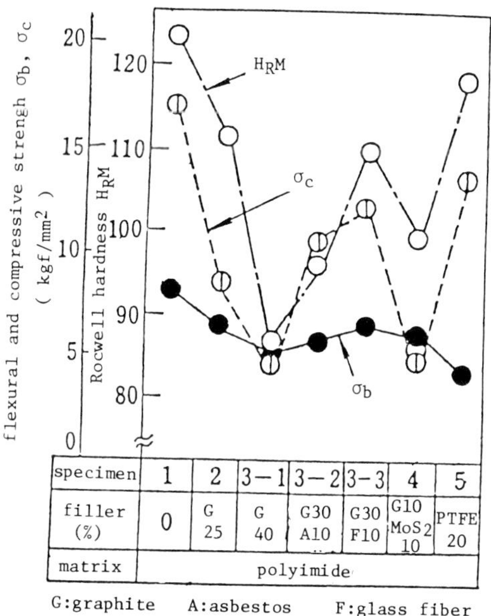

specimen	1	2	3−1	3−2	3−3	4	5
filler (%)	0	G 25	G 40	G30 A10	G30 F10	G10 MoS2 10	PTFE 20
matrix	polyimide						

G:graphite A:asbestos F:glass fiber

Fig. 3.24 Flexural strength, σ_b, compressive strength, σ_c, and hardness of compression mouldings of polyimide composites

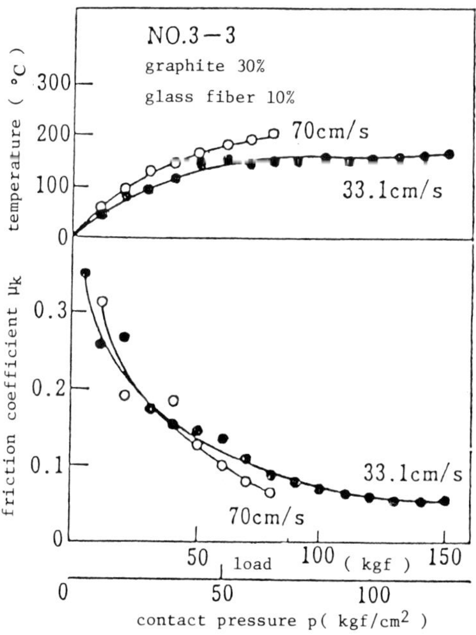

Fig. 3.25 Relationships between surface temperature, friction coefficient, μ_k, and contact pressure at two speeds

TABLE 3.11
Values of limiting pv for various plastics (continuous sliding friction)

	Specimen	pv_{max} (kgf/cm²·cm/s)
Polyimide:		
1	No Filler	-
2	Graphite 25%	>5740
3-1	Graphite 40%	>4592
3-2	Graphite 30%, Asbestos 10%	
3-3	Graphite 30%, Glass Fibre 10%	>4900
4	Graphite 10%, MoS_2	
4	Polytetrafluoroethylene (PTFE) 20%	>2390
	Polytetrafluoroethylene (PTFE)	10-13.5 (23°C atmosphere)
	Cloth-filled Phenolic	850 (2°C water cooling) 750 (23°C atmosphere) 310 (50°C water cooling)
	Polyacetal (Dellin)	1400 (2°C water cooling) 820 (16°C water cooling) 560 (23°C atmosphere) 240 (50°C water cooling)
	Polycarbonate (PC)	460 (23°C atmosphere)
	ABS Resin	500 (23°C atmosphere)

Figure 3.26 shows the μ_k values of each composite against steel at v = 41 cm/s, p = 8.5 kgf/cm² or v = 10 m/s, p = 50 kgf/cm², and indicates that the value of μ_k in the composites is smaller than that of the unfilled polyimide.

Table 3.11 shows the limiting pv values under continuous sliding for the polyimide composites compared with those for other plastic materials. The limiting pv values of these other plastics are generally lower than 1000 kgf/cm²·cm/s; however, those for the polyimide composites range from 2300 to 5740 kgf/cm²·cm/s. Polyimide composites can therefore be used for very severe service conditions involving high contact pressures and speeds.

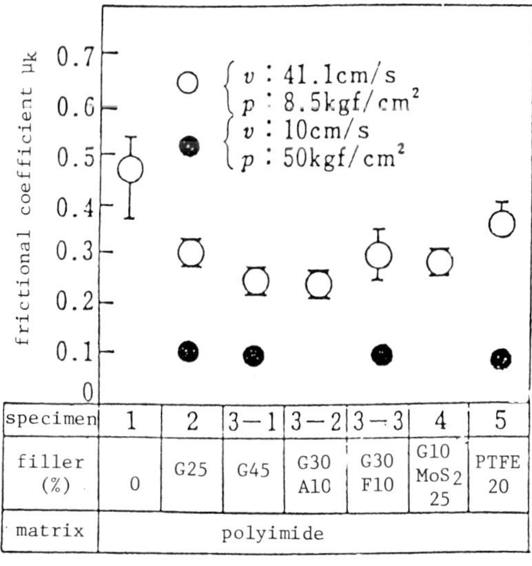

Fig. 3.26 Coefficients of kinetic friction of various polyimide composites at different p and v values (against steel)

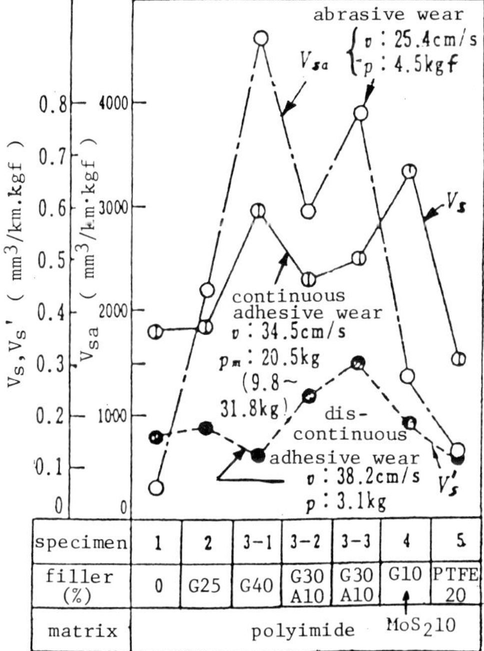

Fig. 3.27 Specific abrasive and adhesive wear rates, V_{sa}, V_s, V'_s of various polyimide composites (against steel)

Figure 3.27 gives values for all the polyimide composites of the specific adhesive wear rates in conforming (continuous) contact V_s, and non-conforming (discontinuous contact V'_s, together with the specific abrasive wear rates V_{sa}. This figure indicates that the values of both V'_s and V_s, for the unfilled polyimide are generally smaller than those of the polyimide composites, except for the one containing PTFE. The V'_s and V_s values of polyimide and its composites are compared with those of other plastic materials in Table 3.12. The values for the polyimide and its composites are generally amongst the lowest.

TABLE 3.12
Specific wear rates of polyimides and other plastics (against steel)

Material	Specific Adhesive Wear Rate Nonconforming contact (V'_s, mm^3/kgf·km)		Material	Specific Adhesive Wear Rate Continuous contact (V_s, mm^3/kgf·km)	
Polyimide:					
No. 1	0.16		No. 1	0.36	
No. 2	0.18		No. 2	0.37	
No. 3-1	0.10		No. 3-1	0.60	
No. 3-2	0.22	v: 38.1	No. 3-2	0.46	v: 34.5
No. 3-3	0.30	(cm/s)	No. 3-3	0.50	(cm/s)
No. 4	0.18		No. 4	0.67	
No. 5	0.08		No. 5	0.30	
Other Plastics:					
Ryton R4	0.66		POM No filler	0.78	
Ryton R14	0.20		CF 25%	0.8-1.3	
POM	0.14		PTFE 10%	0.2	
ABS	0.44				
PC	0.84		PTFE	2.6	
PVC	1.10		Cloth filled phenolic	2.4	
Cloth Filled Phenolic	0.24				
Epoxy Resin	5.18		ABS	2.0	
DAP Resin	14.24		PC	12.5	
UP	11.24		Ekonol	0.098	

3.2.5 *Polyacetal composites containing carbon fibres and PTFE*

Polyacetal (polyoxymethylene, POM) is a highly crystalline polymer having a comparatively low elongation, a high tensile strength and excellent sliding properties, as shown in Tables 1.15 and 3.5. It has already been shown that additions of PTFE or carbon fibres to POM are very effective in improving the frictional resistance and the limiting pv value. This section discusses polyacetal copolymer (Duracon M90) composites, filled with low modulus carbon fibres of various lengths (as shown in Table 3.13) or with low-molecular-weight PTFE powders, with respect to the improvement of their sliding properties. The specimens were produced by compression moulding, as shown in Table 3.13.

TABLE 3.13
Types and moulding conditions of specimens

Filler Type	Filler Length (mm)	Filler Content %
	0	0
		5
	1	10
		25
Carbon Fibre		5
	3	10
		25
		5
	5	10
		25
		5
PTFE	Powder	10
		25

Matrix: Polyacetal (P.Co., Duracon M90)
Moulding: Compression moulding (temperature: 200°C; pressure: 130 kgf/cm^2; rapid cooling after moulding)

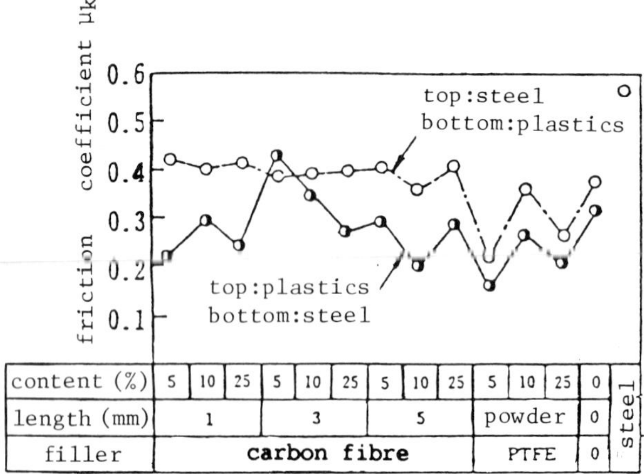

speed:6.2cm/s contact pressure:1.4kgf/cm^2

Fig. 3.28 Coefficients of kinetic friction, μ_k, of polyacetal composites

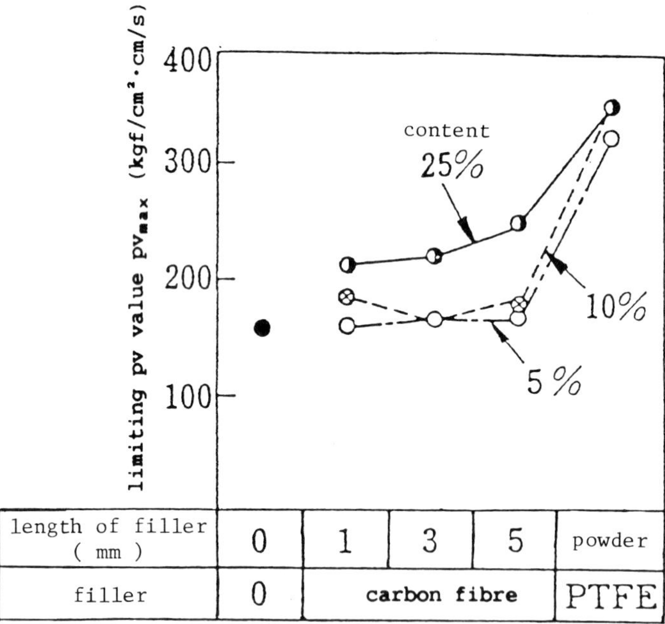

Fig. 3.29 Limiting pv values of polyacetal composites (against steel)

The μ_k values were measured under continuous sliding between the end surfaces of two opposing hollow cylinders, as shown in Fig. 2.1(a). Figure 3.28 shows these values for all the composites in the two specimen arrangements of top/bottom, steel top/plastic bottom and vice versa. It is clear from Fig. 3.28 that the composites containing 5% PTFE have a minimum μ_k value of 0.2 or 0.15, which is around 0.35 or 0.3 times that of the matrix itself. However, the μk values of the composites filled with carbon fibres are not very different from those of the matrix itself, regardless of the length of content of carbon fibres. Table 3.14 shows the continuous specific adhesive wear rate, V_s, of the various POM composites and for unfilled POM and PTFE. The V_s value of unfilled POM is 0.78mm^3/kgf·km, and that of the composite filled with carbon fibres is 0.83 - 1.39 mm^3/kgf·km; the wear resistance therefore decreases. For the composites filled with PTFE powders, V_s is 0.2 - 0.32 mm^3/kgf·km, i.e. wear resistance is improved by three to four times. The improvement in the adhesive wear resistance over the range of 5% to 10% PTFE is particularly significant.

TABLE 3.14
Specific adhesive wear rate

Matrix	Filler Type	Filler Length (mm)	Filler Content(%)	Specific Continuous Adhesive Wear Rate (V_s, mm^3/kgf·km)
POLYACETAL (POM)		0	0	0.78
		1	5	1.21
			10	1.23
			25	0.97
	Carbon Fibre	3	5	1.12
			10	1.23
			25	1.16
		5	5	1.18
			10	0.83
			25	0.93
	PTFE	Powder	5	0.21
			10	0.20
			25	0.32
PTFE				2.26

Figure 3.29 shows the limiting pv value at which the adhesive wear rates increase suddenly for the various POM composites. The value for the POM matrix itself is about 160 kgf/cm²·cm/s, and that for the composites filled with 5 - 10% carbon fibre is nearly the same. However, for the composites filled with 25% carbon fibres (5 mm in length) or with PTFE powders, the pv is about 250 or 350 kgf/cm²·cm/s.

3.2.6 *Composites of polyvinylchloride (PVC) [29]*

Polyvinylchloride (PVC), an amorphous high polymer, is not suitable for severe sliding conditions because it is high in frictional resistance and low in heat resistance, as compared with semi-crystalline polymers, such as polyacetal, nylon and polytetrafluoroethylene. However, PVC is very useful for floor materials, pipes, furniture and various construction applications because of its relatively low cost. This section examines the effects of fillers and of coupling agents which increase the bonding between the matrix and fillers in order to improve the abrasive wear resistance for floor materials, and the adhesive wear resistance in general.

The composites used for the experiments consisted of PVC as the matrix and four types of fillers: glass fibres, carbon fibres, asbestos fibres and PTFE powder; details are given in Table 3.15.

TABLE 3.15
Types of filler

Type	Details	Maker
Glass Fibre	Glasslon Yarn, dia.: about 6 μm	Asahi Glass Fibre Co.
Carbon Fibre	Lignin Graphite (CLL) dia.: about 14 μm	Nippon Kayaku Co.
Asbestos	Crysotile Fibre W-30,length: below 840 μm	Nippon Asbestos Co.
PTFE	Unon P 310, dia.: below 10 μm	Nippon Valker Co.

The glass fibres, 3 mm and 6 mm in length, were treated with a silane or chrome coupling agent. The type and chemical structure of the coupling agents used are shown in Table 3.17. The PVC composite specimens, containing each filler in the amounts shown in Table 3.16, were produced by compression moulding. Abrasive and adhesive wear properties were measured using an Elmendorf impact abrasive wear tester, as shown in Fig. 2.39, or a non-conforming contact (plate-ring) wear tester (Ogoshi-type) as shown in Fig. 2.1(b). The contact angle, θ, between the molten polymer and a glass surface, as shown in Fig. 3.30, was measured to estimate the wettability of the glass. Figure 3.30 shows each contact angle, θ, for each coupling agent with which the glass surface was treated. Figure 3.30 also shows that the minimum contact angle was obtained following treatment with the coupling agent, A-187. Composite specimens containing glass fibres treated with this particular coupling agent were thus used for the wear tests.

TABLE 3.16
Specimen

FILLER				
	Type		Length (mm)	Content (%)
Glass Fibres:				0
	No treatment		3	10
				20
			6	10
				20
	Silane	A-187	3	10
	Coupling Agent			20
			6	10
				20
		A-172	6	10
		A-174	6	10
	Chrome Coupling Agent	MCC	6	10
Carbon Fibres				5
			3	10
				20
				5
			6	10
Asbestos				5
				10
				20
PTFE			Powder	10
				20

Matrix: PVC
Moulding: Compression: temperature: 190°C; pressure: 10 kgf/cm^2

TABLE 3.17
Types of coupling agents used for surface treatment

Type		Chemical Structure	Maker
Silane Type	A-172	$CH_2{=}CHSi(OCH_2CH_2OCH_3)_3$	Union Carbide Co.
	A-174	$CH_2{=}C(CH_3){-}C({=}O){-}O(CH_2)_3Si(OCH_3)_3$	
	A-187	CH_2-OH-CH_2-O-$(CH_2)_3Si(OCH_3)_3$ (with O bridging the first CH_2 and OH)	
Chrome Type	MCC	CH_3 / $CH_2{=}C{-}C$ bonded to O—$CrCl_2$ and O—$CrCl_2$, both $CrCl_2$ bridged by OH	Du Pont Co.

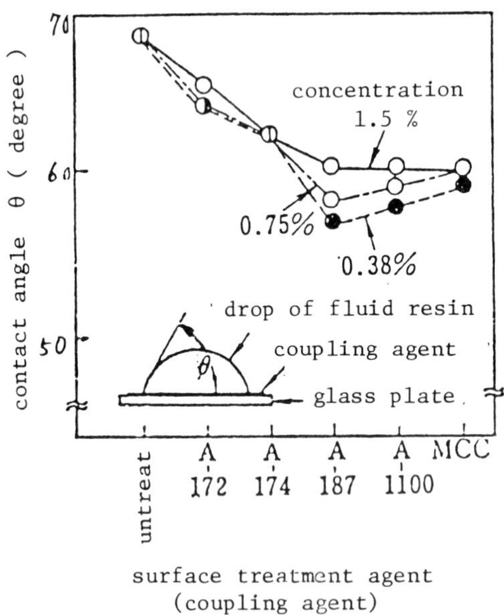

Fig. 3.30 Contact angle θ, between resin and a glass plate treated with various coupling agents

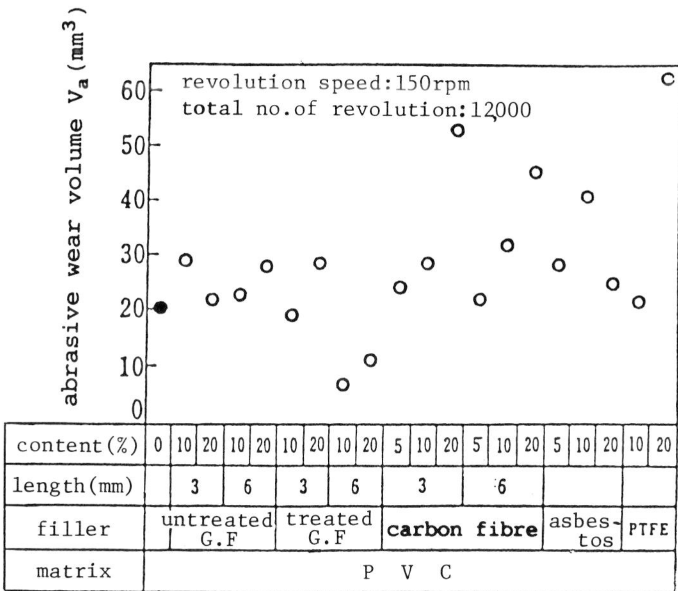

Fig. 3.31 Abrasive wear volumes for PVC and its composites

Figure 3.31 shows the abrasive wear volume V_a, after 1200 revolutions at 150 rpm for PVC alone and all the PVC composites. It indicates that the abrasive wear volume of the composites is generally equal to or greater than that of the PVC alone. However, for the composites containing 10% or 20% glass fibres treated by the coupling agent A-187, V_a decreases slightly from 1/1.5 to 1/2.5 of the value for PVC itself. It should be noted that the abrasive wear volume is proportional to the square of the rotational speed in this test.

Figure 3.32 shows the adhesive wear volumes V' obtained for each composite in a non-conforming (plate-ring) tester under a 3.1 kgf load, at a speed of 30 cm/s and after a distance of sliding of 200 metres. It is clear from this figure that the adhesive wear volumes for the composites are smaller than that for PVC itself. The composite containing 10% PTFE has the lowest value, and that containing glass fibres treated with the silane coupling agent is about 1/3 that for PVC itself. In terms of specific wear rates, the V_s' value for PVC itself is 0.92 mm^3/kgf·km and for the PVC composites, ranges from 0.09 to 0.64 mm^3/kgf·km; the fillers thus improve the adhesive wear resistance.

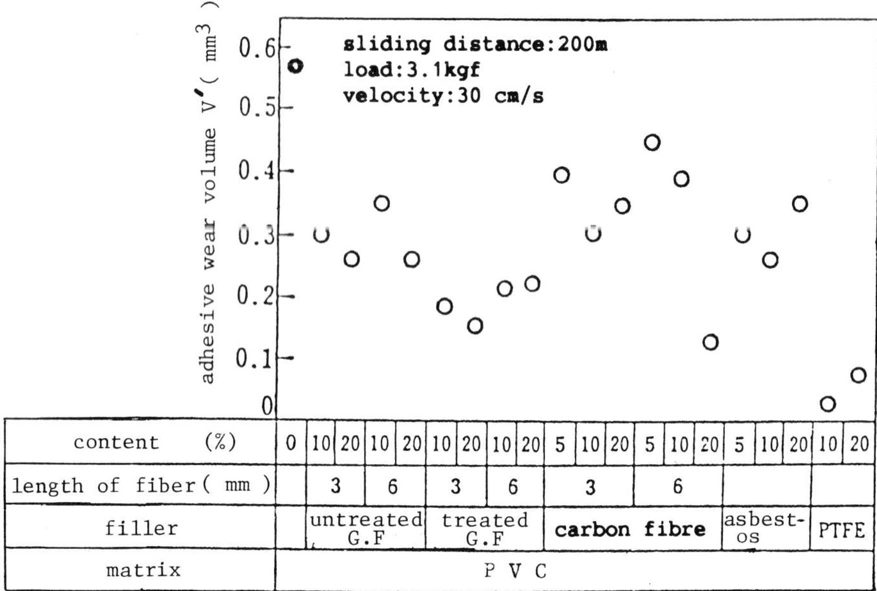

Fig. 3.32 Adhesive wear volumes for PVC and its composites

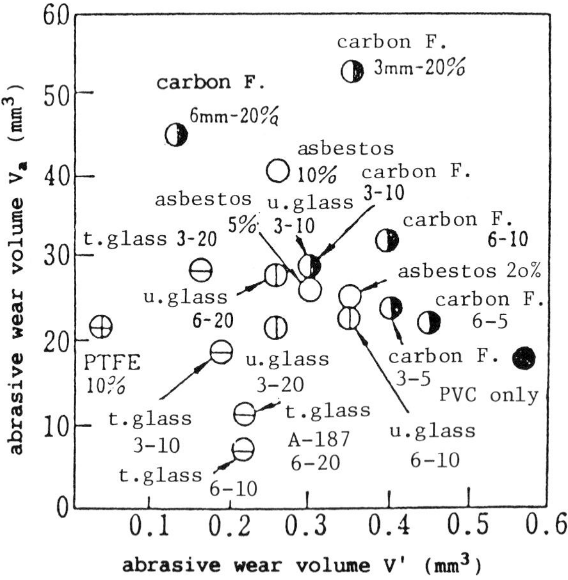

Fig. 3.33 Relationship between abrasive wear volume and adhesive wear volume

Figure 3.33 shows the relationship between the abrasive wear volume, V_a (ordinate), and the adhesive wear volume, V' (abscissa), for each PVC composite. It is very clear from the figure that there is no definite relationship between the two kinds of wear.

3.2.7 *PTFE composites filled with carbon fibres [30,31]*

Polytetrafluoroethylene (PTFE) has excellent frictional properties, and its coefficient of kinetic friction, μ_k, against other materials is often less than 0.1, as shown in Table 1.5. However, its adhesive wear rate increases appreciably above a critical contact pressure. In an attempt to increase the limiting pv value for this plastic, some composites with carbon fibres have been examined and one example carried out in the author's laboratory [30,31] is described in this section. There are many kinds of carbon fibres, as a result of varied manufacturing methods. For these particular tests, both low modulus fibres, CLF and GLF, and high modulus fibres, Sermolon, WH and S, were used as fillers, as shown in Table 3.18. The PTFE matrix was Polyflon moulding powder M-1.1, manufactured by the Daikin Company. The specimens were produced by sintering after mixing the matrix with the fillers prior to the preforming process.

TABLE 3.18
Characteristics of carbon fibres

Type	Diameter (μm)	Specific Weight	Tensile Strength (kgf/mm^2)	Elongation (%)	Modulus (kgf/mm^2)
CLF (Carbonized)	23.9-6.2 (12-14)	(1.6)	30.5 (60)	0.64 (1.0)	1.66x10^3 (3-6x10^3)
GLF (Graphite)	18.1-7.4 (12-14)	(1.8)	36.9 (60)	0.86 (0.8)	1.43x10^3 (3-6x10^3)
Sermolon WH	18.4-10.5 (13-15)	(1.77-1.78)	54.4 (60-90)	0.94 (7.0)	1.84x10^3 (30-70x10^2)
Sermolon S	11.4-6.2 (10-12)	(2.0)	177.1 (182)	0.0 (0.0)	- (42x10^3)

( ): from the manufacturer's catalogue

TABLE 3.19
Properties of carbon fibre-filled PTFE

	Length of fibre (mm)	Content (wt%)	Pre-forming pressure	Modulus (kgf· mm^{-2})	Tensile Strength (kgf· mm^{-2})	Yield Stress (kgf· mm^{-2})	Elong-ation (%)	*Hardness		Specific Weight		**Wear Vol. (mm^3)	Static Friction Coefficient μ_s
								$\overline{H}_R$	H_R	Theoretical	Actual		
No filler	-	-	300	63.8	2.32	0.89	365	-20.1	52.3	-	2.16	3.59	0.18
GLF	3	5	300	113	1.69	1.69	198	1.0	64.4	2.14	2.01	0.87	0.21
Sermolon	3	5	300	191	1.98	1.17	287	19.5	71.5	2.14	2.08	0.72	0.11
		5	300	77.0	1.24	0.95	83.6	5.6	70.8	2.12	2.08	1.30	0.16
	3	15	700	78.0	0.97	0.97	7.1	-37.6	13.6	2.05	1.71	1.42	0.14
		30	1000	41.9	0.30	0.30	3.9	-33.7	-1.9	1.95	2.06	1.28	0.14
		5	300	52.5	0.85	0.85	16.4	-21.1	43.9	2.12	2.05	0.68	0.15
CLF	9	15	700	73.7	0.67	0.67	6.9	-41.2	10.3	2.05	2.05	0.18	0.12
		30	1000	78.0	0.39	0.39	4.8	-41.8	1.2	1.95	2.01	0.98	0.12
		5	300	60.8	1.24	0.98	71.4	-21.6	45.6	2.12	1.99	0.62	0.12
	15	15	700	80.8	0.81	0.81	7.9	-37.5	7.7	2.05	1.99	1.06	0.14
		30	1000	93.1	0.41	0.41	6.8	-43.1	1.6	1.95	2.01	0.98	0.12
		5	300	146	2.24	2.24	12.5	-28.5	87.4	2.14	2.10	0.46	0.15
	3	15	700	115	0.97	0.97	7.1	-37.6	-0.8	2.09	1.61	1.02	0.18
		30	1000	85.5	0.55	0.55	6.7	-19.5	25.3	2.03	1.54	0.69	0.17
		5	300	135	1.19	1.19	7.6	-12.4	55.7	2.14	2.07	0.86	0.12
	9	15	700	133	0.99	0.99	6.9	-42.3	-9.0	2.09	1.51	1.10	0.12
		30	1000	82.5	0.65	0.65	5.8	-34.9	8.3	2.03	1.33	0.98	0.11
Sermolon		5	300	78.6	0.83	0.83	12.6	-12.9	51.2	2.14	1.95	0.76	0.16
WH	15	15	700	126	0.98	0.98	7.0	-12.8	28.6	2.09	1.50	0.52	0.14
		30	1000	111	0.60	0.60	6.8	-35.7	5.7	2.03	1.38	1.37	0.15

* Rockwell R scale: $\overline{H}_R$ at 30 s after loading; H_R: at 30 s after unloading
** by Ogoshi's wear tester, sliding distance 200 m

Mechanical properties of these PTFE composites with a range of carbon fibre contents are shown in Table 3.19. An example of the relationships between fibre content and the tensile modulus of elasticity, tensile strength, yield stress, and maximum elongation for each fibre length is shown in Fig. 3.34.

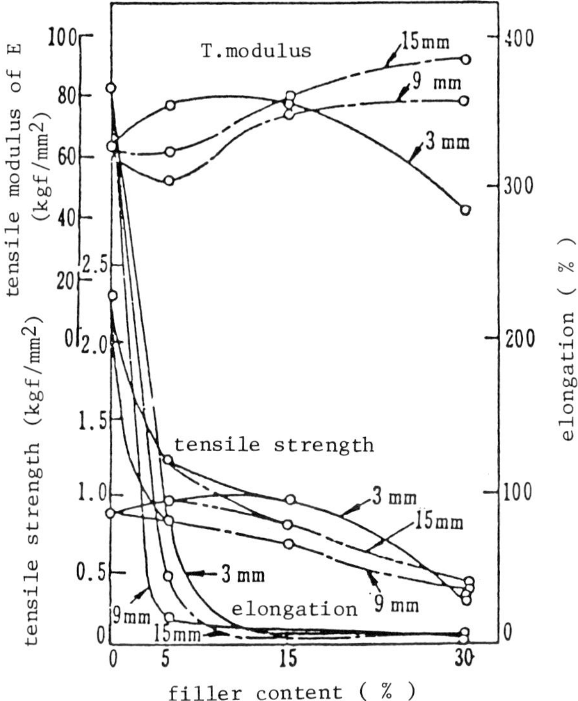

Fig. 3.34 Relationships between tensile characteristics and filler content of CLF-filled PTFE mouldings

It can be seen that as the fibre content increases the tensile modulus of elasticity becomes slightly higher, the tensile strength becomes slightly lower, and the maximum elongation decreases. The same trend occurs for each length of fibre. It is also shown in Table 3.19 that the μ_s values of PTFE itself against steel is 0.18, whereas the values for most of the composites are lower and can even decrease to 0.11. The adhesive wear volume, V, of PTFE generally decreases with carbon fibre content, and becomes 1/8 to 1/3 of itself with a 5% filler content. There is, however, a slight increase with fibre contents beyond 15%. Table 3.20 gives the continuous (conforming contact) specific adhesive wear rate, V_s, for each of the PTFE composites against steel, and it can be seen that the values range from 0.57

to 1.88 mm^3/kgf·km compared with that for PTFE alone of 2.61 mm^3/kgf·km. The value of V_s for the composites with 5% fibres of 1-3 mm length is approximately 1/4 of that for PTFE alone. Figure 3.35 shows the limiting p-v relationships for PTFE and composites filled with carbon fibres, CLFR, of 3 mm length, and indicates that the limiting pv value increases only very slightly with a 5% fibre content, and decreases with high fibre contents.

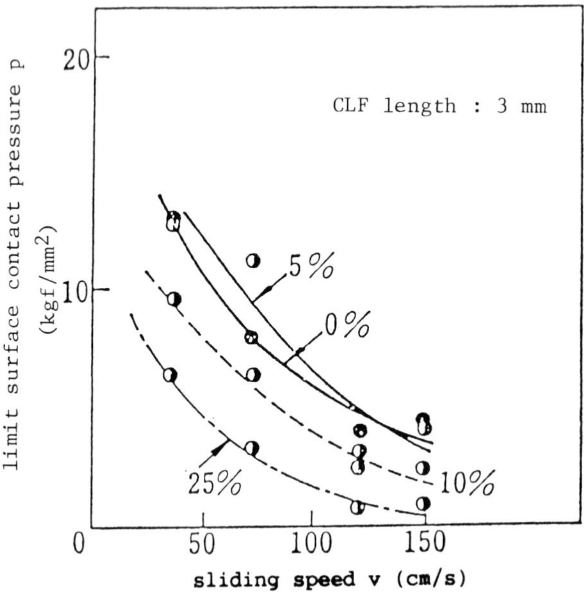

Fig. 3.35 Relationships between limiting contact pressure and speed of PTFE composites filled with carbon fibres of 3mm length

3.2.8 *PPS composites [32]*

Polyphenylene sulfide (PPS, Ryton), developed by the Philips Pet. Co., U.S.A., is now becoming recognized as a useful engineering plastic with excellent heat resistance (288°C melting point), high chemical resistance, and low flammability. In particular, the addition of approximately 40% glass fibres results in marked improvements in the mouldability and mechanical strength of PPS. In this section, some experimental results are described aimed at improving the sliding behaviour of PPS by the addition of various fillers.

TABLE 3.20
Specific wear rate V_s of carbon fibre-filled PTFE (against steel)

Matrix: PTFE*
Filler: Carbon fibres (CLF)
Sliding Speed: 36 cm/s

Filler (CLF) Length (mm)	Content (%)	Specific Wear Rate V_s (mm^3/kgf·km)
	0	2.61
	5	0.83
1	10	0.60
	25	1.40
	5	0.68
3	10	0.57
	25	1.36
	5	0.63
5	10	0.81
	25	1.88

* Continuous adhesive wear - conforming contact

The compositions of the various PPS composites tested are shown in Table 3.21. Friction and wear tests of these materials against steel (S45C) were made using a continuous sliding friction tester between the end surfaces of opposing hollow cylinders and a non-conforming (plate-ring) tester (Ogoshi-type), as shown in Figs. 2.1(a) and (b) respectively.

Figure 3.36 shows the relationship between μ_k and the temperature of the specimen surface for various PPS composites; R4 (40% glass fibre), RFC_f (30% carbon fibre, 15% PTFE) and $RMSC_f$ (17% MoS_2, 20% Sb_2O_3, 18% carbon fibre. It can be seen that the μ_k value increases significantly at a surface temperature of approximately 180°C and is comparatively low between 50-100°C; in particular, that of the $RMSC_f$ becomes approximately 0.09. Table 3.22 shows the values of μ_k, limiting pv, specific adhesive wear rate during sliding in a non-conforming (plate-ring) arrangement V_s', and two other characteristics; μ_{kb} and the limiting pv_b, for journal bearings of PPS composites and other plastic materials.

TABLE 3.21
Composition of specimens

No.	Code	Filler (wt%)
0	R6	None
1	R4	Glass Fibre (G): 40%
2	R4F	Glass Fibre (G): 40% PTFE (F): 10%
3	RC_f	Carbon Fibre (C_f): 30% PTFE (F): 15%
4	RCA	Graphite (C): 10% Asbestos (A) 40%
5	RFC_f	Carbon Fibre (C_f): 30% PTFE (F): 15%
6	RMS	MoS_2 (M): 33% Sb_2O_3 (S): 27%
7	$RMSC_f$	MoS_2 (M): 17% Sb_2O_3 (S): 20 Carbon Fibre (C_f): 18%
8	RMSG	MoS_2 (M): 20% Sb_2O_3 (S): 17% Glass Fibre (G): 20%

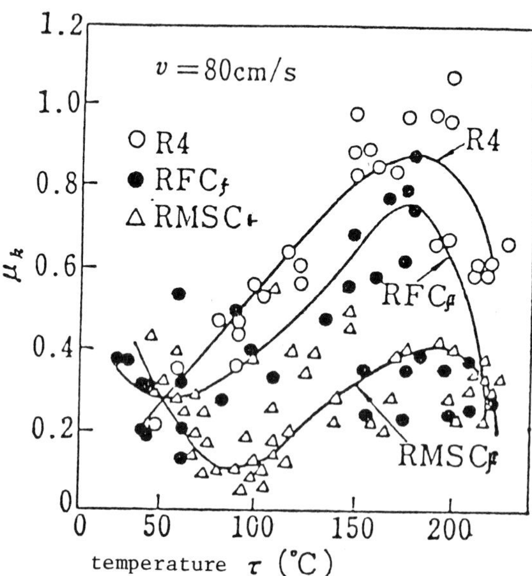

Fig. 3.36 Relationships between μ_k and temperature for various PPS composites

TABLE 3.22
Sliding and dry-bearing characteristics for plastics against steel

Material	Continuous sliding friction			Dry sliding bearing characteristics		Specific adhesive wear rate (V'_s) (Ogoshi's tester) ($mm^3/km \cdot kgf$)
	μ_s R/steel	μ_k	Limiting pv ($kgf/cm^2 \cdot cm/s$)	μ_{kb}	Limiting pv ($kgf/cm^2 \cdot cm/s$)	
R6	0.333			0.04-0.21	250	1.30
R4	0.311	0.2-1.10	478	0.18-0.9	457	0.50
R4F		0.14-0.47	843	0.05-0.61	862	0.15
RC_f		0.24-0.72	847	0.18-0.32	420	
RCA			1037	0.04-0.25	515	0.39
RFC_f		0.15-0.61	925	0.04-0.36	715	0.15
RMS		0.25-0.90	630	0.19-0.32	496	0.87
$RMSC_f$		0.09-0.8	913	0.10-0.29	652	0.06
RMSG		0.31-0.80	354	0.06-0.64	311	0.12
POM Delrin		0.2-0.45	510	0.03	210	0.08
POM Duracon			180			
PTFE		0.1-0.2[12)]	80-100	0.02	181	0.18-0.30
Nylon 6		0.09-0.24	590	0.05	89	
Phenolics Cloth		0.15-0.82	750			
Phenolics 10% Graphite					610	
DAP G:20%; A: 10%				0.009	790	
Polyimide (25% graphite)		0.095-0.3	5740			0.14

Experimental ranges of μ_k: $p = 1.13 - 13.5\ kgf/cm^2$, $v = 35.3 - 117.0\ cm/s$
Experimental ranges of μ_{kb}: $p = 1.5 - 10.0\ kgf/cm^2$, $v = 67 - 300\ cm/s$

One example of the relationship between μ_k and contact pressure, p, for these composites is presented in Fig. 3.37. This figure shows that μ_k normally decreases with increasing p but begins to increase abnormally at high pressures and speeds, that is above a limiting pv. The limiting pv values of these composites are shown in Table 3.22, and that of RCA is the highest, at 1037 kgf/cm^2·cm/s. The V_s' values of the composites and other plastic materials are shown in Table 3.22, and this table indicates that V_s' can decrease markedly when using certain types of fillers; in particular, $RMSC_f$ becomes 0.06 mm^3/kgf·km, being 1/30 of R6 (PPS itself) and 1/8 of R4 (PPS containing 40% glass fibres).

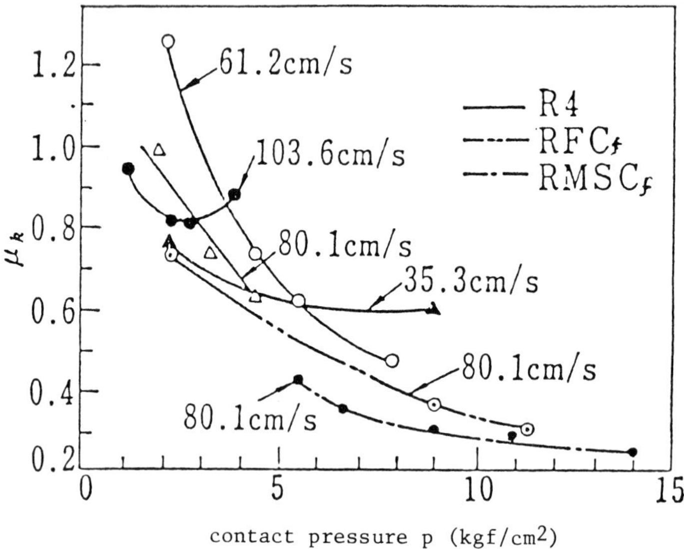

Fig. 3.37 Relationships between μ_k and contact pressure for various PPS composites at different speeds

3.2.9 *Composites of polyarylate (U Polymer) and polyethersulphone (PESF)*

Polyarylate (U Polymer), developed by the UNITIKA Co. in Japan (chemical structure shown in Table 3.23), is an engineering plastic which has excellent heat resistance and mechanical strength and is useful as a sealing material, owing to its wide range of elastic strain. The effects of fillers, 30% MoS_2 and 5% PTFE, on the friction and wear properties of this polymer are described in this section and compared with those of polyethersulfone (PESF), developed by I.C.I. as a high heat-resistance engineering plastic. Table 3.23 shows details of the composition and moulding conditions of these materials.

TABLE 3.23
Specimen

Type	Maker	Mouldings	Note
Polyethsulfone (PESF) 200 P grade	I.C.I. Co.	Injection moulding Pressure: 1140 kgf/cm^2 Temperature: 370°C Mould temperature: 180°C	Heat deflection temperature: 203°C Tensile strength: 8.3-9.5 kgf/cm^2 Tensile modulus: 240-290 kgf/mm^2
U Polymer U-100 3% MoS_2-filled 5% PTFE-filled	Unitica Co.	Injection moulding	Heat deflection temperature: 174°C Tensile strength: 7.5-8.5 kgf/mm^2 Tensile modulus: 220 kgf/mm^2

PESF
Amorphous
T_g: 230°C

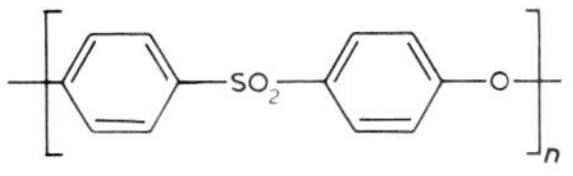

$$\left[-C_6H_4-SO_2-C_6H_4-O- \right]_n$$

U Polymer Polyarylate
Amorphous
T_g: 190°C

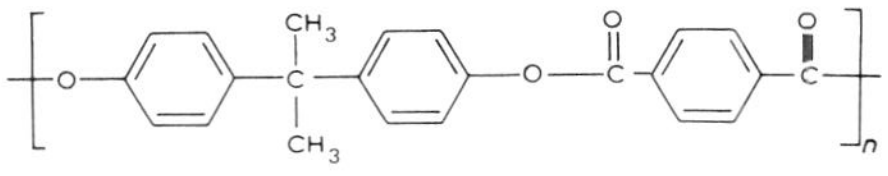

$$\left[-O-C_6H_4-C(CH_3)_2-C_6H_4-O-C(=O)-C_6H_4-C(=O)- \right]_n$$

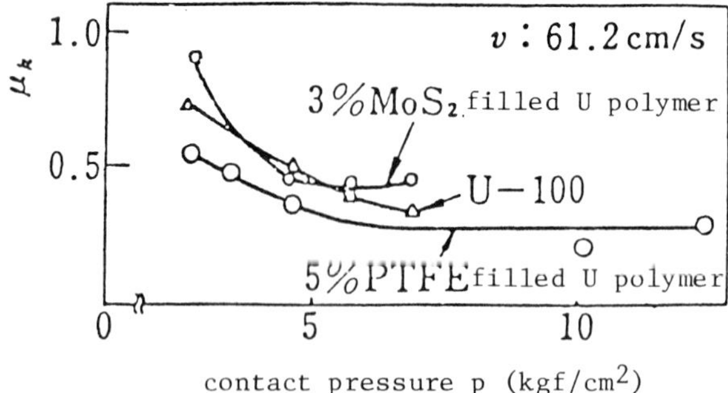

Fig. 3.38 Relationships between μ_k and contact pressure, p, for U polymer composites

Figure 3.38 shows the relationship between the dynamic friction coefficient μ_k and contact pressure p for each composite sliding against steel with conforming contact between the end surfaces of opposing hollow cylinders. Table 3.24 shows the individual values of μ_k. It is clear from the figure and the table that the μ_k decreases with increasing contact pressure and is also lower when fillers are present, especially PTFE.

TABLE 3.24
Average limiting pv value and friction coefficient of some plastics

Material	Average limiting pv value (kgf/cm²·cm/s)	Kinetic coefficient of friction μ_k
Polyethersulfone (PESF)	990	0.14 - 0.53
U Polymer U-100	510	0.33 - 0.9
3% MoS_2 filler	538	0.25 - 0.9
5% PTFE filler	954	0.14 - 0.62

Figure 3.39 shows the relationship between the pv value and temperature of the specimen surface during the friction test, and Fig. 3.40 shows the relationship between p and v for limiting conditions. Table 3.24 shows the limiting pv values of these composites; the value for the U Polymer itself is 510 kgf/cm²·cm/s, but that for the composite containing 5% PTFE has increased markedly to 954 kgf/cm²·cm/s.

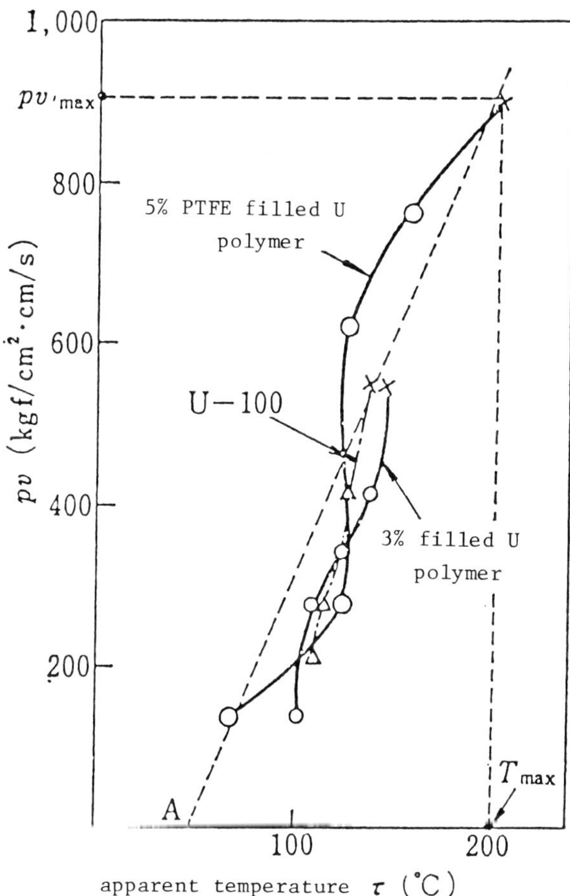

Fig. 3.39 Relationships between pv value and temperature for U Polymer composites

Table 3.25 shows the specific adhesive wear rate V_s' in non-conforming contact conditions (plate-ring), the specific adhesive wear rate V_s in continuous conformal contact, and the specific abrasive wear rate V_{sa}, using JIS-K 7205, of these composites and other plastic materials. It can be seen that the values of V_s' and V_s of the U Polymer composite containing 5% PTFE decrease appreciably to 1/5 - 1/50 of the values for the U Polymer itself. However, the abrasive wear rate V_{sa} greatly increases following the addition of PTFE to the U Polymer. The effect of MoS_2 on the abrasive wear rate, however, appears to be negligible.

TABLE 3.25
Specific wear rate of various plastics [33]

Material	Non-conforming contact Specific adhesive wear rate V'_s (Ogoshi's Tester) $mm^3/km \cdot kgf$	Conforming contact Specific adhesive wear rate V_s $mm^3/km \cdot kgf$	Specific Abrasive Wear rate V_{sa} (JIS-K 7205) $mm^3/km \cdot kgf$
Polyethersulfone (PESF)	4.45		58.8
U Polymer U-100	1.10	9.57	86.5
3% MoS_2-filled	1.21	10.35	94.5
5% PTFE-filled	0.24	0.2	334.1
Polyacetal (POM)	0.14	0.78	
Polycarbonate (PC)	4.84		89.0
ABS Resin	0.44		
Polyvinylchloride (PVC)	1.90		
Polyethylene (PE)	0.24		96.3
Nylon 6		0.45 (0.93)	
PTFE	8.2	2.61	
Graphite-filled polyimide	0.19	0.37	2199
PTFE-filled polyimide	0.10	0.30	766
Unsaturated polyester (UP)	10.24		361
Cloth-filled phenolic	0.24		344
Epoxy resin	5.18		
DAP resin	14.24		

( ): Initial specific wear rate

General rules governing the improvement of friction and wear properties by filler additions to plastics have not yet been elucidated from the above experimental results. However, it seems clear that significant improvements can be achieved by suitable selection of the type of fillers and matrix plastics.

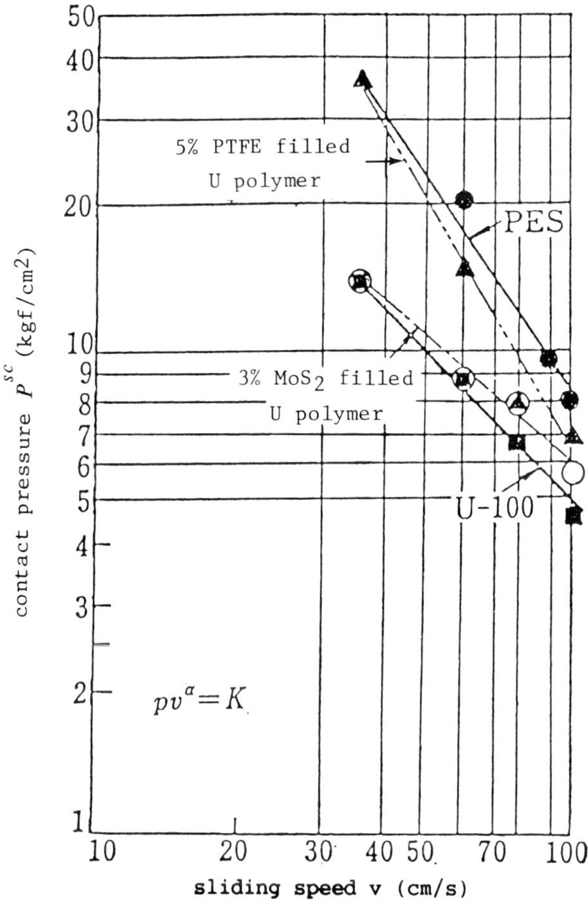

Fig. 3.40 Limiting pv diagrams for PESF and U Polymer

3.2.10 *Plastic coatings on metal surfaces [34]*

Following the development of many new plastic materials suitable for sliding parts and advances in coating process technology, plastic coatings on metals are now well-worth considering for tribological applications. Such coatings may be expected to improve the following five characteristic:

(i) *Improvement of overall mechanical strength and rigidity.* Plastics are in general, inferior in mechanical strength and rigidity to metals, althoug they may have excellent sliding properties due to their self-lubricatin characteristics. Materials consisting of a plastic surface and metallic bas are thus useful, as they can combine excellent sliding properties wit moderate overall rigidity.

(ii) *Reduction in size.* The dimensions of a plastic machine part ar usually greater than that of the corresponding metal one because th mechanical strength and rigidity of plastics are inferior to those of meta However, a component made from plastic-coated metal may thus be reduce in size from one made solely from plastic.

(iii) *Improvement of corrosion resistance.* Metallic surfaces are generall subject to corrosion; plastic-coated ones, however, are generally mor resistant to corrosion at high humidities and air temperatures.

(iv) *Cost reductions.* The new, high-performance plastics such a polyimide, polyoxybenzylene and polyphenylenesulfide are very expensive b costs may be reduced appreciably by using these materials as thin films o a metallic base.

(v) *Improved sliding properties.* Improved sliding properties may b achieved if the plastic film has suitable friction and wear characteristic:

TABLE 3.26
Details of coating materials

Code	Material	Details	Maker
A	Ekonol	EPDX-200	Sumitomo Kagaku Co.
B	Ekonol (with primer)	EPDX-200 EK-1883 GB	Sumitomo Kagaku Co. Daikan Kogyo Co.
C	PPS	Ryton	Philips Pet. Co.
D	PPS PTFE (mix)	Ryton	Philips Pet. Co.
E	PTFE (with primer)	Polyflone enamel (E-4105 GN)	Daikin Kogyo Co.
F	Primer	EK-1883 GB	Daikin Kogyo Co.

Some experimental examples of the performance of plastic coatings on metals, carbon steel and brass, are described in this section, taking into consideration the above-mentioned characteristics. Table 3.26 shows the details of four kinds of coating materials: polyoxybenzylene (Ekonol), having high chemical and heat resistance and excellent sliding properties; polyphenylene sulfide (PPS, Ryton or Susteel), having high chemical and heat resistance, excellent sliding properties and high rigidity; polytetrafluoroethylene (PTFE), having the highest chemical and heat resistance and excellent sliding properties; and a primer produced for PTFE coating.

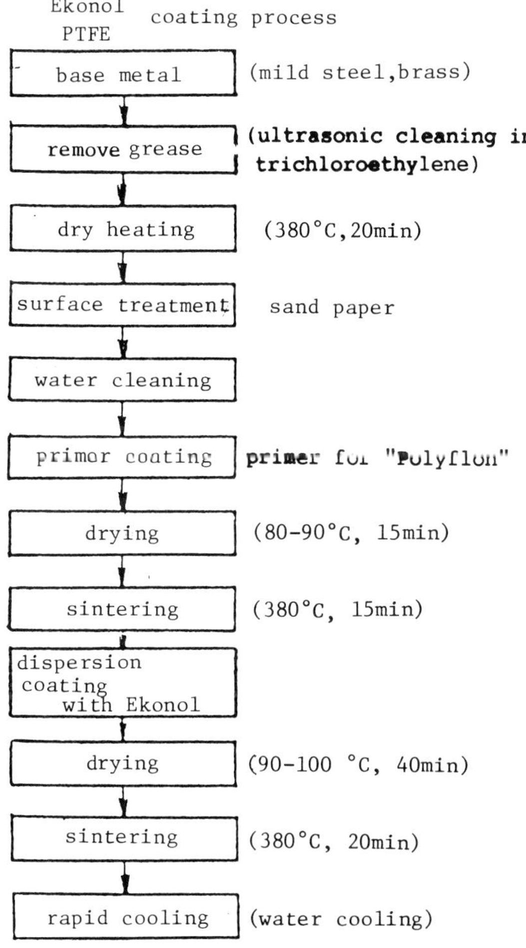

Fig. 3.41 Coating process for C and D

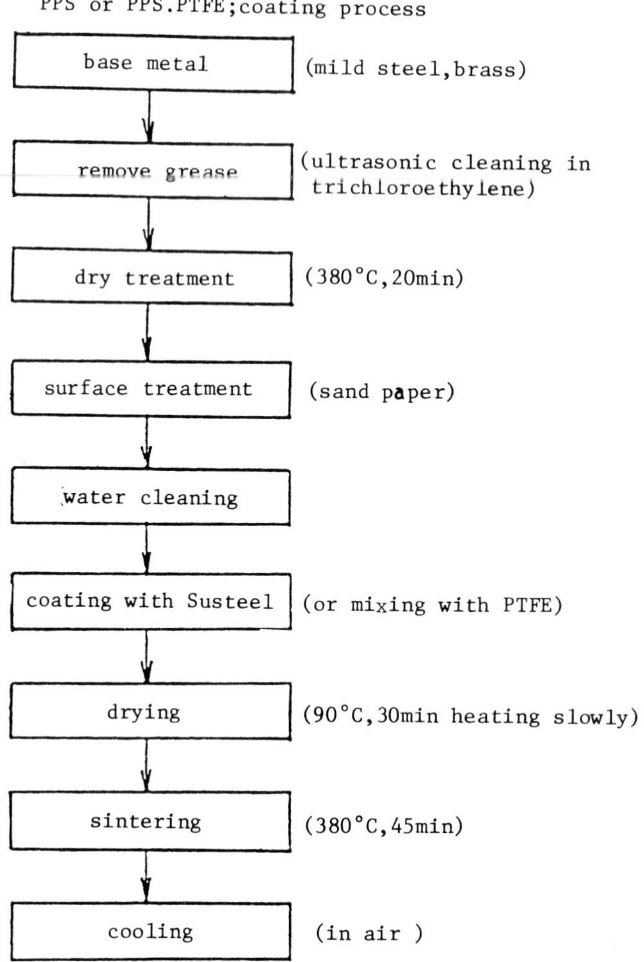

Fig. 3.42 Coating process for B, E and D

Figure 3.41 shows the coating process for materials, A (Ekonol), C (PPS) and D (mixture of PPS and PTFE), whilst Fig. 3.42 shows this process for materials, B, E (with primer) and F (primer only). Sliding tests were carried out using a rotating hollow steel cylinder-end surface as the upper specimen and a plastic-coated surface as the lower one, as shown in Fig. 2.1(a), with a velocity of 50 cm/s and at 20°C and 60% R.H. The temperature of sliding surface, the μ_k value, the limiting pv and the value of V_s were measured experimentally. Figure 3.43 shows the relationship between μ_k and contact pressure p for the four plastic coatings, B, D, E and F. Figure 3.44 shows the variation of surface temperature, τ and μ_k with time of sliding for coating material B (Ekonol with primer) at different contact pressures.

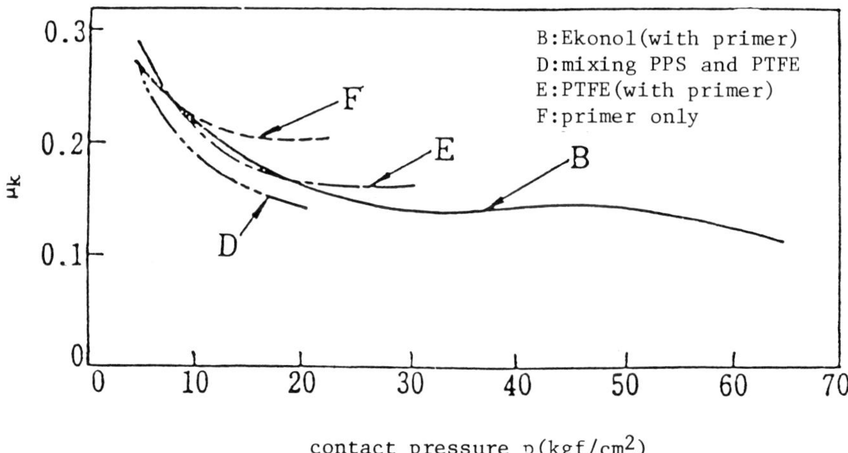

Fig. 3.43 Relationship between μ_k and contact pressure, p, for various surface coatings (v = 50 cm/s)

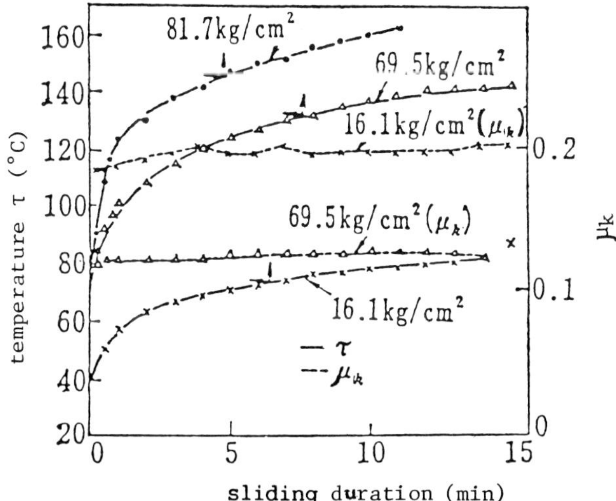

Fig. 3.44 Relationships between μ_k, counterface temperature and sliding duration, t, for Ekonol coating surface with poymer at different contact pressures

TABLE 3.27(a)
Sliding characteristics of plastic-coated surfaces [34]

Code	Coating Materials	Base metal	Film thick- (μm)	Coefficient of friction, μ_k (μm): On each base metal	Coefficient of friction, μ_k (μm): Average	Limiting pv ($kgf/cm^2 \cdot cm/s$): On each base metal	Limiting pv ($kgf/cm^2 \cdot cm/s$): Average	Specific adhesive wear rate in conforming contact V_s ($mm^3/km \cdot kgf$): On each base metal	Specific adhesive wear rate in conforming contact V_s ($mm^3/km \cdot kgf$): Average
A	Ekonol	SS41	48.0	-	-	<224	<224	-	-
		Brass	-	-		<224		-	
B	Ekonol (with primer)	SS41	62.5	0.114-0.276	0.107-0.271	3996	4147	-	-
		Brass	69.7	0.109-0.266		4299		-	
C	PPS	SS41	20.2	-	-	<224	<224	-	-
		Brass	3.8	-		<224		-	
D	Mix of PPS and PTFE	SS41	35.0	0.141-0.274	0.137-0.263	1471	1471	0.100	0.096
		Brass	19.5	0.134-0.253		1471		0.092	
E	PTFE (with primer)	SS41	72.7	0.151-0.304	0.162-0.299	1870	1419	0.315	0.353
		Brass	36.3	0.173-0.291		968		0.392	
F	Primer	SS41	15.0	0.190-0.288	0.196-0.301	1370	1924	0.250	0.319
		Brass	22.0	0.203-0.315		1218		0.389	

TABLE 3.27(b)
Sliding characteristics of various plastics [34]

Code	Coating Materials	Coefficient of friction, μ_k (μm)	Limiting pv ($kgf/cm^2 \cdot cm/s$)	Specific adhesive wear rate in conforming contact V_s ($mm^3/km \cdot kgf$)
		Average	Average	Average
PLASTIC	Phenolic (cloth filled)	0.49	750	2.4
	Polyimide (graphite filled)	0.47	>4542	0.37
	ABS	0.37	500	2.0
	Polycarbonate	0.33	460	12.5
	Polyacetal (POM)	0.15	560	0.78
	PTFE	0.10	12.5	2.6

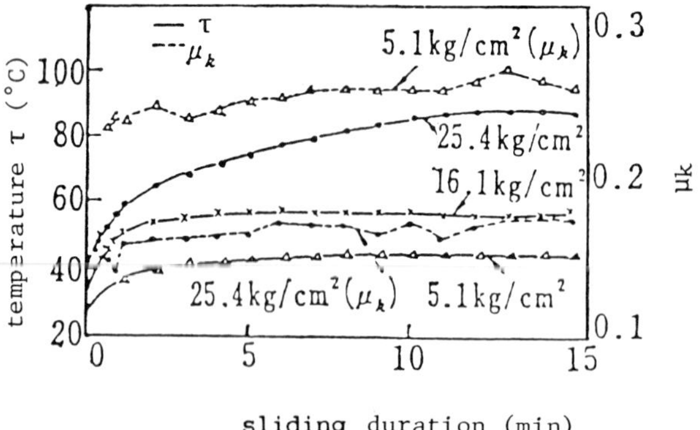

Fig. 3.45 Relationships between μ_k, counterface temperature and sliding duration, t, at various contact pressures for surfaces with a mixture of PPS and PTFE

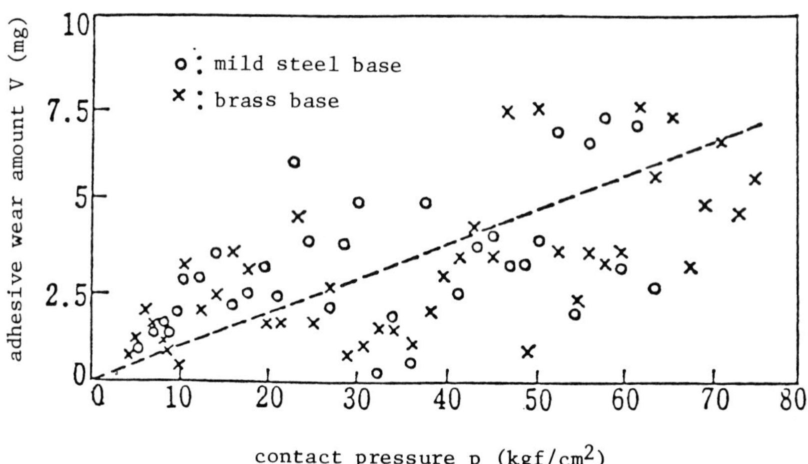

Fig. 3.46 Relationship between the amount of adhesive wear and contact pressure, p, for an Ekonol coated surface (for 5 minutes)

Figure 3.45 shows similar relationships for the other coating materials, C (PPS) and D (mixture of PPS and PTFE). Table 3.27 gives the average values of the coefficient of kinetic friction μ_k of the six kinds of coating materials on each base metal (steel and brass) and indicates that the values range from 0.109 and 0.315 and are generally smaller than those for the other plastics shown in the lower part of the same table.

The limiting pv values, pv_{max}, of the plastic-coated surfaces for unlubricated continuous sliding are between 224 and 4299 kgf/cm^2·cm/s, as also shown in Table 3.27. That of a surface coated with B (Ekonol with primer) is the highest, and these values are, in general, higher than those for most of the other plastics shown in the lower part of the table. It may be noted that the pv_{max} for coating B is almost the same as that for the graphite-filled polyimide.

As an example of the adhesive wear characteristics, Fig. 3.46 shows the relationships between the adhesive wear volume, V, during a 15-minute test and the contact pressure, p, for the surface coated with B (Ekonol with primer). The wear volume is almost directly proportional to the contact pressure, as shown by the dotted straight line, but there is considerable scatter. Finally, it may be noted from Table 3.27 that the lowest wear rate, V_s, of all is obtained with coating D (a mixture of PPS and PTFE).

REFERENCES

1 Yamaguchi, Y., Lubrication of plastic materials, Nikkan Kogyo Press Co. (1981) p. 38-39.
2 Yamaguchi, Y. and I. Sekiguchi, Proceedings of 3rd Int. Solid Lub. Conf. of ASLE (1984) p. 187.
3 Okubo, Y., Y. Yamaguchi and I. Sekiguchi, Research Rep. of Kogakuin Univ., No. 53 (1982) p. 22.
4 Lomax, J.Y. and J.T. O'Rourke, Machine Design, June 23 (1966), p. 158.
5 Play, D., A. Floquet and M. Godet, Proceedings of 3rd Leeds-Lyon Symp. on Tribol. Sess. 1 (1978) p. 32.
6 Giltrow, J.P. and J.K. Lancaster, Wear, 16 (1970) 359.
7 Lancaster, J.K., Wear, 20 (1972) 315.
8 Lancaster, J.K., Wear, 20 (1972) 335.
9 Miyake, S. and Y. Katayama, J. JSLE, 18 (1973) 404.
10 Tanaka, K., Wear of Materials, Proc. Int. Conf. on Wear of Mat., ASME (1977), 510.
11 Briscoe, B.J. and M.D. Steward, ibid. (1977) 5318.
12 Eiss, N.S., N.E. Lewis and C.W. Reed, Wear of Material, Proc. Int. Conf. on Wear of Mat., ASME (1979) 589.
13 Plumb, R.S. and W.A. Glaeser, Wear of Material, Proc. Int. Conf. on Wear of Mat., ASME (1977) 337.
14 Tsukizoe, T. and N. Ohmae, ibid. (1977) 518.
15 Pogosian, A.K. and N.A. Lambarian, ibid. (1977) 547.
16 Halberstadt, M.L., J.A. Mansfield and S.K. Rhee, Wear of Materials, ASTM (1977) 560.
17 Clauss, F.J., Solid Lubricants and Self-Lubricating Solids, Academic Press (1972) 179.
18 Ibid (1972) 206.

19 Dzhanakmedov, A.H. and M.W. Pascoe, Session 1, The wear of non-metallic materials (1978) 60.
20 Yamaguchi, Y. and I. Sekiguchi, Research Rep. of Kogakuin Univ. No. 28 (1970) 21.
21 Yamaguchi, Y., I. Sekiguchi, T. Kondo and H. Sakai, Research Rep. of Kogakuin Univ. No. 30 (1971) 59.
22 Yamaguchi, Y., I. Sekiguchi, S. Matsuda and Y. Matsuda, Research Rep. of Kogakuin Univ. No. 30 (1972) 65.
23 Yamaguchi, Y., I. Sekiguchi, T. Ohtomo and T. Kakiya, J. JSLE, 22 (1977) 58.
24 Lewis, R.B., Lubricated Vespel Bearing, du Pont Accession, No. 12912 (1967).
25 Matsubara, K., M. Watanabe and M. Karasawa, J. JSLE, 14 (1969) 43.
26 Matsubara, K., M. Watanabe and M. Karasawa, Preprint of JSLE (1970).
27 Albariz, R.T. and F.P. Darmorz, Tech. Papers of 29th ANTEC of SPE, Section 11-A (1974) 1.
28 Yamaguchi, Y., I. Sekiguchi, K. Sugiyama and J. Suzuki, J. Soc. Mat. Sci. of Japan, 21 (1972) 826.
29 Sekiguchi, I., Y. Yamaguchi, N. Higuchi and S. Nagasawa, Research Rep. of Kogakuin Univ. No. 35 (1973) 30.
30 Yamaguchi, Y., R. Kawai and H. Goto, The Plastics, JSPT, 17, No. 6 (1971) 49.
31 Yamaguchi, Y. and I. Sekiguchi, The Plastics, JSPT, 17, No. 9 (1971) 49.
32 Yamaguchi, Y., I. Sekiguchi, S. Takane and M. Shibata, J. JSLE, 25 (1980) 451.
33 I. Sekiguchi, Y. Yamaguchi, M. Shibata and S. Takane, Research Rep. of Kogakuin Univ., No. 44 (1978) 34.
34 Yamaguchi, Y., Kogyo-zairyo, 26, No. 10 (1978) 91.
35 For example: J.J. Licari, Trans. by Nagasaka, Plastic Coatings for Electronics, Maki Shoten Co. (1974).

CHAPTER 4

APPLICATION TO SLIDING MACHINE PARTS

4.1 BEARINGS

Plastics materials can be used for the following three types of bearings
(1) lubricated journal bearings
(2) dry journal bearings
(3) ball bearings
The characteristics and performance of these bearings are described in this chapter.

4.1.1 *Lubricated journal bearings*

(i) *History of bearings lubricated with water*

Metallic materials have been used almost exclusively in journal bearings, with a lubricant supplied between the steel shaft and the bearing surface. An exception, however, is lignum vitae, which is used in stern tube bearings with moderate-viscosity grease and oil as lubricants in order to prevent corrosion and avoid direct contact between the shaft and the bearing surface. However, following the development of phenolic resins some fifty years ago, with their characteristics of high shock absorbance and ability to use water as a lubricant, they began to be used as journal bearings, particularly in rolling mills. With recent developments, newer types of plastics have been used as journal bearings lubricated with water in various applications such as rolling mills, stern tubes, guide bearings in water turbines, ball-mills, paper machines and mining machines [1,2]. A water-lubricated plastic journal bearing has the following characteristics.
(1) long life
(2) low friction coefficient
(3) low running costs
(4) clean running
(5) accurate dimensions due to low wear rate
(6) high resistance to large loads owing to its high resistance to impact
(7) high corrosion resistance
(8) low wear rate of the journal

This section describes experiment results on the performance of plastic journal bearings lubricated with water and compares the results with those obtained when using oil lubricants.

There are comparatively few experimental results for water-lubricated journal bearings. A summary will be given, however, of work on phenolic bearings by Sasaki and Sugimoto [3], PTFE bearings lubricated with sea water by Craig [4] and gum bearings by Pieper [5]. Details of various research reports from the author's laboratory [6-13], will also be discussed.

The conclusions from six reports [3] on the characteristics of phenolic journal bearings lubricated with water or oil are as follows.

(1) For a 50% cotton cloth-filled, phenolic journal bearing lubricated with abundant water, the bearing temperature rise is much lower than that of a bearing lubricated with mobile oil, and the limiting pv value can become as high as 400 $kgf/cm^2 \cdot cm/s$.

(2) The limiting pv value for a journal bearing lubricated with mobile oil is the highest, but is only slightly different from that for spindle oil and water when the same amount of lubricant is used. The performance of a bearing lubricated with water was particularly high at low bearing pressures.

(3) The limiting temperature rise of the bearing is about 15°C when using water as a lubricant. The limiting pv value may be increased if the temperature rise is controlled by increasing the amount of water and, hence, the rate of cooling.

(4) The limiting pv value of a bearing soaked in water or oil becomes lower than that of a bearing with the lubricant supplied externally, owing to the lower cooling efficiency.

(5) With respect to a 1% MoS_2-filled phenolic bearing lubricated with water, the limiting pv value is increased by the addition of 1% NaOH into the water.

(6) Cotton cloth or asbestos and 5% Sn, Al, graphite, PbO or Pb filled phenolic journal bearings have better running characteristics with water than with an oil lubricant, particularly the bearing material filled with Sn because it supports the highest load.

Craig [4] has reported friction coefficient values of 0.02 - 0.5 for a spherical bearing surface coated with PTFE fibre oscillating at angles from 5 - 9° in sea water under pressures between 1000 and 2300 kgf/cm^2. These values are greater than those obtained in pure water, 0.015 - 0.45. In general, the frictional resistance in water is less than that in air but the wear rate in water becomes greater than in air.

According to Pieper's report [5], friction coefficients ranged from 0.0025 to 0.2 for a journal bearing with an inner surface coated with soft rubber running with a water lubricant at contact pressures of 0.2 - 9.3 kgf/cm^2 and peripheral speeds of 1-16 m/s. Four kinds of inner surface were used: eight longitudinal grooves, spiral grooves, a regular double pentagonal groove pattern, and smooth. Friction generally decreases with increasing speed and becomes constant in the speed range above 3 cm/s. In this study, three kinds of rubber were used having Shore hardnesses ranging from 33-81.

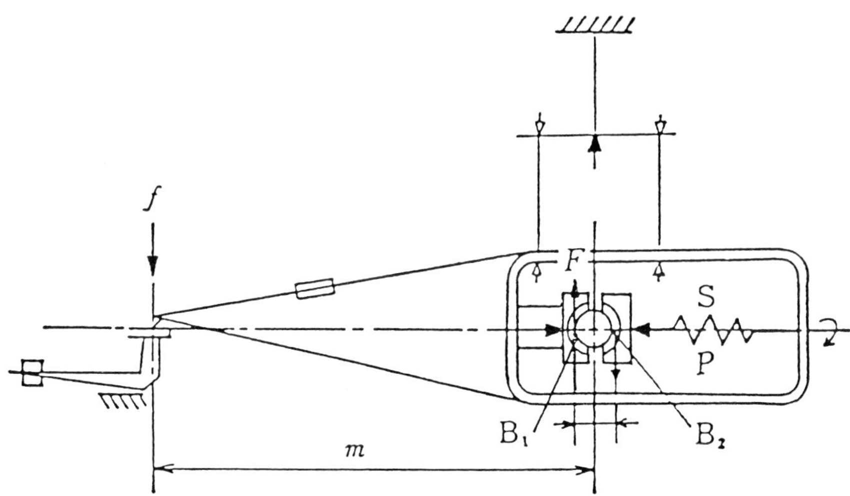

Fig. 4.1 Bearing tester

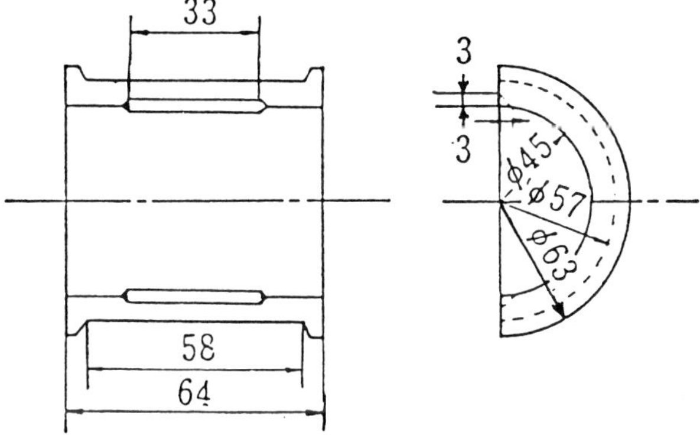

Fig. 4.2 Dimensions of plastic bearings

(ii) *Experimental performance with water lubrication*

(1) *Thermosetting plastics*

(a) *Phenolic journal bearings.* Figures 4.3 to 4.11 show various bearing characteristics obtained by measuring the kinetic friction coefficient, μ_{kb}, between a steel shaft and phenolic bearing surfaces. The apparatus used is shown in Fig. 4.1 and the bearing dimensions in Fig. 4.2. A continuous supply of water or oil lubrication was provided at 300 cc/min. The bearing materials were wood flour-filled phenolics containing graphite, MoS_2 or PTFE powder, as shown in Table 4.1. Experiments were carried out under various conditions to clarify the effects of the fillers and their content.

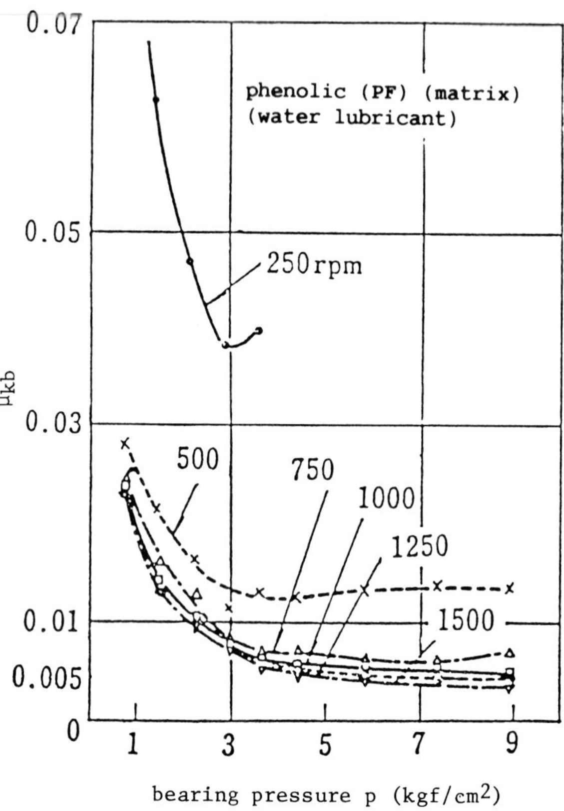

Fig. 4.3 Relationships between μ_{kb} and p for phenolic (PF) bearings (water lubrication)

Figures 4.3 and 4.5 [6] show examples of the relationships between the kinetic friction coefficient, μ_{kb} and bearing pressure, p, for the unfilled phenolic bearing lubricated with water at different speeds. Figure 4.6 shows similar relationships for a cotton-laminated phenolic bearing [12]. It is clear from these figures that the value of μ_{kb} decreases when the rotational speed of the shaft or the bearing pressure is increased. Figure 4.4 shows a similar relationship for the same phenolic bearing lubricated with 100 cc/min of machine oil (poured) and indicates that the μ_{kb} values in this case are generally greater than for those lubricated with water. The μ_{kb} value also increases slightly with an increase in the rotational speed under larger load.

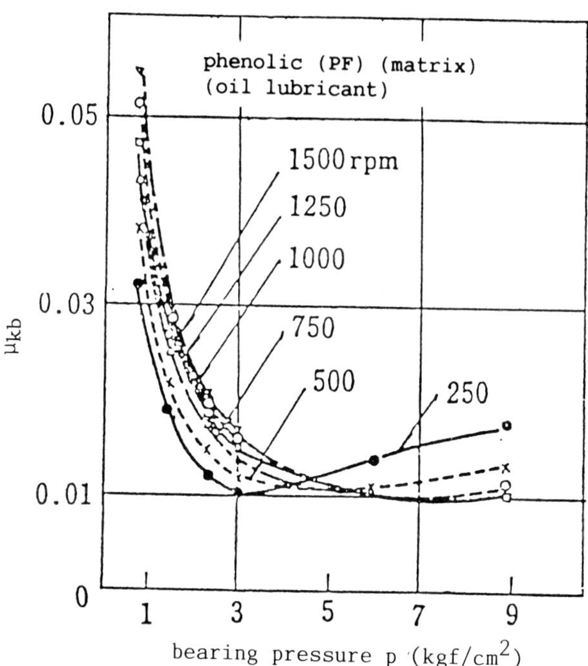

Fig. 4.4 Relationships between μ_{kb} and p for phenolic (PF) bearings (oil lubrication)

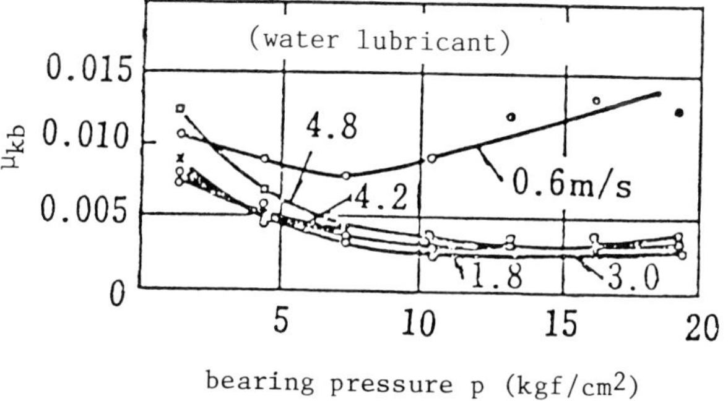

Fig. 4.5 Relationships between μ_{kb} and p for phenolic bearings

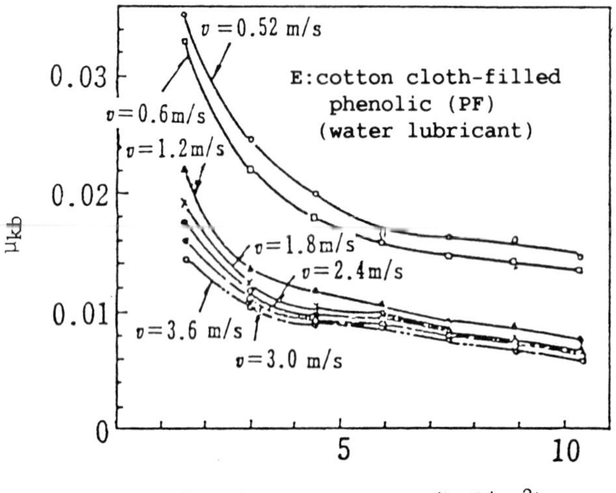

Fig. 4.6 Relationships between μ_{kb} and p for bearing "E"
"E": Table 4.2

TABLE 4.1
Details of phenolic bearing materials

Material: Matrix	Material: Filler	Matrix, Maker	Moulding Conditions
P	3% Graphite		
H	5% Graphite		
E	8% Graphite		
N	3% MoS_2		
O	5% MoS_2	Sumitomo Bakelite	
L	8% MoS_2	PM-40	Moulding temperature: 140°C
I	5% PTFE	F-3	Moulding pressure: 200 kgf/cm^2
C	15% PTFE		Moulding time: 10 minutes
S	30% PTFE		
Cloth filled phenolics		Sumitomo Bakelite PM-700, F-5	
Bronze (BC7)			Casting

Figure 4.7 shows the relationship between μ_{kb} and p for phenolic bearings filled with various materials (shown in Table 4.1), and indicates that the value of μ_{kb} depends on the type of filler, and is particularly low when the filler is PTFE.

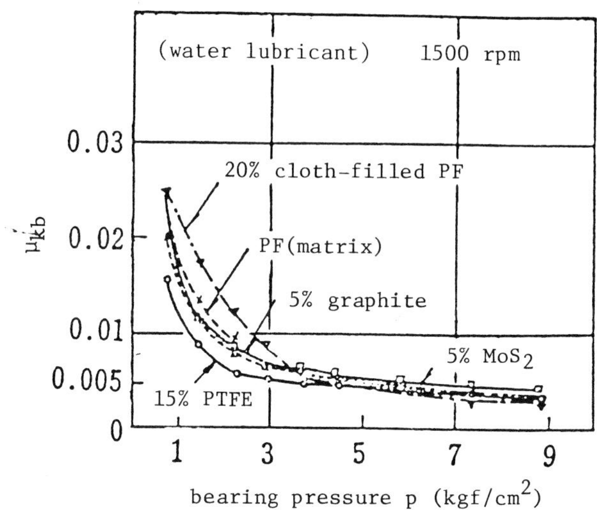

Fig. 4.7 Relationships between μ_{kb} and p for various phenolic bearings

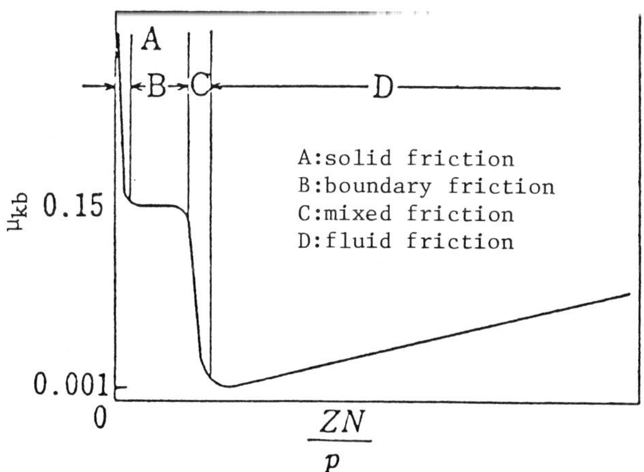

Fig. 4.8 μ_{kb} - ZN/p relationships for journal bearings

The characteristics of a journal bearing are generally presented by the relationship between μ_{kb} and ZN/p, the Sommerfeld variable, as shown in Fig. 4.8 (Z: viscosity of lubricant; N: rotational speed; p: bearing pressure) [14]. There are four regions depending on the value of ZN/p: solid friction, boundary friction, mixed friction and fluid friction.

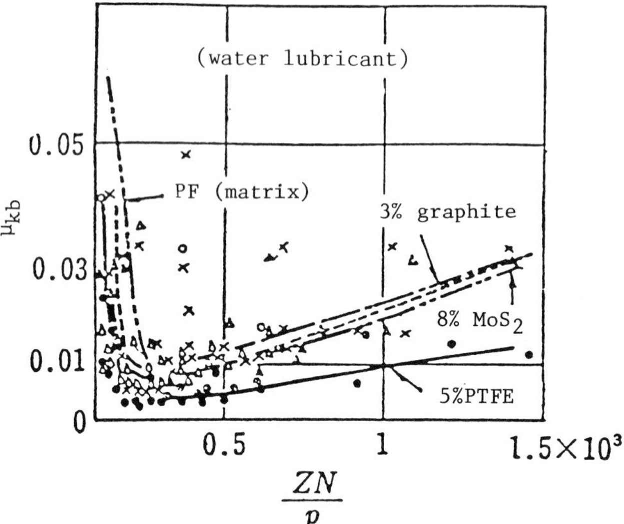

Fig. 4.9 μ_{kb} - ZN/p relationships for various PF bearings

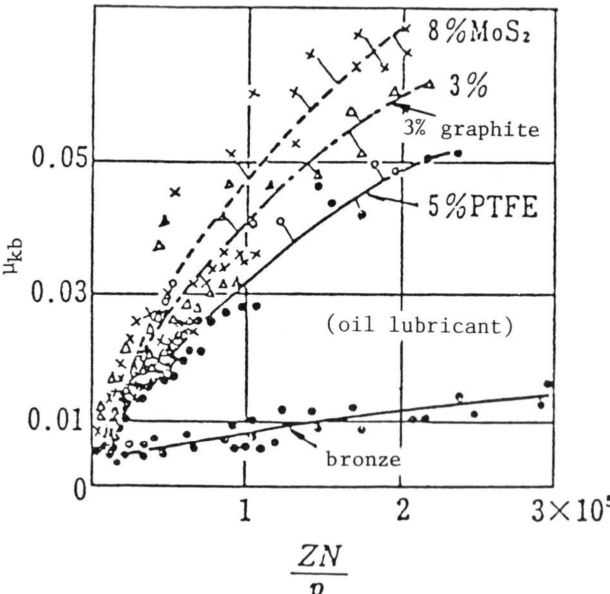

Fig. 4.10 μ_{kb} - ZN/p relationships for various bearings

Figure 4.9 shows the relationships between μ_{kb} and ZN/p for various phenolic journal bearings lubricated with water, and Fig. 4.10 shows similar relationships for phenolic and bronze bearings lubricated with machine oil. In Fig. 4.9, it can be seen that with water lubrication every bearing has a fluid friction range beyond 0.2 x 10^3 of ZN/p. The 5% PTFE filled phenolic bearing shows the smallest value of μ_{kb}; however, there is almost no difference in the μ_{kb} values between the graphite or MoS_2 filled and the unfilled phenolic bearings. When comparing Fig. 4.9 with Fig. 4.10, the μ_{kb} values of phenolic bearings lubricated with machine oil are generally greater than those lubricated with water or for that of a bronze bearing lubricated with machine oil.

Figure 4.11 shows the relationships between μ_{kb} and p for a journal bearing lubricated with water when the bearing material is a composite of wood flour-filled phenolic and 5% carbon fibre (CLF dia. of 12 - 14 μm). The addition of carbon fibre increases μ_{kb} and similar tendency occurs for a DAP composite bearing i.e. the addition of carbon fibre does not improve the lubrication performance of a bearing lubricated with water.

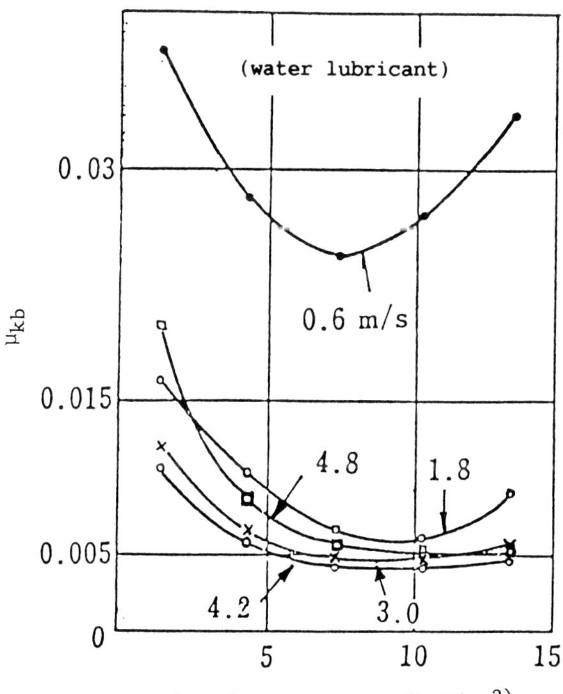

Fig. 4.11 μ_{kb} - p relationships for a 5% carbon fibre-filled PF bearing

(b) DAP composite journal bearing. Figure 4.12 shows the relationships between μ_{kb} and p at various speeds for a 20% glass fibre (G.F.)-filled DAP bearing lubricated with water (similar to that of the phenolic bearing described above).

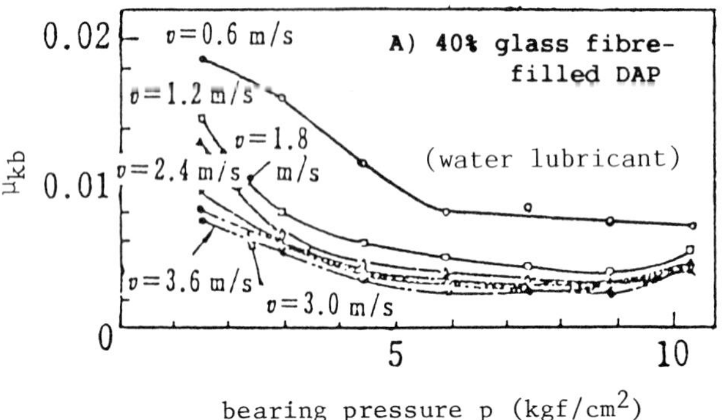

Fig. 4.12 μ_{kb} - p relationships for bearing "A" (40% glass fibre-filled DAP)

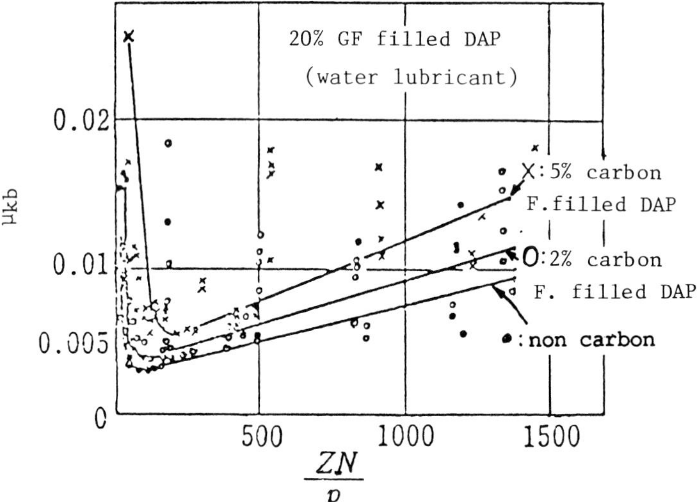

Fig. 4.13 μ_{kb} - ZN/p relationships for DAP composite bearings

Figure 4.13 shows the relationship between μ_{kb} and ZN/p for various DAP composite bearings (20% G.F. filled DAP and 2% or 5% carbon fibre) lubricated with water. Figure 4.14 shows the relationship between μ_{kb} and p for the various DAP composite bearings (20% G.F. filled DAP and 3.5%-10% asbestos as shown in Table 4.3) lubricated with water at a speed of 3.5 m/s; the μ_{kb} values increase slightly when asbestos is added.

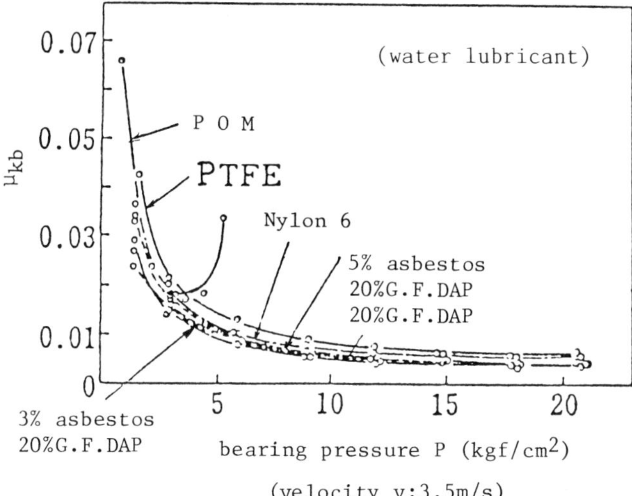

Fig. 4.14 μ_{kb} - p relationships for various plastic bearings lubricated with water

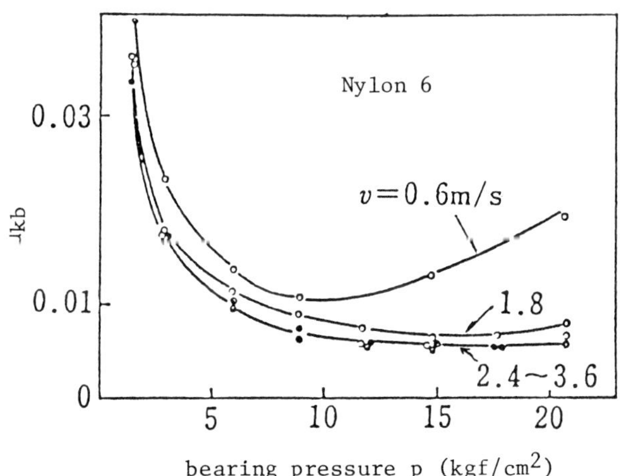

Fig. 4.15 μ_{kb} - p relationships for Nylon 6 bearings lubricated with water

(2) *Thermoplastic journal bearings.* Although phenolics (thermosetting plastics) were first used for water-lubricated journal bearings, interest has been shown more recently in those thermoplastics, e.g. PTFE, Nylon 6 and polyacetal, which are frequently used for unlubricated journal bearings. In this section, some experimental examples will be given to illustrate the characteristics of the above-mentioned thermoplastic journal bearings

lubricated with water, and the results will be compared with those for lignumvitae and thermosetting bearings.

Figure 4.15 shows the relationships between μ_{kb} and p for Nylon 6 journal bearings (shown in Table 4.3), lubricated with water at speeds between 0.6 - 3.6 m/s. Figures 4.16, 4.17, 4.18 and 4.19 show similar relationships for polyacetal (POM - Table 4.3), ultra-high molecular weight polyethylene, UHPE (Table 4.2), PTFE (Table 4.3) and lignumvitae (Table 4.2) respectively.

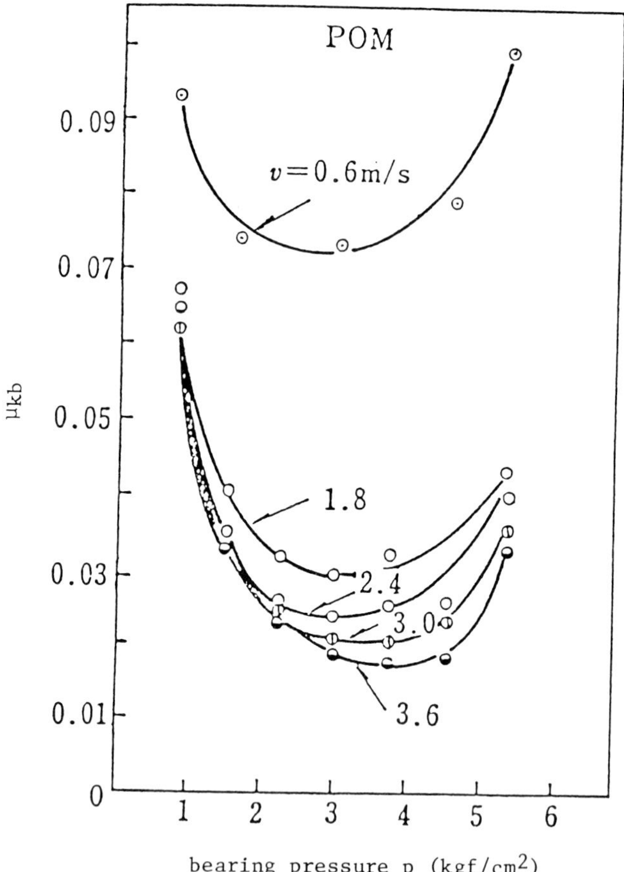

Fig. 4.16 μ_{kb} - p relationships for a POM bearing lubricated with water

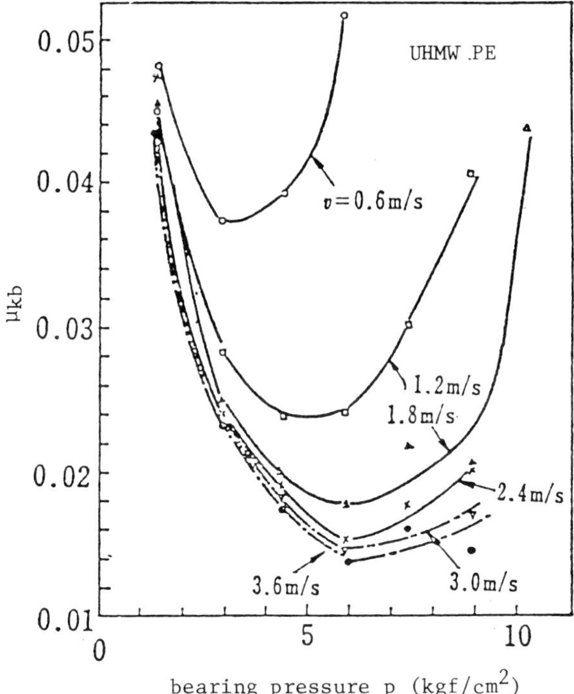

Fig. 4.17 μ_{kb} - p relationships for bearing "F" (UHMWPE) lubricated with water

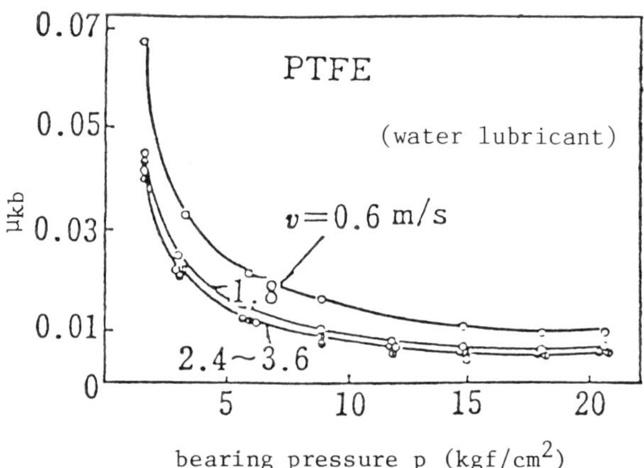

Fig. 4.18 μ_{kb} - p relationships for a PTFE bearing lubricated with water

TABLE 4.2
Bearing materials

Mark	Materials Matrix	Filler	Details of Matrix	Moulding Temperature
A	20%	0	(DAP) resin	Compression moulding Temp: 160°C
B	Glass Fibre	Graft copolymer of MMA and Cellulose (Graft 74%) 10%	Sumikon AM200	Pressure: 200 kgf/cm^2 Time: 15 min
C	Filled DAP	Graft copolymer of MMA and Cellulose (Graft 132%) 10%	Sumitomo Bakelite (KK)	
D		Nylon 4 10%		
E	Cotton cloth-filled phenolic			Compression moulding
F	Ultra high molecular weight polyethylene		Mitsui Pet.Chem. M.W.: over 10^6	Extrusion and machining
G	Lignumvitae		South America	Machining

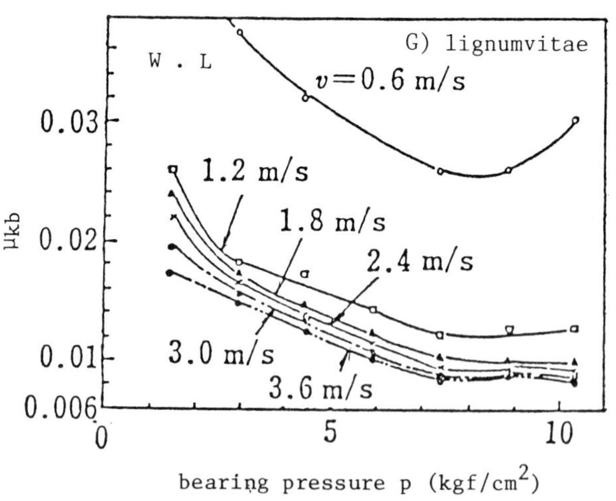

Fig. 4.19 μ_{kb}-p relationships of bearing "G" (lignumvitae) lubricated with water

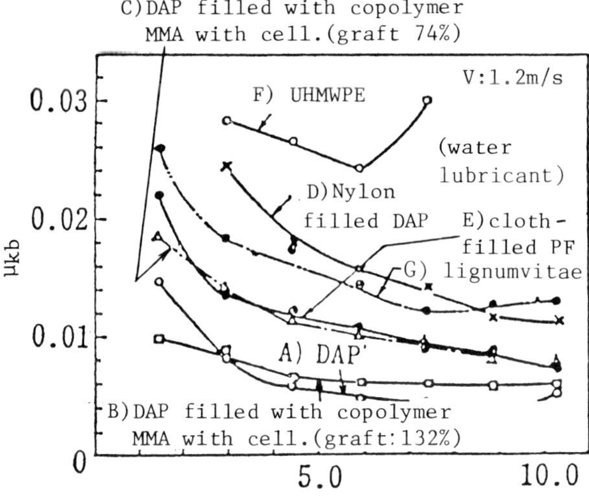

bearing pressur p (kgf/cm²)

Fig. 4.20 μ_{kb} - p relationships for various plastic bearings lubricated with water

TABLE 4.3
Types of moulding methods of bearing materials

Material	Details	Moulding Method
Nylon 6	N.Co. A 1025	Machined from extruded bar
Polyacetal	D.Co. Delrin	
PTFE	D.Co. Polyflon	Preforming pressure: 300kgf/cm^2 Preforming time: 10 mins Sintering time: 60 mins Sintering temperature: 360°C
20% Glass Fibre Filled DAP	0% Asbestos 3% Asbestos 5% Asbestos 10% Asbestos Matrix: Sumikon AM200 filled with asbestos fibres (Crysotile)	Compression pressure: 200kgf/cm^2 Temperature of compression mould: 150°C Compression time: 10 mins

Figure 4.20 compares the relationships between μ_{kb} and p for the various plastic journal bearings, shown in Table 4.2, lubricated with water at a speed of 1.2 m/s. It is clear from these figures that μ_{kb} decreases with an increase in p but can increase suddenly above some limiting pressure. The latter is reduced as the speed decreases. The limiting pressures of various plastic journal bearings may be derived from the figures as follows: Nylon 6 - about 10 kgf/cm^2 at 0.6 m/s (Fig. 4.15); POM - 3 kgf/cm^2 at 0.6 m/s and 4.5 kgf/cm^2 at 3.6 m/s (Fig. 4.16); UHPE - 3 kgf/cm^2 at 0.6 m/s and 3.6 kgf/cm^2 at 3.6 m/s (Fig. 4.17); PTFE - over 20 kgf/cm^2 (Fig. 4.18) and lignumvitae - 8 kgf/cm^2 at 0.6 m/s and over 10 kgf/cm^2 at 1.8 m/s (Fig. 4.19). It is clear from Figs. 4.14 and 4.20 that the μ_{kb} values for thermosetting plastic bearings, and the limiting pressure for the former is greater than that for the latter.

Of the thermoplastic bearings, the μ_{kb} values for PTFE and Nylon 6 are lower, and their limiting pressure is higher than that of lignumvitae. However, the μ_{kb} values for POM and UHPE bearings are greater, and their limiting pressures are lower, than that of lignumvitae.

The various μ_{kb} values for the three thermoplastic and DAP journal bearings unlubricated and lubricated with water, and for a bronze bearing lubricated with machine oil are shown in Table 4.4 [9]. This table indicates that μ_{kb} values for an unlubricated and a water-lubricated PTFE bearing are very low, and that for an unlubricated POM bearing is also comparatively low. However, those for a POM bearing lubricated with water are greater than that of a Nylon 6 bearing lubricated with water.

TABLE 4.4
Range of μ_{kb} values for various journal bearings

Bearing Material	Range of μ_{kb} values Unlubricated Friction	Water Lubrication
Nylon 6	0.043 - 0.18	0.0054 - 0.040
Polyacetal	0.035 - 0.188	0.0172 - 0.10
PTFE	0.022 - 0.015	0.005 - 0.067
20% Glass fibre-filled DAP:		
0% Asbestos	0.0144 - 0.060	0.0035 - 0.026
3% Asbestos	0.0078 - 0.044	0.0038 - 0.043
5% Asbestos	0.0078 - 0.053	0.0031 - 0.033
10% Asbestos	0.0091 - 0.052	0.0031 - 0.034
Bronze bearing with oil lubrication	-	0.0048 - 0.0165

(3) *The effect of the water absorption properties of bearing materials on performance.* It is clear from Table 4.4 that the lubricity of water on Nylon 6 bearings, which have a high hygroscopicity, is greater than that on POM bearings, which have a low hygroscopicity. It was indicated in the previous section that a cotton cloth-filled phenolic bearing having a high hygroscopicity showed excellent lubricity with water. Some experimental examples will now be described showing how the performance of journal bearings lubricated with water can be improved by promoting hygroscopicity in the bearing materials.

Table 4.5 shows the composition and moulding conditions of 20% G.F.-filled DAP composites containing α cellulose, polyvinyl alcohol or wood powder (all with a high hygroscopicity) and silicone resin composites. Figure 4.21 shows the relationships between μ_{kb} and p for an α-cellulose-filled (50%) DAP composite (FRDα50%) journal bearing lubricated with water at various speeds from 0.6 to 3.6 m/s, and indicates that the value of μ_{kb} is generally very small. Figure 4.22 compares the relationships between μ_{kb} and p for the various plastic bearings lubricated with water (shown in Table 4.5) at 3.6 m/s and shows that the μ_{kb} values for DAP composites bearings without α-cellulose (FRD) are generally greater than those of FRD α bearings containing a hygroscopic filler.

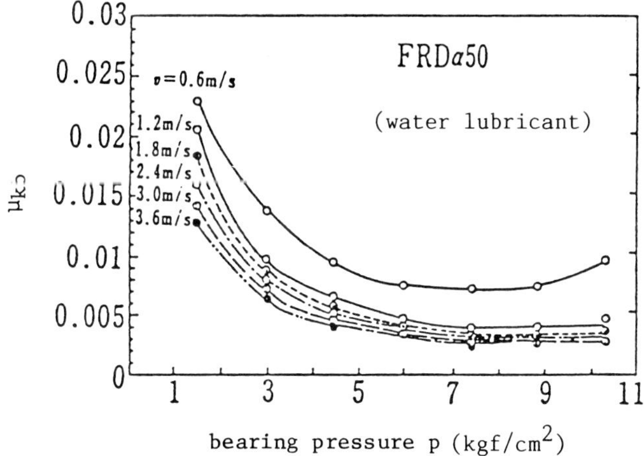

Fig. 4.21 μ_{kb} - p relationships for FRDα50 resin bearings lubricated with water

Figure 4.23 shows the relationships between μ_{kb} and p for MoS_2-filled phenolic, polyacetal (POM), Nylon 6, FRD (DAP) and 50% cellulose-filled FRD (FRDα50) journal bearings lubricated with water. It is clear from this figure that the μ_{kb} values for the hygroscopic plastic bearings are by far the lowest. Figure 4.24 shows the relationships between μ_{kb} and ZN/p for a bronze journal bearing lubricated with machine oil and a 50% cellulose filled FRD DAP (FRα50) lubricated with water. It indicates that the μ_{kb} value of the latter is generally less than that of the former. The reasons for the high lubricity of water with the hygroscopic plastic bearings can be explained by considering the contact angles between water and the plastic surfaces, as shown in Fig. 4.25.

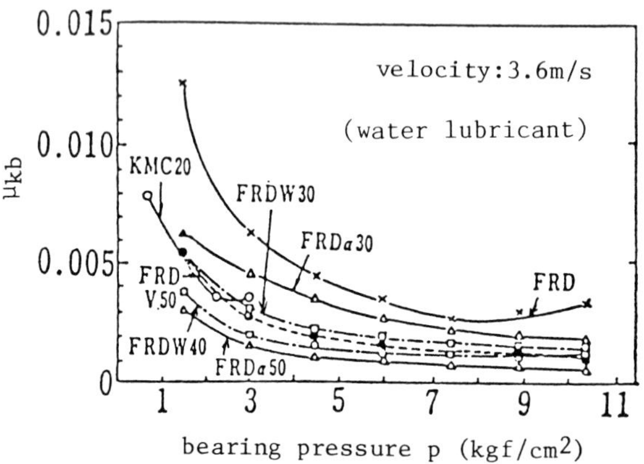

Fig. 4.22 μ_{kb} - p relationships for plastic bearings lubricated with water

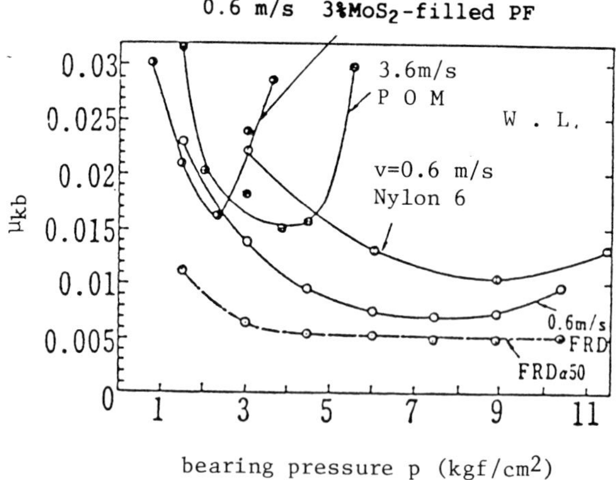

Fig. 4.23 μ_{kb} - p relationships for various plastic bearings lubricated with water

TABLE 4.5
Types and composition of bearing materials containing hygroscopic fillers

Matrix	Filler Type	Filler % Cont.	Mark	Details	Moulding Conditions
20%		0	FRD	Sumikon AM-200	Compression
Glass	α Cellulose	20	FRDα20	(Sumitomo	moulding pressure:
		30	FRDα30	Bakelite)	200 kgf/cm²
Fibre-		50	FRDα50	α Cellulose: filter paper powder	Temperature: 160°C Time: 15 mins
Filled	Polyvinyl Alcohol	30	FRD V30	Wood Powder:	
		50	FRD V50	powder of Lavan wood	
DAP					
	Wood Powder	30	FRD W30		
		40	FRD W40		
Silicone resin filled with glass fibres			KMC 10 KMC 20	Shinetsu Silicone	

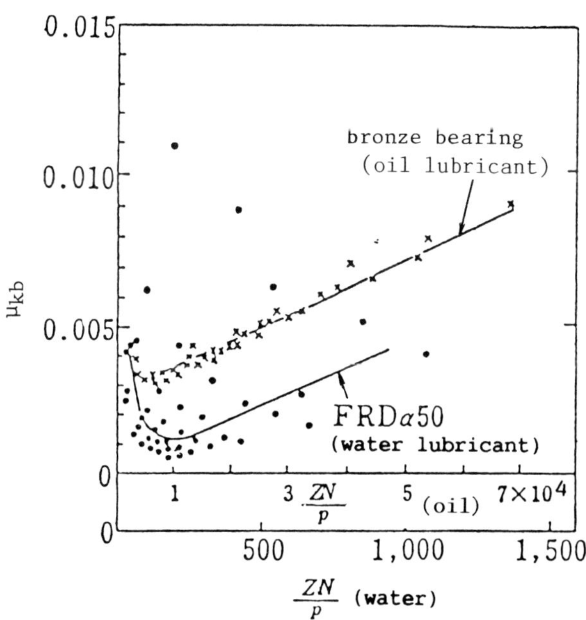

Fig. 4.24 μ_{kb} - ZN/p relationship for a FRDα50 resin bearing

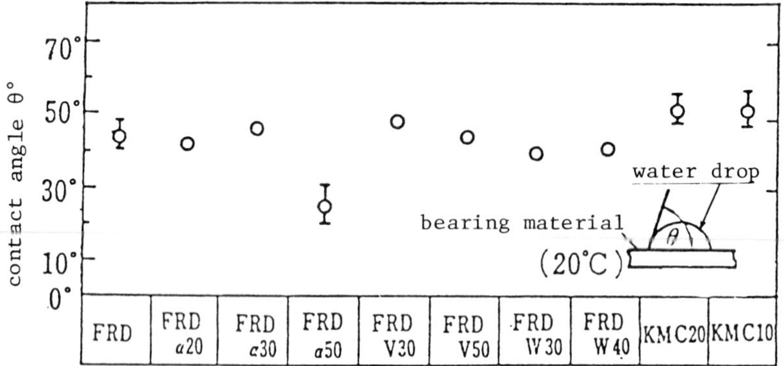

Fig. 4.25 Contact angles of various plastic bearing materials

Figure 4.26 gives the relationship between the minimum μ_{kb} value for each plastic journal bearing lubricated with water and the contact angle, θ, and it can be seen that the smaller the contact angle, the lower is the μ_{kb} value.

Figure 4.27 shows the relationship between the contact angle, θ, for each plastic material and the limiting pressure for each bearing lubricated with water at a speed of 0.6 m/s.

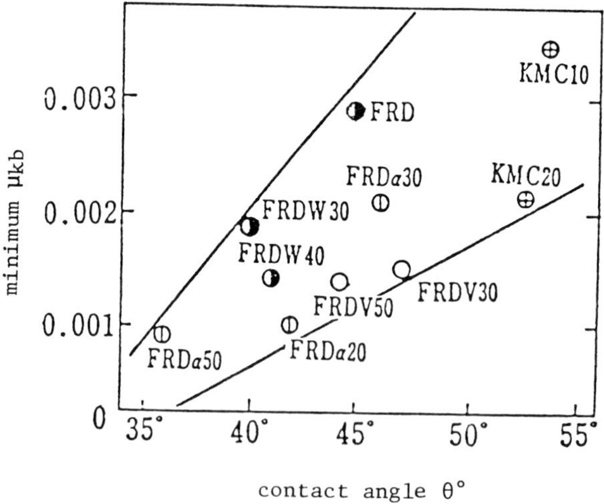

Fig. 4.26 Relationships between minimum μ_{kb} and contact angle, θ, for various plastic bearings

TABLE 4.6
μ_{kb} values of various plastic bearings lubricated by water

Mark	Bearing Material Matrix	and Filler	μ_{kb}
A	40%	0	0.0024 - 0.0188
B	Glass	Graft copolymer MMA with cellulose	0.0039 - 0.0125
	Fibre-	10% (graft 74%)	
C	Filled	Graft copolymer MMA with cellulose	0.0056 - 0.0282
	DAP	10% (graft 132%)	
D		Nylon 4%	0.0092 - 0.0463
E	Cotton cloth filled phenolics		0.0058 - 0.0361
F	Super high molecular weight polyethylene		0.0139 - 0.054
G	Lignumvitae		0.0084 - 0.0455
DAP containing 30% α cellulose			0.0018 - 0.0196
DAP containing 50% α cellulose			0.0009 - 0.011
DAP containing 50% PVA			0.0011 - 0.0219
DAP containing 40% wood powder			0.0011 - 0.0212
Nylon 6			0.0051 - 0.041
Polyacetal (POM)			0.0172 - 0.107
PTFE			0.0052 - 0.069
DAP containing 50% carbon fibres			0.0058 - 0.049
Phenolics containing 15% PTFE			0.0034 - 0.0173
Phenolics containing 5% MoS_2			0.0046 - 0.0282
Phenolics containing 8% graphite			0.0041 - 0.0423

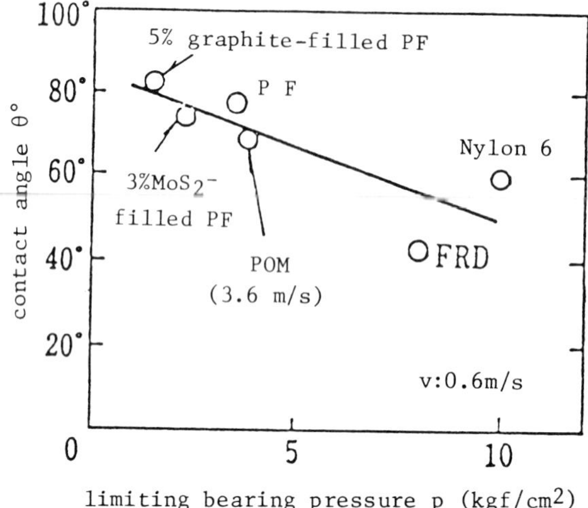

Fig. 4.27 Relationships between contact angle, θ, and limiting bearing pressure for various plastic bearings at a speed of 0.6 m/s and lubricated with water

It appears that the limiting bearing pressure increases with a decrease in the contact angle. It may therefore be generally concluded that the performance of water-lubricated journal bearings can be improved by using plastics incorporating high hygroscopic fillers. Table 4.6 [13] shows a summary of the range of μ_{kb} values for all the various plastic journal bearings lubricated with water.

(iv) *Problems with bearings lubricated with water*

(1) *Bearing material.* Metallic materials for journal bearings are generally too rigid to sustain a fluid film with low viscosity water and are also susceptible to corrosion. However, plastic materials are less rigid and, therefore, the water film is not so easily broken, and they are also resistant to corrosion. Particularly high bearing performances can be obtained by the incorporation of high hygroscopic fillers into plastics.

(2) *Water lubricants.* The water used for lubrication is generally city water or, in some cases, sea water or water containing small amounts of NaOH to inhibit corrosion. Therefore, the composition of water and its additives to improve lubricity is a problem for consideration. When supplying the water as a lubricant, a continuous feed is more effective than immersion, but the amount, temperature and circulation must all be considered.

(3) *Running conditions.* The clearance between the shaft and a plastic bearing surface must be larger than that for metallic bearings; in practice about 0.001 - 0.005 D [2], where D is the diameter of the shaft. However, the exact value will depend on the type of material, the purpose for which it is being used and the bearing dimensions. When the peripheral velocity is so low that the water film is broken even under a comparatively low bearing pressure, the value of μ increases and the frictional heat limits normal running. The peripheral speed of the shaft must therefore be increased to some extent when the bearing pressure in comparatively high. It should be noted that the limiting pv values for dry sliding are not usually relevant to the water-lubricated situation.

TABLE 4.7
Types and manufacturing methods of bearings

Bearing Materials	Details	Manufacturing Method
Nylon 6	N.Co. A 1025	Machined from commercial extruded bar
Polyacetal (POM)	D.Co. Delrin	
PTFE	D.Co. Polyflon M11	Preforming pressure: 300kgf/cm^2 Preforming time: 10 mins Sintering time: 60 mins Sintering temperature: 360°C
DAP containing 20% glass fibre: 0% Asbestos 3% Asbestos 5% Asbestos 10% Asbestos	Matrix: SumikonAM200 Filler: Asbestos fibre (Crysotile)	Compression moulding pressure: 200 kgf/cm^2 Temperature of compression moulding: 150°C Time of compression moulding: 10 mins

4.1.2 *Dry journal bearings*

Plastic materials possessing self-lubricating properties are particularly valuable for unlubricated (dry) journal bearings. There is a long history of the application of plastics to sliding bearings and some of the literature available includes Braithwaite [15], Clauss [16], Kawasaki [17] and Matsubara [25]. Other papers concerning specific types of plastic bearings are: PTFE, MoS_2, graphite or stearine-filled phenolic bearings by Willis [18], PTFE bearings by Twiss [20] and polyacetal (POM) bearings by Pratt [20]. In this section, some experimental results obtained in the author's laboratory will now be described for journal bearings of Nylon 6, POM, PTFE, Diallylphthalate (DAP), polyphenylene sulfide (PPS), polyimide and their composites.

(i) *Bearing materials and test conditions*

The materials used for the journal bearings tested were Nylon 6, polyacetal (POM), PTFE, 20% G.F.-filled DAP and 3%, 5% or 10% asbestos-filled G.F.DAP (details and moulding conditions are show in Table 4.7); polyphenylene sulfide (PPS) and eight kinds of PPS composites containing various fillers (shown in Table 4.8), and polyimide and seven kinds of polyimide composites containing various fillers (shown in Table 4.9).

TABLE 4.8
Composition of PPS bearing specimen

Code	Filler
R6	Nil
R4	Glass fibre: 40%
R4F	R140; Glass fibre: 40%; PTFE (F): 10%
RC_f	Carbon fibre (Cf): 30%
RCA	R150; Graphite (C): 10%; Asbestos (A): 40%
RC_fF	Carbon fibre (C_f): 30%; PTFE (F): 15%
RMS	R1800; MoS_2 (M): 33%; Sb_2O_3 (S): 27%
$FMSC_f$	MoS_2 (M): 17%; Sb_2O_3 (S): 20%; Carbon fibre (C_f): 18%
RMSG	M: 20%; S: 17%; Glass fibre (G): 20%

TABLE 4.9
Bearing characteristics of polyimide materials [23]

Materials		Range of μ_{kb}	μ_{kb} (average)	Limiting pv_b value ($kgf/cm^2 \cdot cm/s$)
Polyimide composites:				
No.1	No filler	0.132 - 0.828	0.361	209
No.2	Graphite 25wt%	0.151 - 0.504	0.264	334
No.3-1	Graphite 40wt%	0.229 - 0.569	0.366	311
No.3-2	Graphite 30wt% Asbestos 10wt%	0.137 - 0.163	0.366	938
No.3-3	Graphite 30wt% Glass fibre 10wt%	0.117 - 0.862	0.369	1251
No.4	Graphite 10wt% MoS_2 25wt%	0.161 - 0.760	0.438	348
No.5	PTFE 20%	0.450 - 0.909	0.662	209
Nylon 6		0.059 - 0.440	0.249	188
DAP		0.113 - 0.375	0.249	209
Bronze (oil lubricated)		0.0037- 0.016	0.008	1217

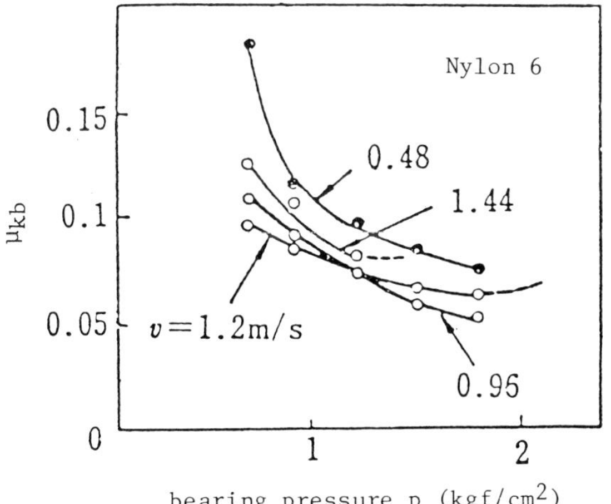

Fig. 4.28 μ_{kb} - p relationships for unlubricated Nylon 6 bearings

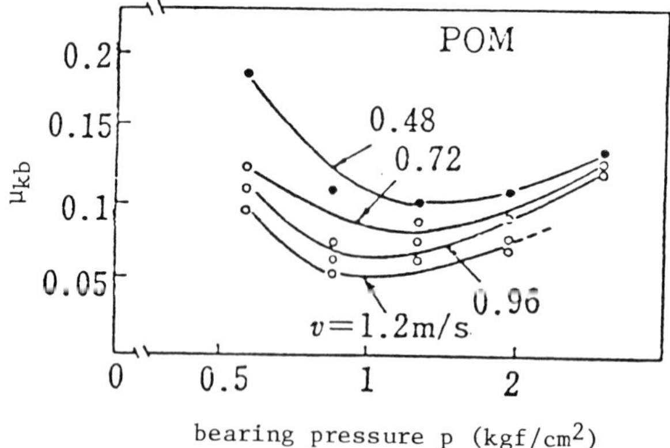

Fig. 4.29 μ_{kb} - p relationships for unlubricated POM bearings

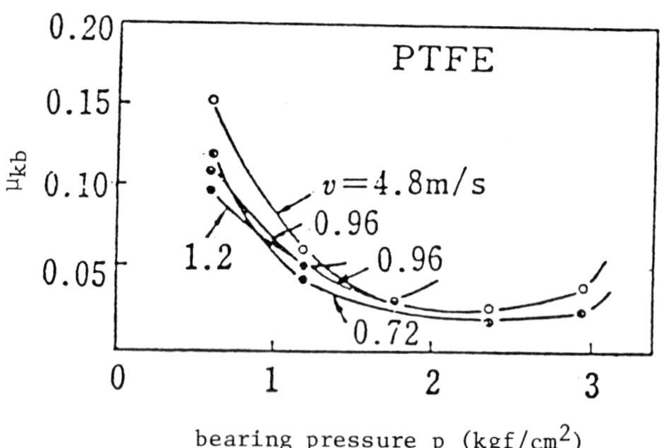

Fig. 4.30 μ_{kb} - p relationships for unlubricated PTFE bearings

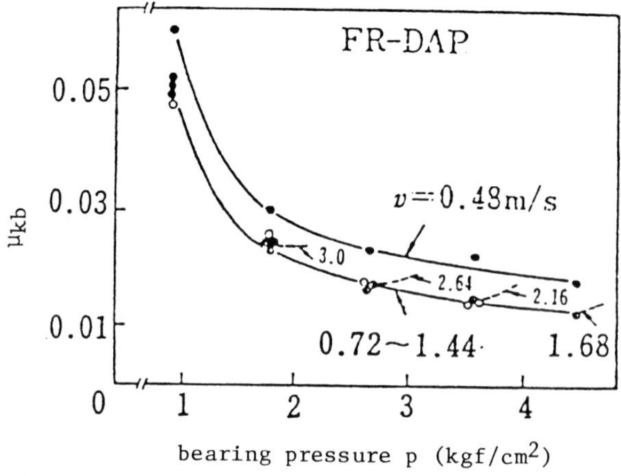

Fig. 4.31 μ_{kb} - p relationships for unlubricated GF-filled DAP bearings

The bearing tester, as shown in Figs. 4.1 and 4.2, consisted of a split-type journal bearing with a 45 mm inner diameter, a 64 mm length and a steel shaft. The tests were carried out under the following conditions; dry (no lubricant supplied), peripheral speeds of 0.43 - 3.5 m/s, bearing pressures of 0.6 - 4.5 kgf/cm² and environment temperatures of 15 - 25°C.

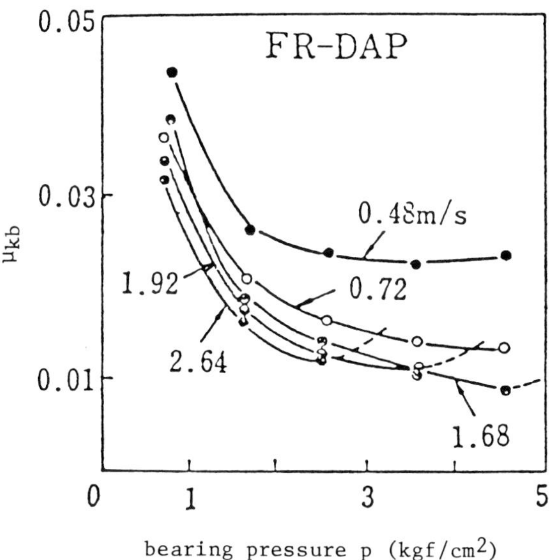

Fig. 4.32 μ_{kb} - p relationships for unlubricated 3% asbestos, 20% GF-filled DAP bearings

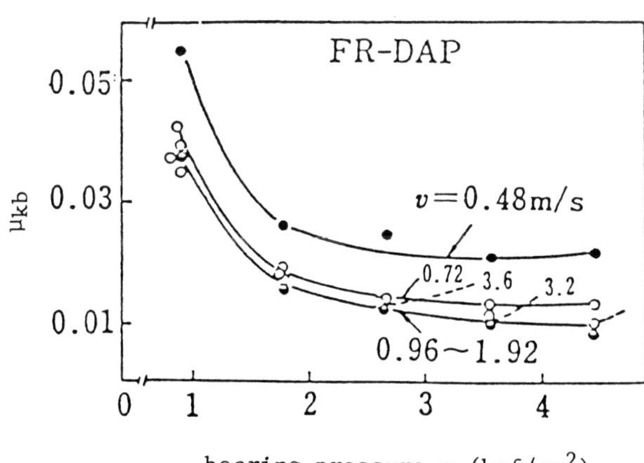

Fig. 4.33 μ_{kb} - p relationships for unlubricated 5% asbestos, 20% GF-filled DAP bearings

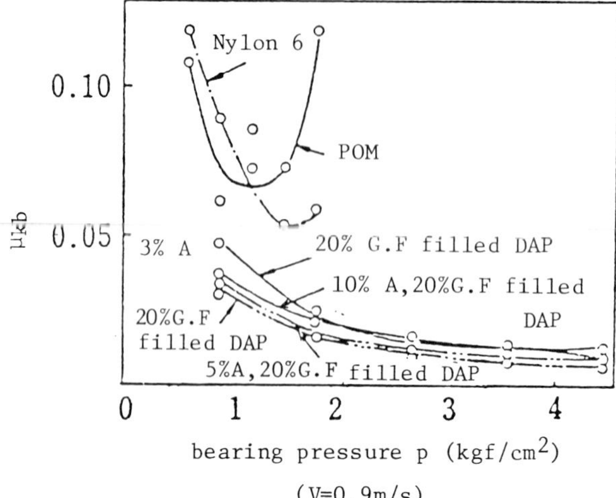

Fig. 4.34 μ_{kb} - p relationships for various unlubricated bearings

(ii) *Relationship between* μ_{kb} *and* p *or* v

The kinetic friction coefficient, μ_{kb}, was measured for each constant bearing pressure, p, and peripheral speed, v, when a stable condition has been reached after 5 - 10 minutes. Figures 4.28 to 4.34 show the relationships between μ_{kb} and p, and Figs. 4.37 to 4.42 show the relationships between μ_{kb} and peripheral speed, v, for each bearing. Figure 4.28 shows the μ_{kb}-p relationship for Nylon 6, Fig. 4.29 for POM, Fig. 4.30 for PTFE, Fig. 4.31 for the 20% G.F.-filled DAP, and Figs. 4.32 and 4.33 for the 3% and 5% asbestos-filled FR DAP respectively. Figure 4.34 compares the relationships between μ_{kb} and p for all the above-mentioned bearings at a peripheral speed of 0.9 m/s. Figure 4.35 shows the relationships between μ_{kb} and p at each speed for three kinds of bearings made of PPS composite materials, R4, RFC_f and $RMSC_f$ (details are shown in Table 4.8). Figure 4.36 shows similar relationships at speed of 3.4 m/s for polyimide composite bearings, details of which are shown in Table 4.9. Figure 4.37 shows the relationship between μ_{kb} and peripheral speed, v, at various bearing pressures for a Nylon 6 bearing, Fig. 4.38 for a POM bearing, Fig. 4.39 for a PTFE bearing, Fig. 4.40 for a 20% G.F.-filled DAP bearing and Figs. 4.41 and 4.42 for various polyimide composite bearings at pressures of 0.295 kgf/cm^2 and 1.47 kgf/cm^2, respectively.

It is clear from all the μ_{kb}-p and μ_{kb}-v relationships that the value of μ_{kb} generally decreases with an increase in p or v to a minimum and then increases again. The range and average value of μ_{kb} for the above-mentioned experimental conditions are shown in Table 4.9. It is clear from Table 4.10 and the above figures, particularly Fig. 4.34, that the values of

μ_{kb} for the unfilled PTFE bearings are smaller than those for Nylon or POM bearings, and those for the 3% or 10% asbestos-filled FR DAP bearings are even smaller, at 0.007 - 0.04.

TABLE 4.10
Range of μ_{kb} values for unlubricated journal bearings [21]

Bearing Material	Range of μ_{kb} Values
Nylon 6	0.043 - 0.18
Polyacetal (POM)	0.035 - 0.188
PTFE	0.022 - 0.150
DAP containing 20% glass fibre:	
0% Filler	0.0144 - 0.060
3% Asbestos	0.0078 - 0.044
5% Asbestos	0.0078 - 0.053
10% Asbestos	0.0091 - 0.052

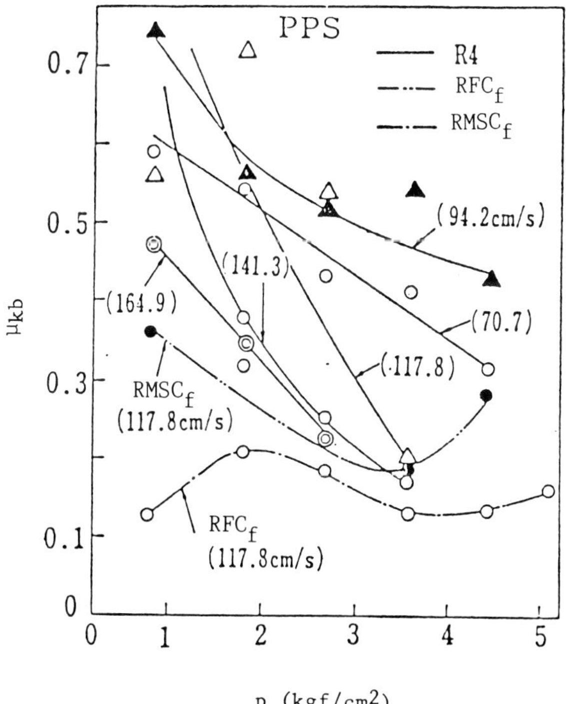

Fig. 4.35 μ_{kb} - p relationships for unlubricated PPS composite bearings

TABLE 4.11
Sliding and journal bearing properties of PPS composites and other materials

Material	Continuous Sliding Friction Properties μ_s (R/Steel)	μ_k	Limiting pv kgf/cm^2·cm/s	Unlubricated Bearing Properties μ_{kb}	Limiting pv_b kgf/cm^2·cm/s
R6	0.333	0.215 (avg)		0.04-0.21	250
R4	0.311	0. 2-1.10	478	0.18-0.9	457
R4F		0.14-0.47	843	0.05-0.61	862
RC_f		0.24-0.72	847	0.18-0.32	420
RCA			1037	0.04-0.25	515
RC_fF		0.15-0.61	925	0.04-0.36	715
RMS		0.25-0.90	630	0.09-0.32	496
$RMSC_f$		0.09-0.80	913	0.10-0.29	652
RMSG		0.31-0.80	354	0.16-0.64	311
POM (Duracon)		0.2 -0.45	189-510	0.03-0.18	210
PTFE		0.1 -0.2	80-100	0.02-0.15	181
Nylon 6		0.09-0.24	590	0.04-0.18	89
Phenolic containing 10% graphite		0.15-0.82	750		610
DAP (G:20%,A:10%)				0.0091-0.05	790

TABLE 4.12
Limiting pv_b values [21]

Bearing Material		(kgf/cm^2·cm/s)
Nylon 6		213
Polyacetal (POM)		214
PTFE		181
DAP containing 20% glass fibre:	0% Asbestos filled	476
	3% Asbestos filled	650
	5% Asbestos filled	714
	10% Asbestos filled	790

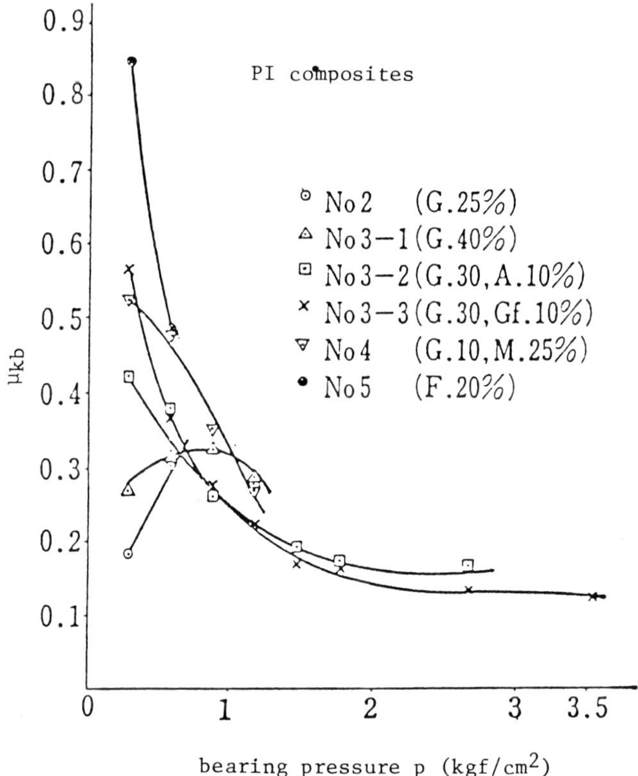

Fig. 4.36 μ_{kb} - p relationships for unlubricated polyimide composite bearings at a speed of 3.4 m/s

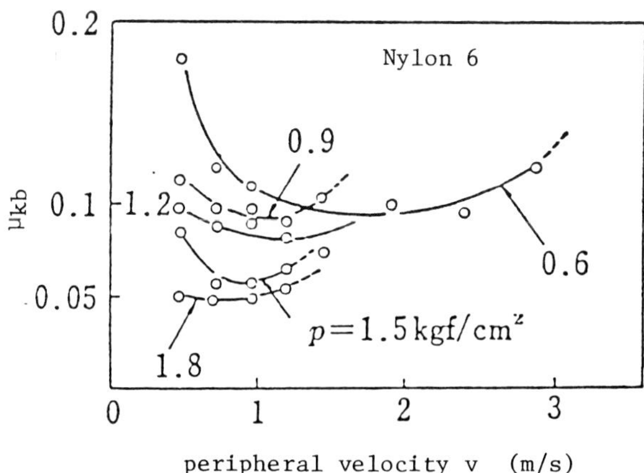

Fig. 4.37 μ_{kb} - v (peripheral velocity) relationships for unlubricated Nylon 6 bearings

It is shown in Fig. 4.35 and Table 4.11 that the values of μ_{kb} in PPS composites are reduced to 0.02 - 0.3 by using PTFE, graphite, Sb_2O_3 or MoS_2 as fillers. Figures 4.36, 4.41, 4.42 and Table 4.9 show that the μ_{kb} values for polyimide composite bearings range between 0.11 and 0.9, with an average of between 0.26 and 0.66, and do not decrease significantly with the addition of PTFE, graphite or asbestos.

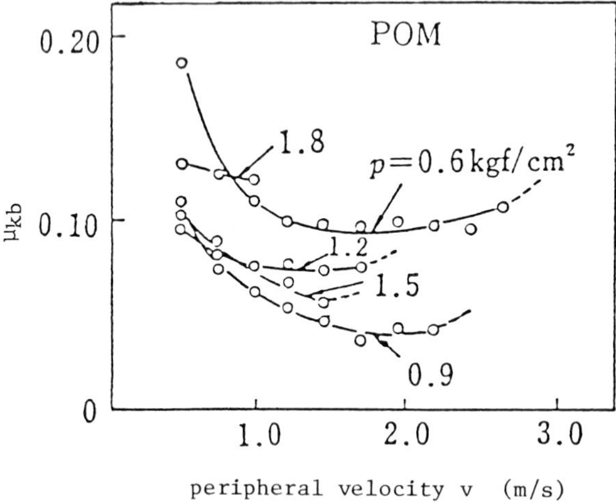

Fig. 4.38 μ_{kb} - v relationships for unlubricated POM bearings

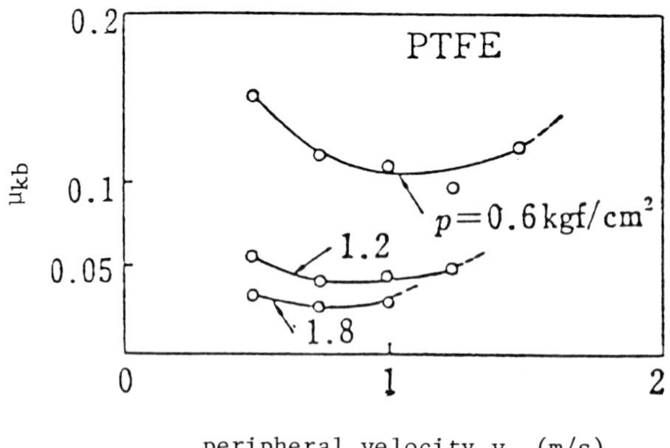

Fig. 4.39 μ_{kb} - v relationships for unlubricated PTFE bearings

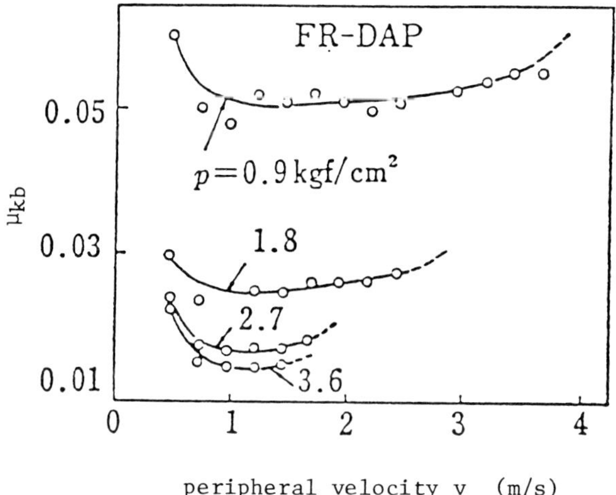

Fig. 4.40 μ_{kb} - v relationships for unlubricated 20% GF-filled DAP bearings at various pressures

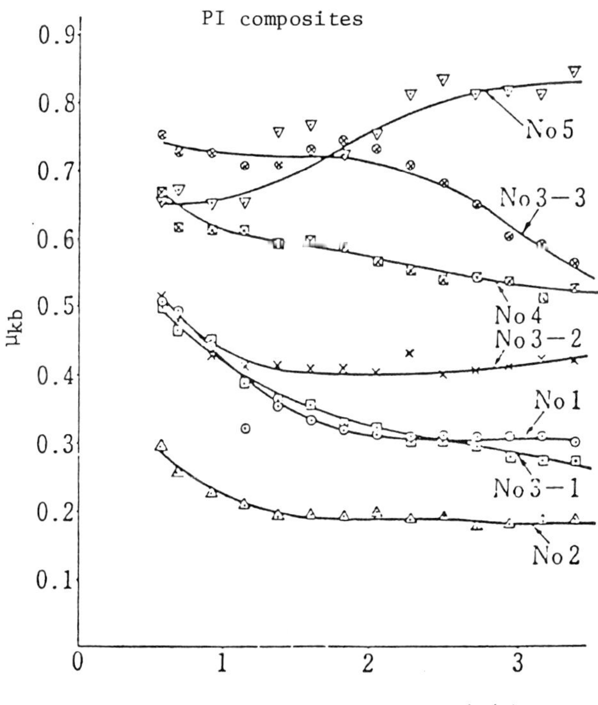

Fig. 4.41 μ_{kb} - v relationships for unlubricated PI composite bearings at p = 0.295 kgf/cm²

(iii) *Limiting* pv_b value of journal bearings

In the above μ_{kb}-p of μ_{kb}-v relationships, there is a point where the μ_{kb} value increases gradually after reaching a minimum as shown, for example, in Figs. 4.30 and 4.38. These points, obtained under various conditions, are given by the coordinates, (p,v), as shown in Figs. 4.43 and 4.44. The p-v curve defined by these points is similar to a right-angled hyperbola and represented by the equation: pv = constant. The value of this constant may be nominated as the limiting p_{vb} value of the dry journal bearing and gives the maximum range for stable running. Experimental values of pv_b for the various dry bearings are shown in Table 4.12; those for PPS composite bearings in Table 4.11 and those for polyimide composite bearings in Table 4.9. It is clear that the pv_{bmax} value is small for unfilled plastic bearings but increases for mixtures of DAP with asbestos, PPS, and PTFE with carbon fibres, and polyimide with graphite. With one polyimide composite bearing, a pv_{bmax} of 1251 kgf/cm²·cm/s is obtained. For reference, another author's data on the pv_b values of plastic bearings are shown in Tables 4.13 and 4.14 [17].

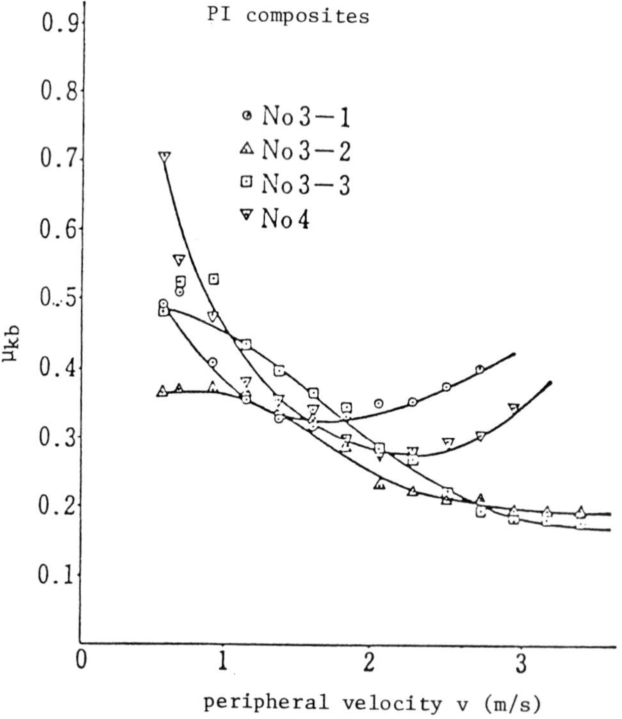

Fig. 4.42 μ_{kb} - v relationships for unlubricated PI composite bearings at p = 1.47 kgf/cm²

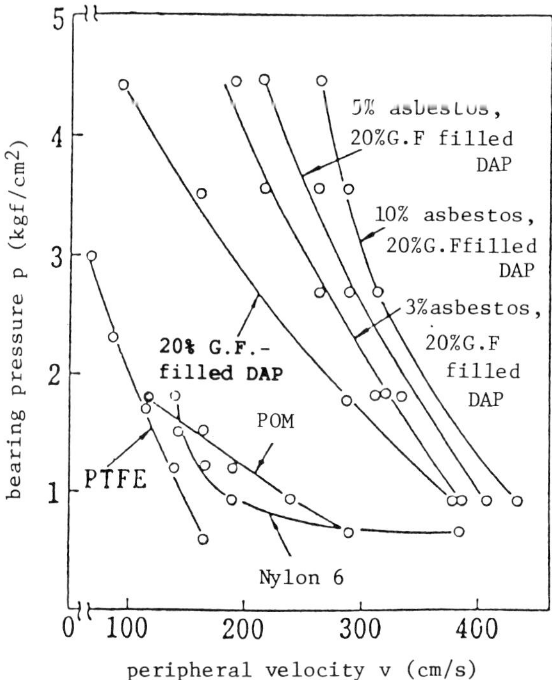

Fig. 4.43 Limiting p - v relationships for various unlubricated plastic bearings

TABLE 4.13
Limiting conditions for unlubricated journal bearings [17]

Bearing Material	Limiting pv_b (kgf/cm²·cm/s)	Limiting Temperature (°C)	Limiting Velocity (cm/s)
PTFE	35	250	650
Polyacetal (POM)	54	100	500
Carbon-graphite	104	350-500	100
PTFE+15% glass fibre	104	250	-
PTFE+25% glass fibre	174	250	-
Metallic carbon	174	150-350	100
Nylon 6 + MoS_2	174	150	250
Cloth-filled phenolic	533	120	1260
Sintered bronze containing PTFE and Pb	418-1750	250	2500

TABLE 4.14
Performance of various journal bearings [17]

Material	Wear Rate ($\times 10^{-10}$ $\frac{cm^3 \cdot s}{kgf \cdot m \cdot h}$)	Limiting pv_b (Velocity 5m/s) ($kgf/cm^2 \cdot cm/s$)
POM filled with oil	7	650
POM copolymer (Duracon)	100	180
POM containing 15% PTFE	40	270
POM homopolymer	120	100
POM containing 15% PTFE fibre	40	290
Nylon 66	400	70
Nylon 66 containing 15% PTFE	27	250
Nylon 66 containing 15% PTFE and 30% glass fibre	32	360
Nylon 66 containing 2.5% MoS_2	400	70
Nylon 6	400	70
Polycarbonate (PC)	5050	150
PC containing 15% PTFE	150	250
PC containing 15% PTFE and 30% glass fibre	60	180
Polyurethane containing 15% PTFE	200	180
Sintered bronze containing oil	203	300
Wood powder-filled phenolic containing 15% PTFE	30-70	630
PTFE containing 25% glass fibre	17	410

Sliding conditions: $v = 0.5$ m/s; $pv = 400$ $kgf/cm^2 \cdot cm/s$

4.1.3 *Ball bearings*

There are three situations where plastic materials can be used in rolling element bearings; the rolling elements themselves, the bearing raceways and finally rolling element retainer.

For each of the above, plastics are effective for preventing noise and vibration because of their low elastic moduli. However, plastics are generally unsuitable for rolling elements because of their limited strengths and the high stress levels involved. They are rather more useful, however, for bearing raceways.

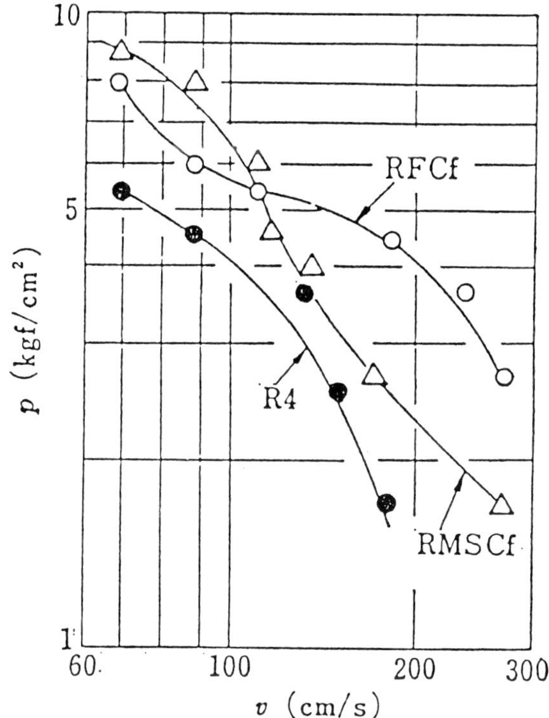

Fig. 4.44 Limiting p - v relationships for unlubricated PPS composite bearings

An example of the application to a retainer has been given in a paper by Wada [24], demonstrating the effectiveness of plastics for wear resistance. An introductory paper by Matsubara [25] and papers by Gremer [26], Montalbano [27] and Boes [28] are also concerned with the application of plastics to rolling bearings. The theories of rolling friction of plastic materials were described earlier in Section 1.2. In this section, some experimental results obtained in the author's laboratory are described, relating to the application of plastics for races in a thrust ball bearing.

(i) *Materials used for thrust ball bearings*

Figure 4.45 shows the thrust ball bearing test apparatus used for the experiments. Twelve steel balls of 4.77 mm diameter were used as rolling elements between three bearing races (inner, centre and outer), and the frictional resistance of the bearings was obtained from the amount of strain at the plate spring measured with a differential transformer. The shape and dimensions of the bearing race are shown in Fig. 4.46, and details of the materials, manufacturing methods, hardness and surface roughness are given in Table 4.15.

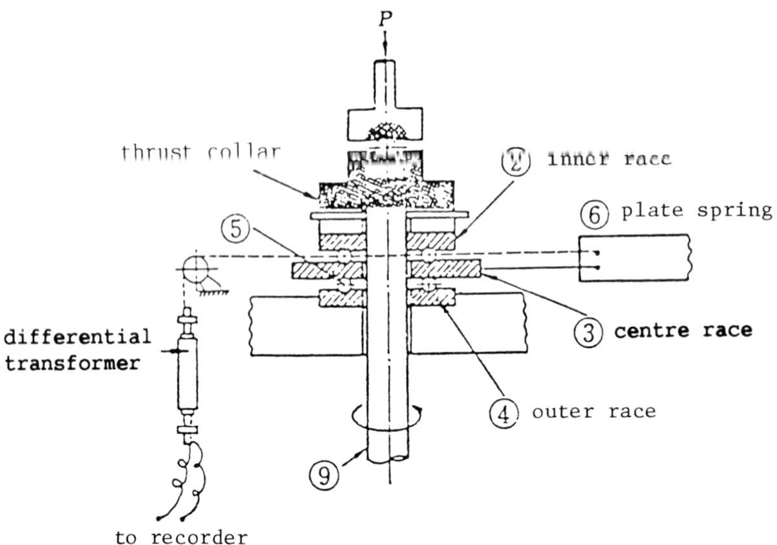

Fig. 4.45 Test apparatus

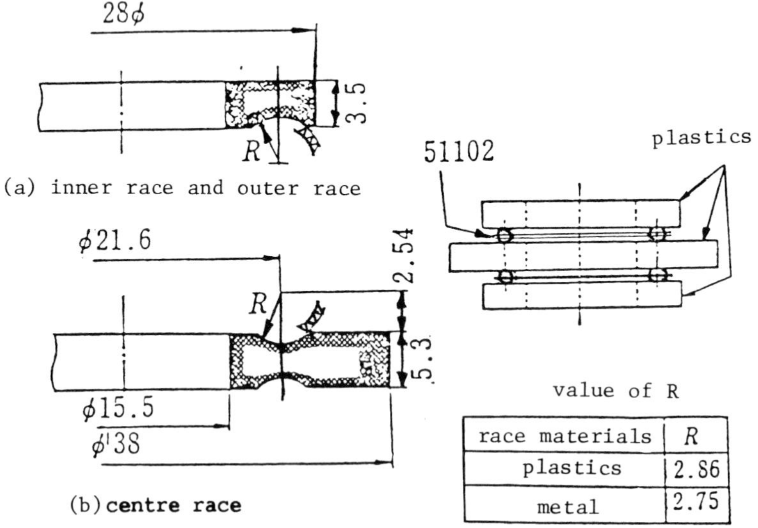

race materials	R
plastics	2.86
metal	2.75

Fig. 4.46 Dimensions of races

TABLE 4.15
Race material

Race Material		Details	Moulding Process	Hardness (Hv)	Surface Roughness μm[1] (max)
Thermosetting Plastic	Melamine (MI)	Sumitomo Bakelite: MM53		26.5	0.28[2]
	Epoxy (ED)	Chiba Co.: Araldite		18.2	
Thermo-plastic Materials	Polyacetal	du Pont Co.: Delrin	Commercial extrusion	8.7	4.1
	Nylon 6	Nichi Ray: A1025	Commercial extrusion	5.3	2.7
	Nylon 11	Lirson: BMN	Compression moulding, temp. 210°C	4.2	1.1[2]
	PTFE	Daikin: Polyflon	Commercial bar	2.2	5.5
	PPO	GE: No. 534	Commercial extrusion	7.4	
	Polyurethane	Elastran: VX7	Injection moulding	3.2	12.3
Metal	Bearing steel	SUJ 2	Commercially available	287	2.6
	Bronze	BC-7	Commercially available	117	6.0
	Standard bearing	51102		613	2.3

1) Roughness of race groove
2) Roughness of centreline μm(Ha)

The race materials were: two kinds of thermosetting plastics; melamine resin (MI) and Epoxy resin (EP); six kinds of thermoplastics; polyimide and five of its composites (shown in Table 4.16); and bearing steel and bronze for reference purposes. The surfaces were machine-finished by milling and a commercially-available steel thrust ball bearing, 51102, was also used in the experiments. Tests of the various thrust ball bearings were carried out at rotational speeds of 200 - 2000 rpm (22 - 220 cm/s rolling velocity) and vertical contact loads of 2 - 30 kgf.

TABLE 4.16
Race materials

Code	Material
A	Polyimide 75%, Graphite 25%
B	Polyimide 60%, Graphite 30%, Asbestos 10%
C	Polyimide 65%, Graphite 10%, MoS_2 25%
D	Polyimide 80%, PTFE 20%
E	Polyacetal (Duracon)
F	Commercially-available thrust ball-bearing

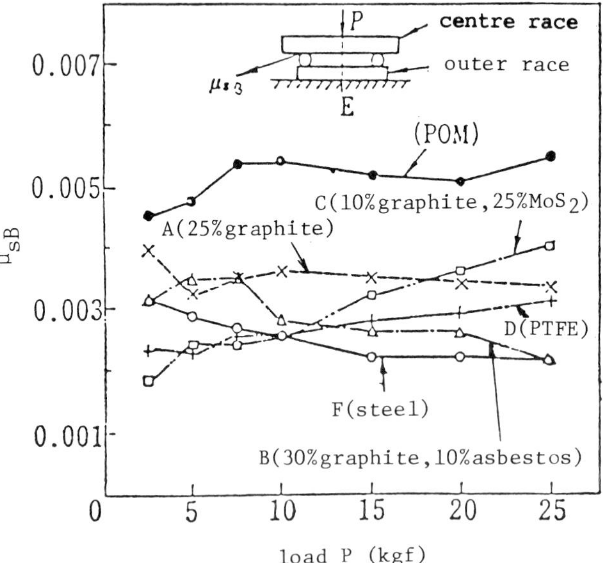

Fig. 4.47 Relationships between μ_{sB} and load for various races

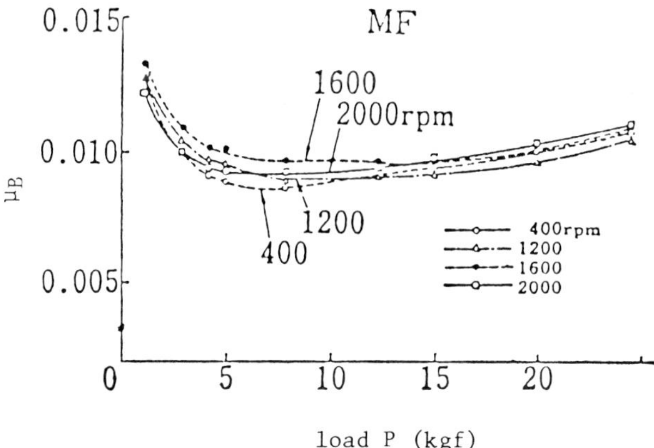

Fig. 4.48 Relationships between μ_B (rolling friction coefficient) and load for MF resin ball bearings

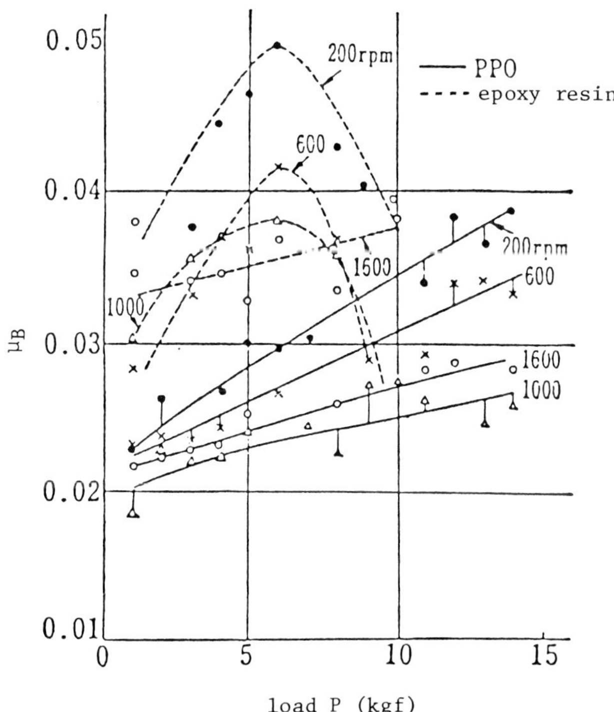

Fig. 4.49 Relationships between μ_B and load P for PPO and Epoxy resin ball bearings

(ii) *Relationships between* μ_b and bearing load P

Figure 4.47 shows the relationship between the static rolling-friction coefficient μ_{sb} and the vertical load P for each plastic bearing race, shown in Table 4.16 when steel balls were placed between a centre bearing race and another bearing race. The values of μ_{sb} range from 0.002 to 0.0056, and those for a POM race are generally greater than those for a PI composite race.

Figures 4.48 to 4.54 show the relationship between the kinetic rolling friction coefficient μ_B and vertical load P for each bearing race material, melamine resin (MF), polyphenylene oxide (PPO), epoxy resin (EP), polyurethane (PU), Nylon 6, Nylon 11, polytetrafluoroethylene (PTFE), polyacetal (POM) and bronze or bearing steel. Figure 4.55 compares the relationships between μ_B and P at a rotational speed of 1000 rpm for each of the six kinds of bearing race material and a steel thrust ball bearing (51102). Figure 4.56 shows similar relationships for each of the various polyimide bearing races.

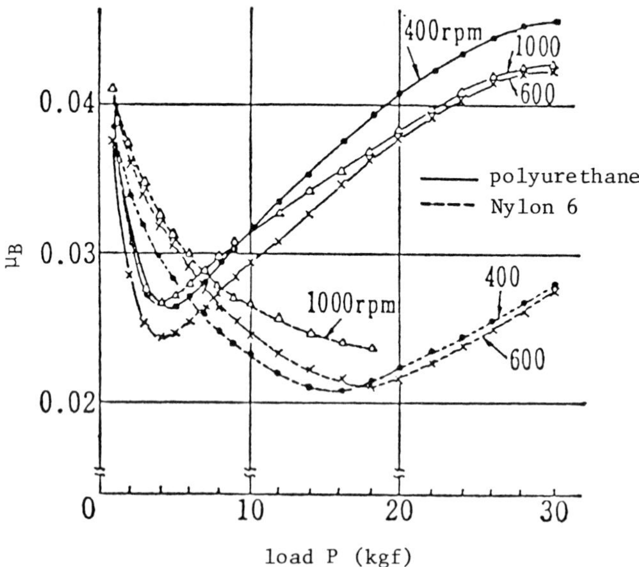

Fig. 4.50 Relationships between μ_B and P for polyurethane and Nylon 6 ball bearings

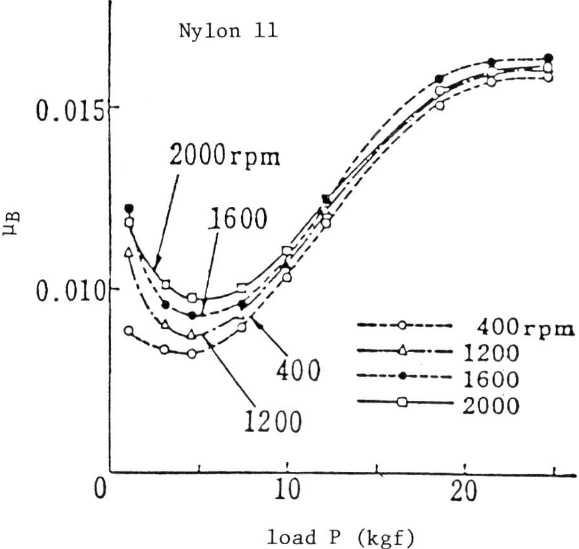

Fig. 4.51 Relationships between μ_B and P for Nylon 11 ball bearings

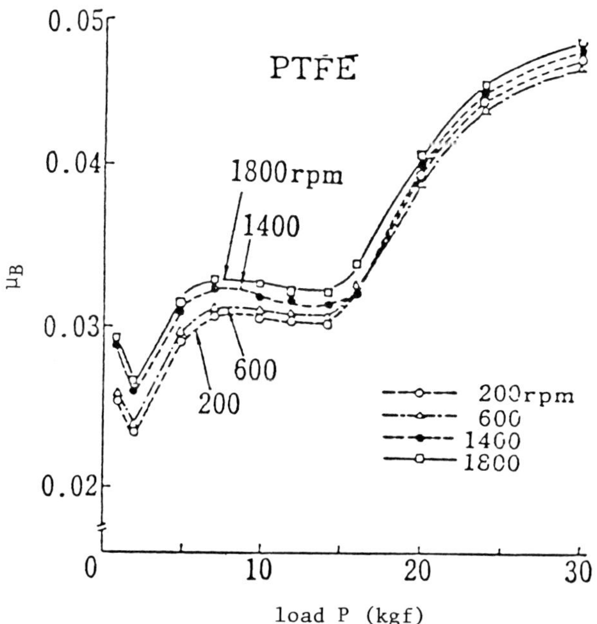

Fig. 4.52 Relationships between μ_B and P for PTFE ball bearings

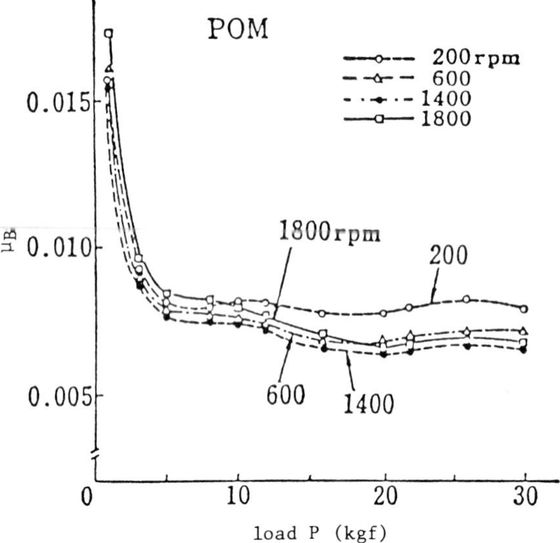

Fig. 4.53 Relationships between μ_B and P for POM ball bearings at various speeds

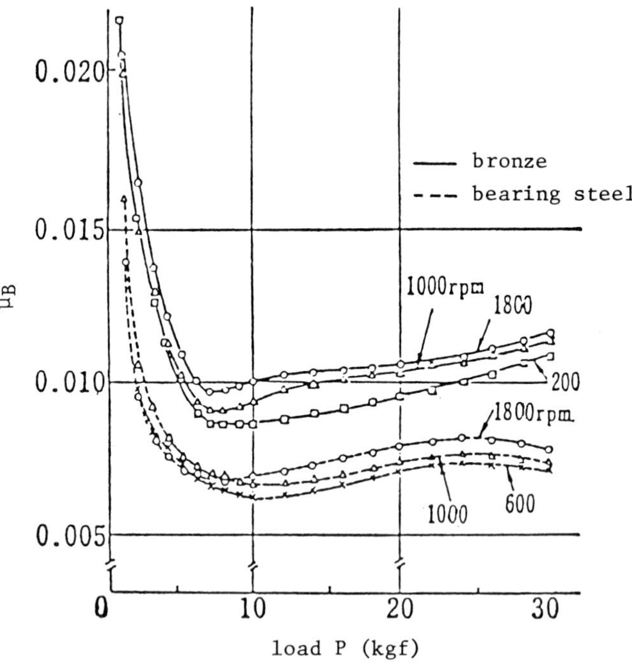

Fig. 4.54 Relationships between μ_B and P for bronze and bearing steel ball bearings at various speeds

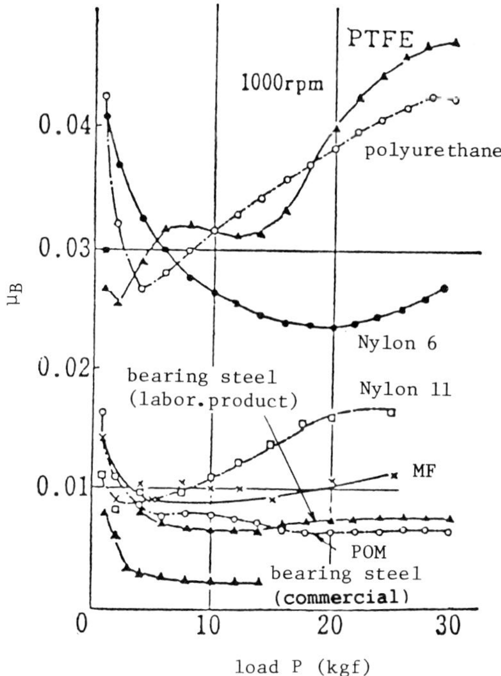

Fig. 4.55 Relationships between μ_B and P for various ball bearings at 1000 rpm

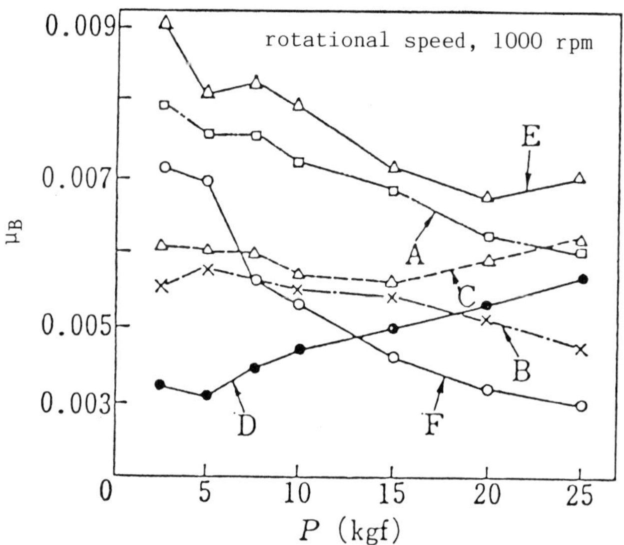

Fig. 4.56 Relationships between μ_B and P for PI composite ball bearings

It is clear from these figures that for the soft plastic bearing races, such as PU, PTFE, Nylon 6 and Nylon 11, μ_B decreases to a minimum under a load of 1 - 3 kgf, and then increases with a further increase in load P. The softer the race material, the smaller the load corresponding to the minimum μ_B. μ_B values for the thermosetting plastic bearing races are comparatively large for the epoxy resin (EP) but for the melamine (MF) and polyimide races they are generally small and nearly equal to those of a steel race.

The μ_B values for the thermoplastic races, such as PTFE, PU, Nylon 6, Nylon 11 and POM, are relatively high except, however, for POM. Table 4.17 shows the μ_B values for each of the various materials of the bearing races.

TABLE 4.17
μ_B values of plastic race thrust ball bearing

Race Material	Rolling Friction Coefficient μ_B
Thermosetting Plastics:	
Melamine resin, Sumitomo Bakelite; MM53	0.0082-0.011
Epoxy resin, Araldite B	0.021 -0.038
Polyimide A Graphite 25% filled	0.006 -0.0084
Polyimide B Graphite 30%, Asbestos 10% filled	0.0046-0.0059
Polyimide C Graphite 10%, MoS_2 25% filled	0.0055-0.0062
Polyimide D PTFE 20% filled	0.0038-0.0058
Thermoplastic Materials:	
PPO (GE:NO 534)	0.028 -0.048
POM (Delrin)	0.0067-0.0093
PTFE (Polyflon)	0.025 -0.046
Nylon 6 (Nichi Ray: A1025)	0.021 -0.035
Nylon 11 (Lirsan:BMN)	0.026 -0.016
Polyurethane (Elastran Vx7)	0.024 -0.043
Metals:	
Bronze (BC-7)	0.0084-0.013
Bearing steel (SUJ 2)	0.004 -0.0076
Standard bearing (51102)	0.002 -0.008

(iii) *Effect of rotational speed*

Figures 4.57, 4.58 and 4.59 show some examples illustrating the effect of rotational speed on μ_B. Each figure gives the relationship between μ_B and rotational speed, N, at different vertical loads, P, for each plastic bearing race; 20% PTFE-filled polyimide D, melamine resin (MF) and polyacetal (POM), respectively. It is clear from these figures that the effect of rotational speed N on μ_B is insignificant for the races of materials D and MF. However, the value of μ_B for a POM race decreases slightly with an increase in N.

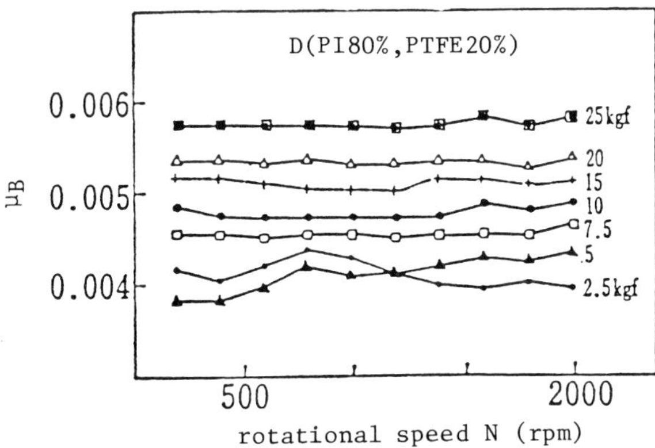

Fig. 4.57 Relationships between μ_B and rotational speed for ball bearings with D race

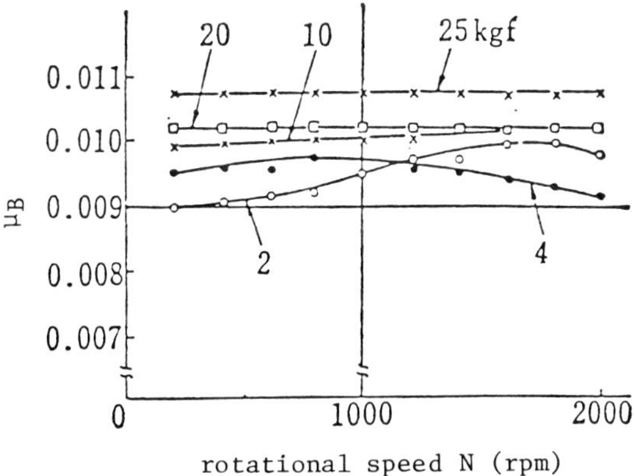

Fig. 4.58 Relationships between μ_B and rotational speed for MF resin ball bearings

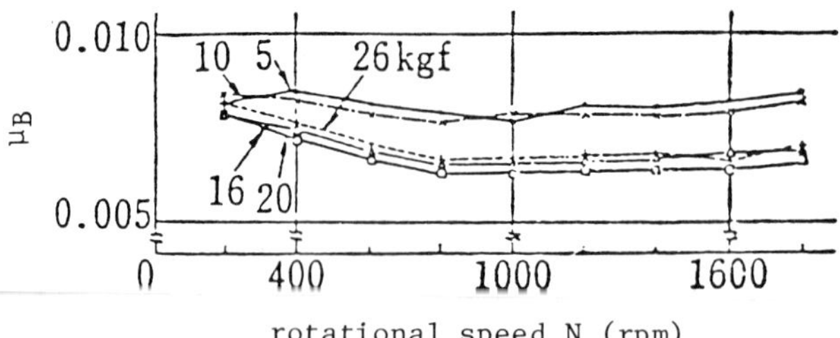

Fig. 4.59 Relationships between μ_B and rotational speed for POM ball bearings

In the rolling friction theories of Section 1.2, the value of μ_B was represented by the following equation (1.30), based on hysteresis loss

$$\mu_B = \alpha_e \frac{3}{16} \cdot \frac{a}{R} \tag{1.30}$$

where a is the radius of the contact circle between the ball and the race, and R is the ball radius. According to Meyer's law,

$$a = cP^{1/n} \tag{1.15'}$$

where c is a constant and n = 2.1 - 2.8 for plastics. Thus

$$\mu_B = c'P^{1/n} \tag{a}$$

It is clear from the data presented in Figs. 4.48 to 4.56 that the experimental results are not as simple as equation (a) predicts.

According to viscoelastic theory, the value of μ_B is presented according to the three regions (1), (2) and (3) as follows:

(1) $\beta \ll \phi$ (2) $\beta = \phi$ (3) $\beta \gg \phi$

where $\beta = \tau S/R$, $\phi = \ell/r$

and τ: retardation time or η/G (η: viscosity, G: shear modulus), S: rolling speed therefore

$\mu_B = \beta$ for region (1)

$\mu_B = k_2 \left(\frac{W}{GR^2}\right)^{1/3}$ for region (2)

$\mu_B = k_3 \left(\frac{W}{G\beta}\right)^{1/2}$ for region (3)

It is also clear from the experimental results presented by Figs. 4.57 to 4.59 that the experimental μ_B-v relationships are not as complex as the above equations and Fig. 1.16 predict.

(iv) *Life of a plastic thrust ball bearing*

(1) *Standard load capacity.* If we define the bearing life as the maximum number of rotations possible without problems such as melting of the bearing race, the relationship between the ball bearing life L and bearing load are show in Figs. 4.60 to 4.62 for, respectively: a melamine (MF) race; a polyacetal (POM) race; and a Nylon 11 race.

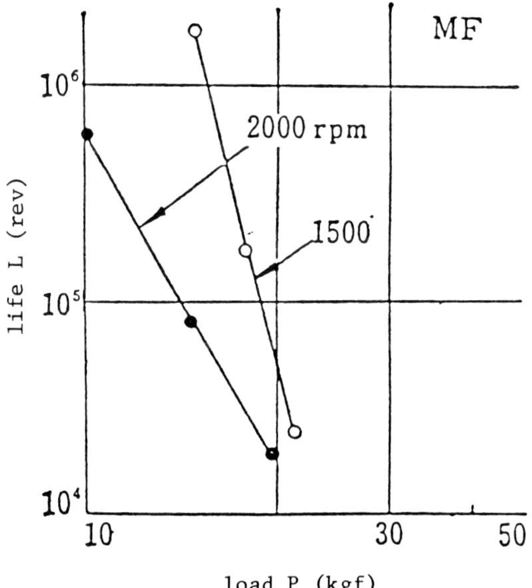

Fig. 4.60 Relationships between life and load for MF resin ball bearings

It is assumed from Fig. 4.60 that the ball bearing with MF races is capable of operating up to 2 x 10^4 - 7.8 x 10^6 rotations under loads from 10 - 25 kgf, although the experimental points defining the relationship are only few in number. If the standard load capacity of a plastic ball bearing is defined as the load capable of achieving up to 10^6 rotations (similar to that for a steel ball bearing), the standard load capacity of the bearings with MF races can be estimated from Fig. 4.60 as 12 kgf. This is equivalent to 1/65 of the load capacity (785 kgf) of the steel thrust ball bearing (51102) of similar dimensions. It is clear from Fig. 4.61 that the life of the bearing with POM races is 1.2 x 10^4 - 3 x 10^4 rotations at bearing loads of 10 - 25 kgf, and the estimated standard load capacity is thus 17.32 kgf, which is equivalent to 1/4 that of the steel bearing (51102) of similar dimensions. The life of the bearing with Nylon 6 races is estimated from Fig. 4.62 as 7 x 10^3 - 5 x 10^6 rotations under bearing loads of 10 - 25 kgf, and the estimated standard load capacity is 14 kgf, equivalent to 1/56 that of the steel bearing.

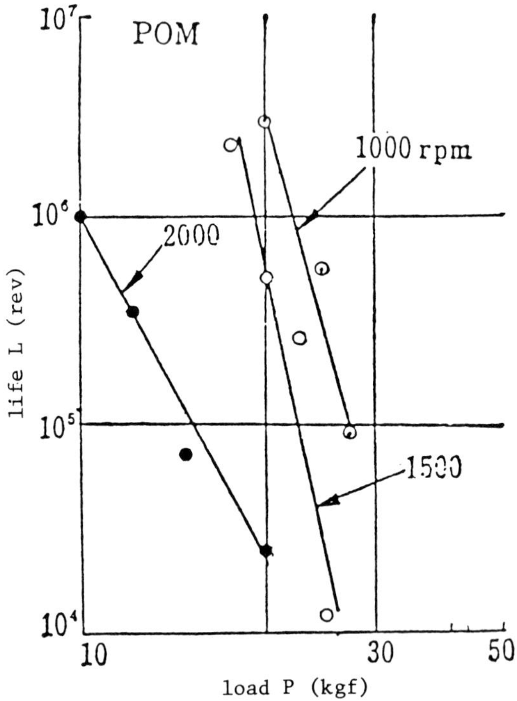

Fig. 4.61 Relationships between life and load for POM ball bearings

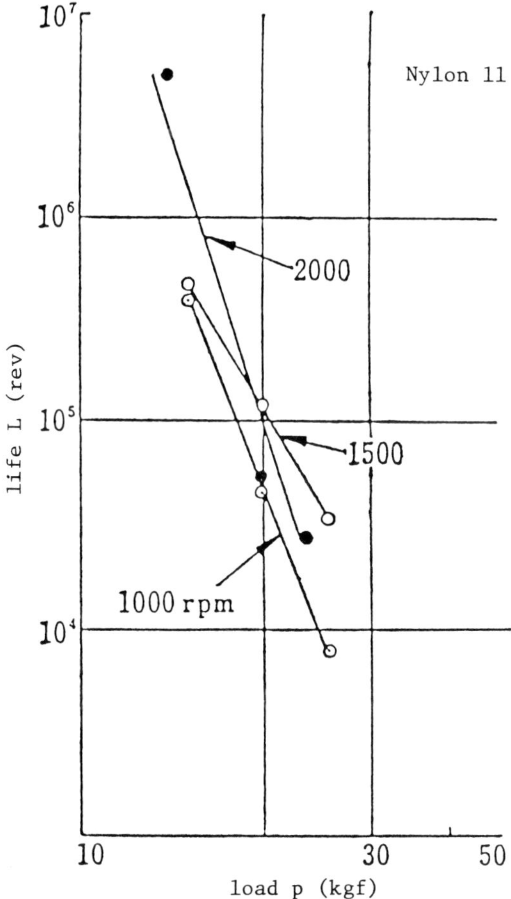

Fig. 4.62 Relationships between life and load for Nylon 11 ball bearings

(2) *Relationship between bearing life and load.* The relationships between the life of plastic thrust ball bearings, L, and the bearing load, P, are presented in Figs. 4.60, 4.61 and 4.62, using the following equation:

$$L = aP^{-m} \tag{4.1}$$

where a and m are constants. The values of m are obtained from the above-mentioned figures (at each rotational speed and for each plastic race as shown in Table 4.18) and range from 5 to 17.0. These values are very large and equivalent to 1.5 - 6.5 times that for a similar steel thrust ball bearing.

TABLE 4.18
Values of m

	Rotational Speed (rpm)		
Race Material	1000	1500	2000
Melamine Resin	-	13.6	5.0
Polyacetal (POM)	10.6	17.0	5.8
Nylon 11	7.8	4.8	5.1

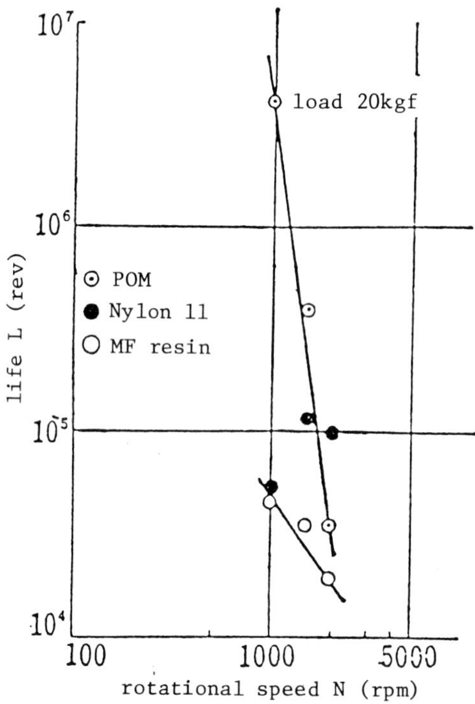

Fig. 4.63 Relationships between life and rotation speed of various ball bearings

(3) *Relationship between L and N.* Figure 4.63 compares the relationships between the life of a plastic ball bearing, L, and the rotational speed, N, under a 20 kgf bearing load for three kinds of plastic races, MF, POM and Nylon 11. The equation representing these relationships is as follows:

$$L = b\ N^{-n} \tag{4.2}$$

where b and n are constants. The value of n is 1.3 for the MF race and 6.7 for the POM race, and these are again much larger than that for a similar steel ball bearing by 3.9 - 20 times.

The main characteristics of the ball bearings with plastic races are freedom from vibration and the ability to operate under unlubricated conditions. The friction coefficients of bearings with suitable plastic races are nearly equal to that of a steel ball bearing. However, the standard load capacity of these plastic ball bearings is very low compared with that of steel bearings. For example, the ratio of the former to the latter is 1/65 for the MF race and 1/45 for the POM race; however, the ratio for a polyimide race is much larger.

4.2 BRAKE BLOCKS

A brake is a component which controls the speed of a running mchine. The most common type is the friction brake, which operates by absorbing kinetic energy by friction. In the past, brake shoe materials were often made from cast iron, soft steel, brass, bronze, wood, asbestos textiles, textile fibres or leather. More recently, however, as a result of increases in the speed and severity of operation of machinery, some sintered metals or plastic composites containing inorganic and metallic materials have been replacing the above materials.

There are many research reports on brake shoe materials; for example, on general aspects by Kanzaki [33], on resin brakes by Wise [34] and Banarad [35], on asbestos-resin brakes by Tanaka [36], on various types of brakes by Fukuoka [37] and Yamaguchi [38], and on filler material by Murata [39]. There are also various research reports on specific applications of resin brake shoes, such as that on railway car brakes by Nishimura [40] and Idemura [41-43], automobile brakes by Hoshino [44,46], Hashikura [45] and Fujii [47] and industrial car brakes by Yoshikawa [48].

In this section, various experimental results [49-53] obtained in the author's laboratory are described and compared with the above-mentioned work.

4.2.1 *Classification and brake performance*

As already mentioned, friction brakes are generally used to reduce or stop the running speed of a machine part by converting the kinetic energy into frictional work. Friction brakes can be classified, depending on their shape and structure, as follows.

(1) Block brake
(2) Band brake
(3) Disc brake
(4) Cone brake
(5) Specific brakes known as "worm", "screw", "coil", "cam" or "centrifugal" types

In this section, the properties and performance of materials are described for a brake shoe contacting a steel drum to provide information relevant to block, disc, and cone brakes. The relationship between the braking torque and the kinetic energy to be absorbed by the frictional work is as follows:

$$\mu_m \cdot P \cdot v_m \cdot t = \frac{1}{2} \cdot \frac{W}{g} \cdot v_o^2 \tag{4.3}$$

where W: weight of rotating countersurface material (kgf); v_o: initial velocity at the radius of gyration of the counter surface; g: acceleration of gravity; v_m: average peripheral velocity of brake drum (m/s); t: braking time (s); P: contact load (kgf); and μ_m: average friction coefficient. If K_r is the radius of gyration and R is the radius of the brake drum, $K_r^2 = 1/2 \cdot R^2$, and v is the initial peripheral velocity of brake drum in m/s, then $v_o^2 = 1/2 \cdot v^2$. When $v_m = 1/2 \cdot v$, the braking time, t, is obtained from equation (4.3) as follows:

$$t = \frac{1}{2} \cdot \frac{W}{g} \cdot \frac{v}{\mu_m} \cdot P^{-1} \tag{4.4}$$

By letting

$$a = \frac{1}{2} \cdot \frac{W}{g} \cdot \frac{v}{\mu_m}$$

then

$$t = a \cdot P^{-1} \tag{4.5}$$

Thus the braking time t to stop motion is inversely proportional to contact load P.

4.2.2 *Required characteristics*

The following six types of brake characteristics are required for brake shoes.

(1) suitable friction coefficient
(2) stable friction coefficient under various conditions
(3) low wear rate
(4) minimum damage or wear to the opposing material
(5) long endurance under heavy loads and high speeds
(6) no brake noise

Cast iron and other materials, as mentioned above, were formerly used as brake shoe materials, and largely satisfy these requirements. Recently, however, new technology for the application of plastic composites to brake shoes, following the development of synthetic high polymers, has led to resin brakes in railway cars and other industrial machines. The main feature of resin or plastic composite brake shoes is that their friction coefficients can be controlled so as to decrease the wear rates of both themselves and the opposing material by suitable choice of the plastic matrix and the type and proportion of filler. Table 4.19 [40] gives an example of the compositions of resin brake shoe materials for a railway car, and indicates that by suitable choice of plastics and fillers it is possible to produce brake shoes with either a high friction coefficient (H) or a low one (L), similar to that obtained with cast iron.

TABLE 4.19
Composition of resin brake for railway car [40]

Material	H (high μ)	L (low μ)	Note
Resin	13-15%	15-30%	Various kinds of resin
Carbon	0-20%	10-40%	" " "
Metal powder	20-40%	15-30%	" " "
Inorganic material	10-20%	10-20%	" " "
Stabilizer	5-15%	5-15%	" " "

Table 4.20 [40] shows the characteristic values of four types of brake shoes: high (H) and low (L) friction coefficient (μ) railway car brakes, a cast iron brake and a special brake shoe. Figure 4.64 [43] shows the relationships between the wear rate of the brake shoes and the initial speed for each type. This figure indicates that the wear rate of a resin brake is much less than that of a cast iron one. Table 4.21 [48] shows the application and friction coefficients of the brakes and clutch as used in industrial vehicles and illustrates the recent increase in the application of plastic composites for brake materials.

TABLE 4.20
General characteristics of various sliding materials [40]

	H High μ resin brake shoe	L Low μ resin brake shoe	Cast iron brake shoe	Special brake shoe
Coefficient of friction	0.25-0.37	0.10-0.25	0.10-0.25	0.07-0.15
Ratio of wear volume to that of cast iron	0.2 -0.3	0.2 -0.3	1	0.2
Thermal conductivity (cal/cm^2·°C·s)	3-7x10^{-3}	4-8x10^{-3}	1.1-1.5x10^{-1}	2-4x10^{-3}
Pressure (kgf/cm^2)	700-1500	700-2000	-	1000-2500
Hardness (Hs) (Shore)	35-60	35-60	40-60	40-70

TABLE 4.21
Types and kinds of brakes and clutches for industrial vehicles

Kind	Type	Application	Sliding Material	Friction Coefficient
DRY:				
Brake	Disc	Running, power control	semimetallic resin mouldings	0.3-0.45
Brake	Drum	Running, turning, winding	(same as above)	"
Brake	Band	Running, turning, winding, power control	(same as above)	"
Clutch	Disc	Main, driving	metallic, semimetallic	0.3-0.5
Clutch	Drum	Main, driving, turning	resin mouldings	0.3-0.5
Clutch	Band	Running, turning, winding	resin mouldings	0.3-0.5
WET:				
Brake	Band	Running, power control	woven lying, resin mouldings	0.08-0.12
Clutch	Disc	Main use, driving	metallic, paper type	0.08 0.12-0.14

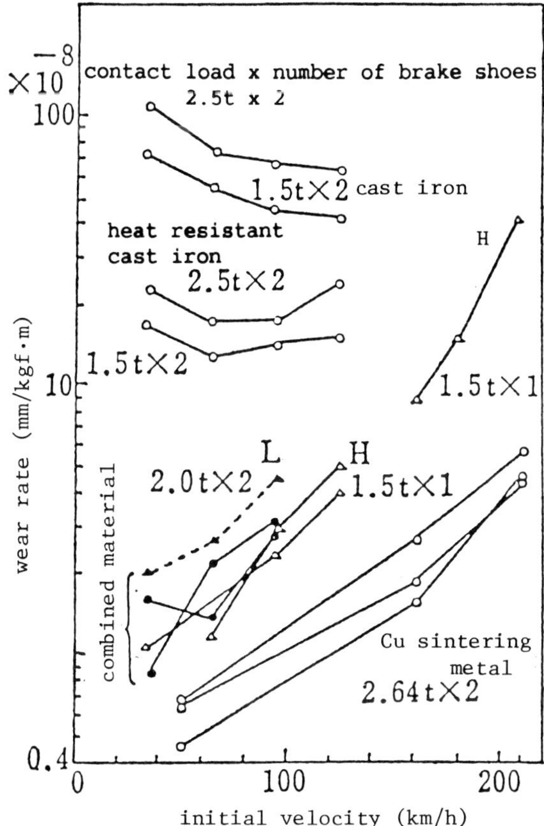

Fig. 4.64 Relationships between wear rate per unit absorbed work, W, and initial velocity for a railway brake

The following section describes various characteristics of plastic composite brake shoes obtained experimentally in the author's laboratory.

4.2.3 *Experimental results [49]*

(i) *PF or DAP composite brake shoes*

(1) *Performance testing.* The brake tests were carried out using the apparatus shown in Fig. 4.65 and Table 4.22. The shape and dimensions of the brake shoes are shown in Fig. 4.66, and the compositions of fourteen types of plastic composite materials, D - L, are given in Table 4.23. The Rockwell hardnesses are given in Fig. 4.68, and Fig. 4.69 shows the compressive strengths. The brake testing conditions are listed in Table 4.24.

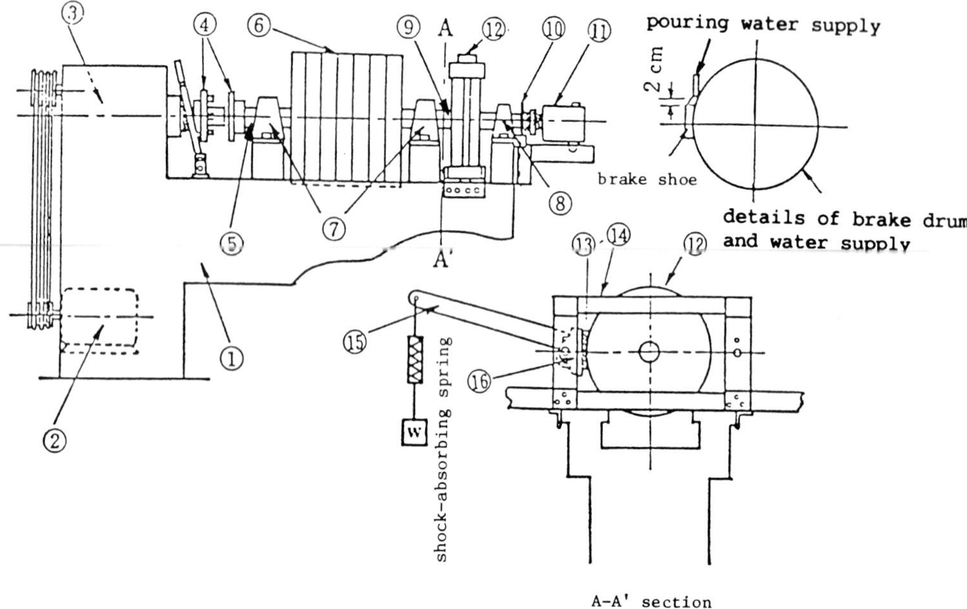

Fig. 4.65 Brake testing apparatus

TABLE 4.22
Details of brake testing machine

No.	Name
1	Bed
2	Motor
3	Speed Reduction Unit
4	Clutch
5	Main shaft
6	Inertial mass
7	Bearing
8	Bearing
9	Gauge for measuring torque
10	Tachometer
11	Slip ring
12	Brake disc
13	Resin brake shoe
14	Frame
15	Weight level
16	Brake shoe base

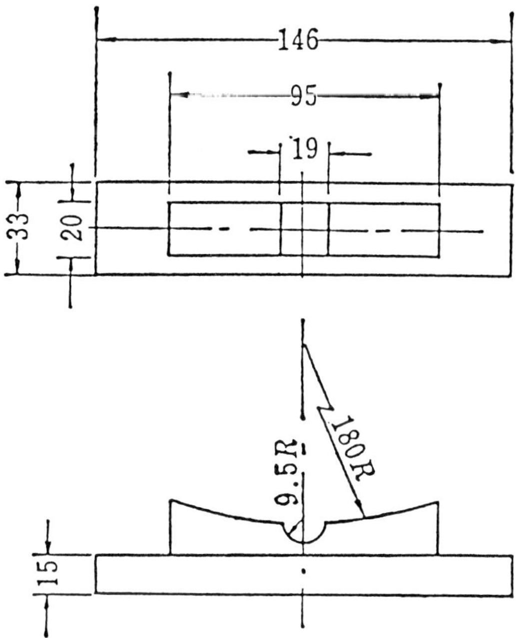

Fig. 4.66 Shape and dimensions of brake shoes

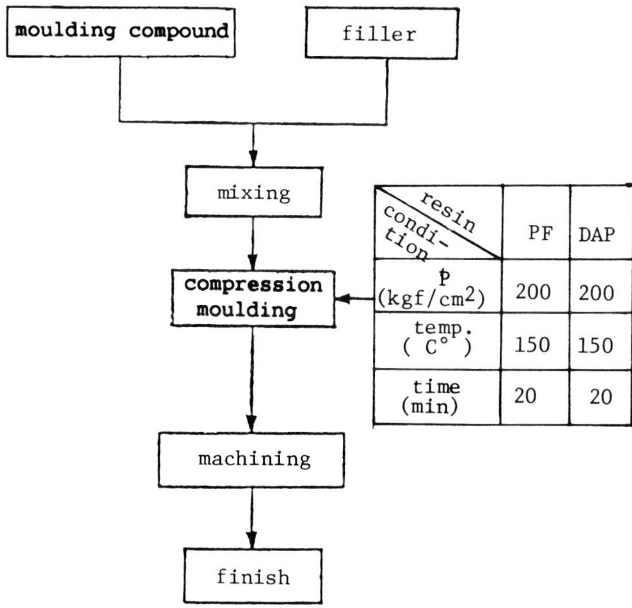

Fig. 4.67 Brake shoe manufacturing process

TABLE 4.23
Types of brake shoes

Code	Iron	Asbestos	Graphite	Filler (wt%)					
				Pb	Cu	Al	Porcelain	Gum	Silica
DAP resin:									
D1	40	5	5	5	5			5	
D2	40	5	5						
D3	40	5	5			5			
D4	40	5	5				5		10
PF, Phenolics:									
A1	40	5	5	5	5			5	
A2	40	5	5						
A3	40	5	5			5			
A4	40	5	5				5		10
Rubber-modified Phenolics:									
B1	40	5	5	5	5			5	
B2	40	5	5						
B3	40	5	5			5			
B4	40	5	5				5		10
H[1)]	10-45	2-5	3-5	2-10	3-5			2-5	
L[2)]	5-25	2-5	25-60	2-15	3-5	2-5			

[1)] High μ brake shoe for railway car
[2)] Low μ brake shoe for railway car

TABLE 4.24
Test conditions

	Brake Testing	
	Continuous wear characteristics	Brake characteristics
Velocity (km/h)	26, 42, 69	26, 42, 69, 110
Contact load (kgf)	20, 30, 40, 50, 60	30, 60, 70, 120
Sliding distance (km)	4, 8, 12, 16	-
Moment of inertia (kgf·m·s^2)	-	0.75, 1.0

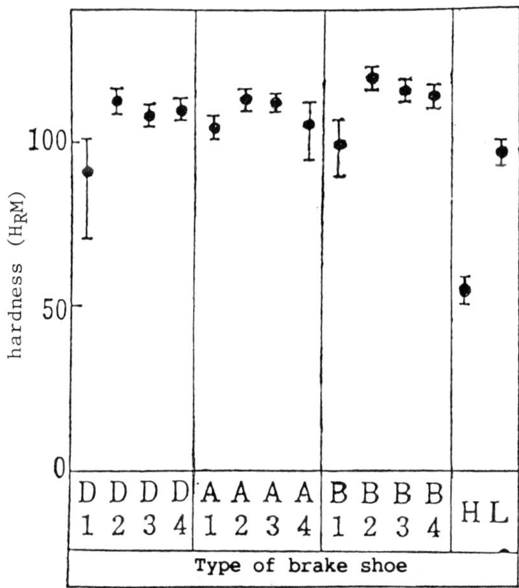

Fig. 4.68 Hardness of brake shoes

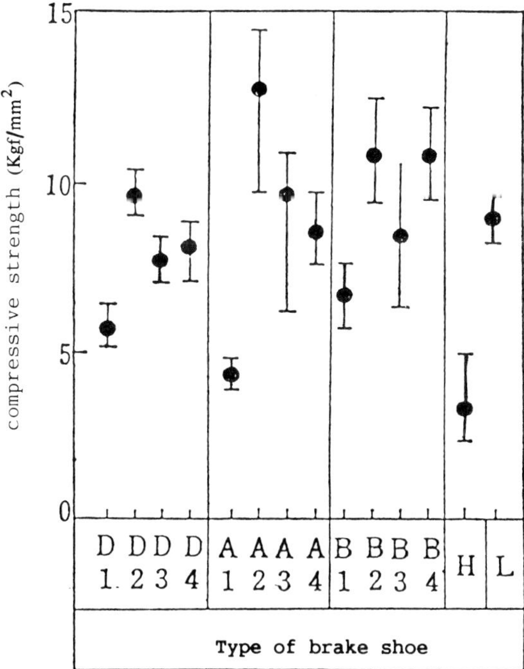

Fig. 4.69 Compressive strengths of brake shoes

In this brake-testing apparatus, the braking torque T (kgf-m) is measured using a strain gauge attached to the shaft with a rotating mass, and the relationship between T and the friction coefficient μ is:

$$T = \frac{D}{2} \cdot \mu P \tag{4.6}$$

therefore,

$$\mu = \frac{2T}{DP} \tag{4.7}$$

where D is the outside diameter of the brake drum (m) and P is the contact load (kgf).

The relationship between μ and time t during a continuous friction test is shown in Fig. 4.70(a). The relationship between T and braking time t, or that between μ (from equation (4.7)) and t, is shown in Fig. 4.70(b). Figure 4.71 shows the relationships between braking time, t, and contact load, P, at different peripheral speeds for the plastic composite shoes, A1, B1 or D1. Figure 4.72 shows similar relationships for brake shoes, B1, D1, H or L, at an initial sliding speed, v_o, of 69 km/h. These relationships, as described above, are consistent with the equation:

$$t = a \cdot P^{-n} \tag{4.8}$$

Values of n, obtained from the experiments, are shown in Table 4.25 and range from 0.87 to 1.01; that is, they are almost equal to 1, as expected from equations (4.4) and (4.5).

TABLE 4.25
Value of n

Brake Shoe	Initial Velocity v_o (km/h) 26	42	69	110
D1	0.96	0.94	0.94	0.92
D2	0.92	0.92	0.93	0.96
D3	0.89	0.93	0.88	0.99
D4	0.86	0.92	0.90	0.96
A1	1.01	0.96	0.96	0.90
B1	0.89	0.88	0.97	0.92
H	0.93	0.97	0.92	0.87

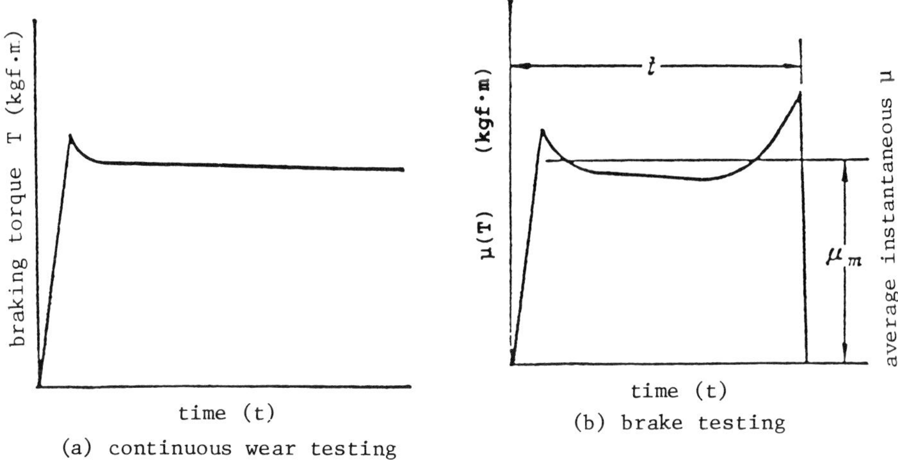

Fig. 4.70 Relationships between braking torque T and friction coefficient μ_T and braking time

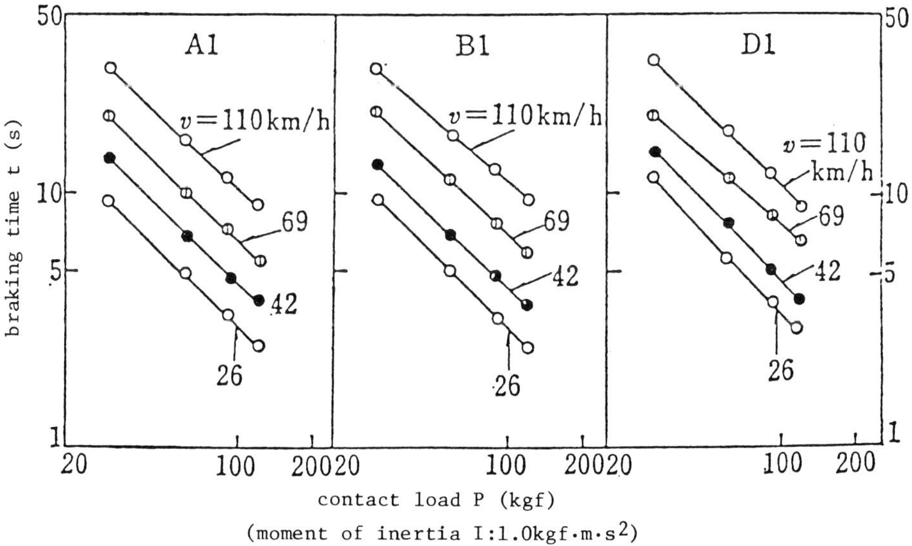

Fig. 4.71 Relationships between braking time and contact load for A1, B1 and D1 brake shoes

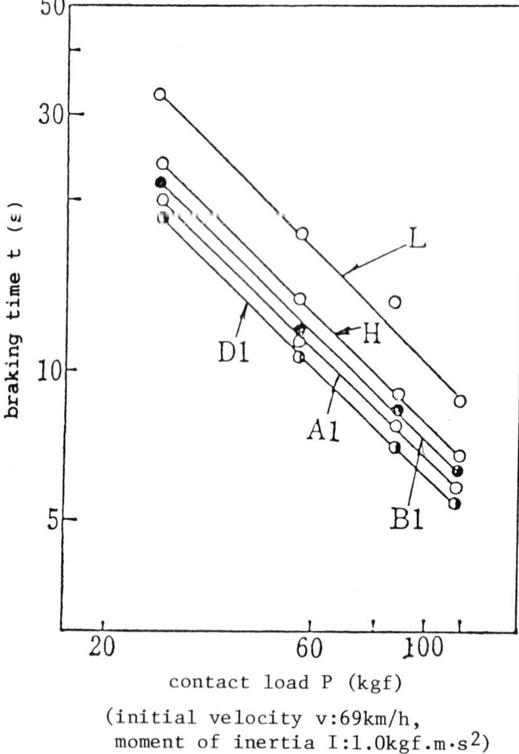

(initial velocity v:69km/h,
moment of inertia I:1.0kgf.m·s^2)

Fig. 4.72 Relationships between braking time and contact load for A1, B1, D1, H, and L brake shoes

TABLE 4.26
Calculated and experimental values of μ

	Initial Velocity V_o					
	42 km/h		69 km/h		100 km/h	
Type	Theo.v[1]	Exp.v[2]	Theo.v[1]	Exp.v[2]	Theo.v[1]	Exp.v[2]
D1	0.411	0.323	0.444	0.354	0.441	0.338
D2	0.411	0.320	0.450	0.335	0.434	0.317
D3	0.352	0.292	0.432	0.319	0.423	0.328
D4	0.352	0.307	0.438	0.346	0.427	0.363

Contact load P = 90 kgf
1) μ from eqn. (4.4)
2) μ from eqn. (4.7) and experiment

Table 4.26 shows the experimental values of friction coefficients and the average instantaneous friction coefficient μ_m obtained from Fig. 4.70(b) and equation (4.7), together with the theoretical values obtained from equation (4.3) or (4.4) for the brake shoes, D1, D2, D3 or D4. The results indicate that the experimental values of μ are all slightly less than the theoretical ones.

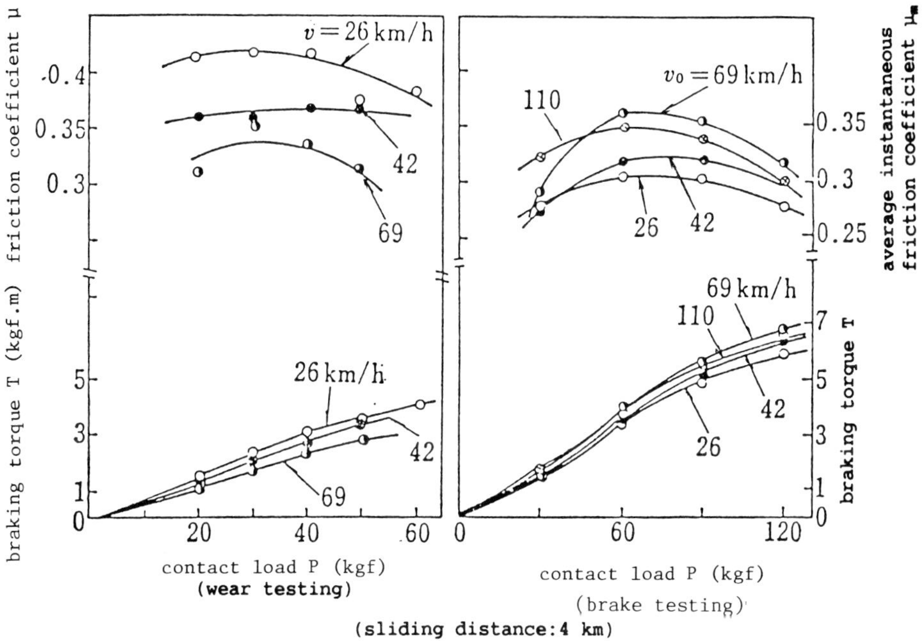

Fig. 4.73 Relationships between braking torque, friction coefficient and contact load for D1 brake shoes

(2) *Friction test.* For both the continuous sliding wear test and the brake test shown in Fig. 4.70 and Table 4.24, respectively, the relationships between friction coefficient μ, average instantaneous friction coefficient μ_m (or braking torque T) and contact load P for each test condition for each brake shoe are shown in Figs. 4.73 to 4.75. Figure 4.73(a) shows the relationships between μ (or T) and P for different speeds, v, or initial speeds, v_o, using brake shoe D1. Figure 4.73(b) shows the relationships between μ_m (or T) and P for similar conditions. Figure 4.74(a) shows the relationships between μ (or T) and P at a peripheral speed, v, of 26 km/h,

for a sliding distance of 4 km with various brake shoes, A1, B1, D1, H and L. Figure 4.74(b) shows the relationship between μ_m (or T) and P at an initial speed, v_o, of 26 km/h using the same brake shoes. Figure 4.75 shows the relationships between μ_m (or T) and P at v_o = 69 km/h and a moment of inertia of 1.0 kgf·m·s^2 for brake shoes, D1, D2, D3 and D4 in the brake test.

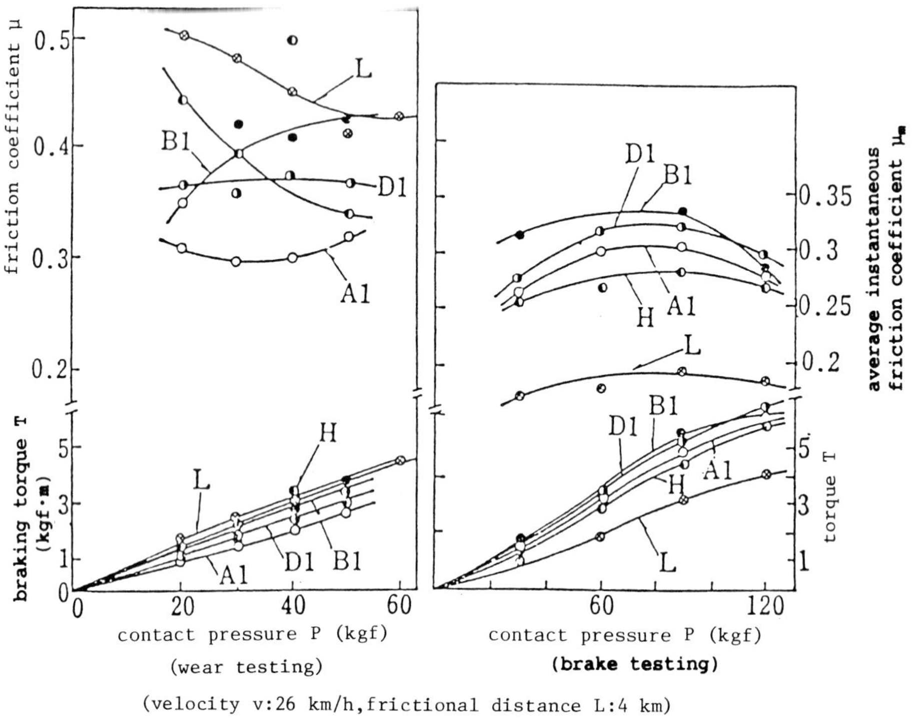

Fig. 4.74 Relationships between braking torque, friction coefficient and contact pressure for various shoes

It is clear from these figures that the braking torque T is almost directly proportional to the contact load P, and that μ is only slightly affected by the type of brake material and the magnitude of v or P. The values of μ range from 0.28 to 0.5 and those of μ_m from 0.18 to 0.34; μ is generally larger than μ_m. The value of μ_m for brake shoe H is 1.4 to 1.5 times that for brake shoe L. It is indicated in Fig. 4.75 that the addition of gum increases the value of μ_m whilst Al additions decrease it.

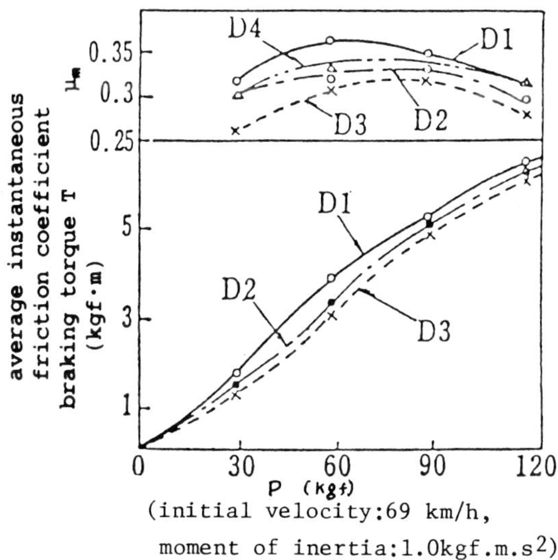

Fig. 4.75 Relationships between braking torque, average instantaneous friction coefficient and contact load for D1, D2, D3 and D4 brake shoes

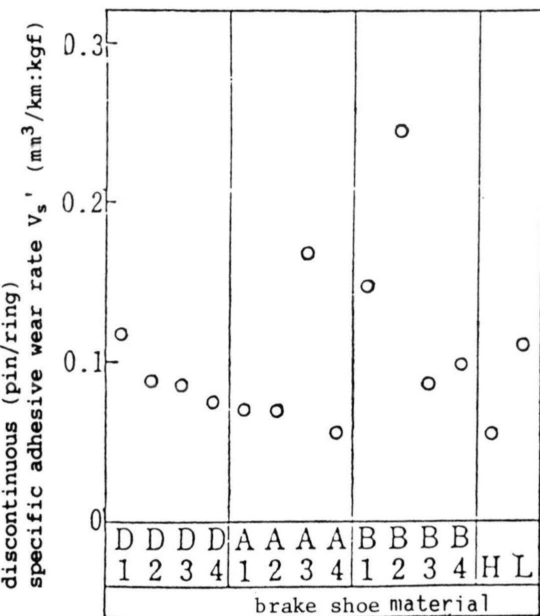

Fig. 4.76 Specific adhesive wear rate, V_s' of various brake shoe materials, measured using an Ogoshi-type, wear testing machine

(3) *Sliding wear characteristics.* Figure 4.76 shows the specific adhesive wear rate V'_s in non-conforming contact for each of the various brake shoes, measured using an Ogoshi-type wear tester. This figure indicates that the V'_s values for the brake shoes A4, B4 and D4 filled with silica sand are generally small, and those for D1, D2, D3 and D4, with a DAP matrix are about 0.092 mm³/kgf·km and are stable. The filler has less effect on the value of V'_s for DAP composites than for those based on phenolic or gum-modified phenolic matrices.

The experimental results obtained from continuous sliding wear tests using the brake-testing apparatus are shown in Figs. 4.77, 4.78 and 4.79. Figure 4.77 shows that the relationships between the adhesive wear volume V and sliding distance ℓ at various contact loads P for brake shoe D1 are linear.

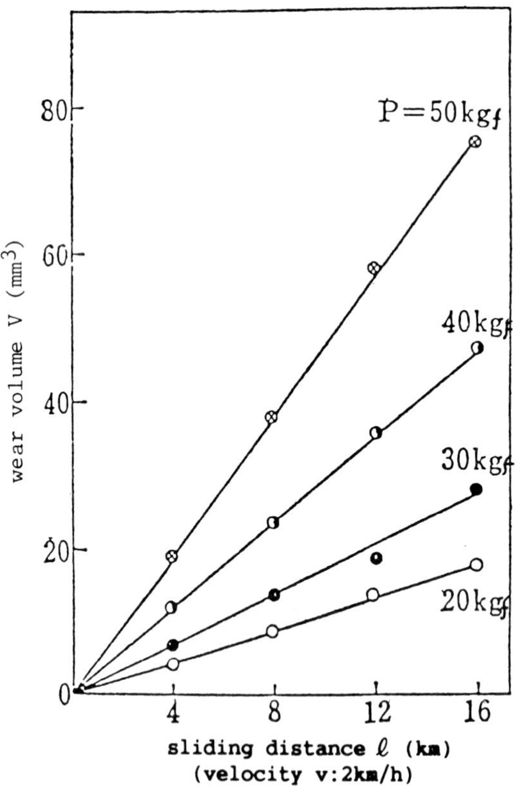

Fig. 4.77 Relationships between wear volume and sliding distance in a continuous friction test for D1 brake shoe material

Figure 4.78 shows the relationships between the adhesive wear volume V and contact load P for each brake shoe, and it can be seen that V is almost directly proportional to P. It is therefore, justifiable to express the sliding wear properties of brake shoes in terms of the specific adhesive wear rates V_s or V'_s.

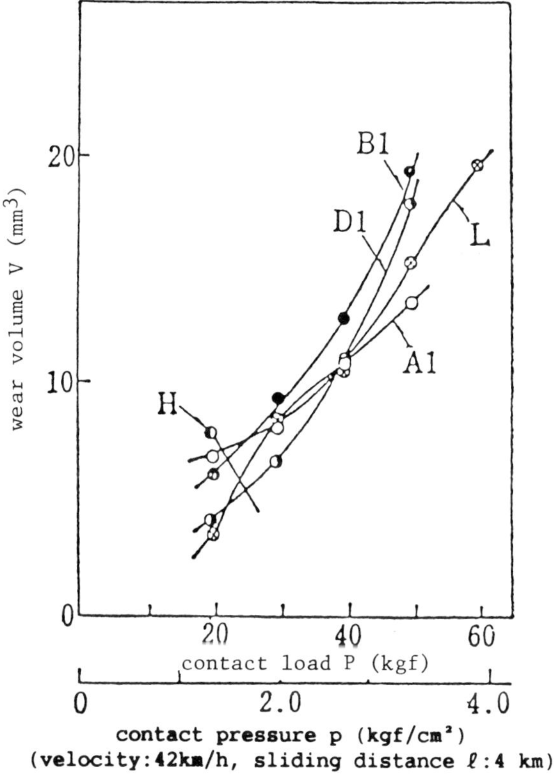

Fig. 4.78 Relationships between wear volume and contact load for A1, B1, Di, H and L brake shoes

Figure 4.79 shows the specific adhesive wear rates V''_s for each brake shoe at a speed of 69 km/h, obtained using the brake-testing apparatus. In this figure, "x" indicates a negative value, resulting from the transfer of worn iron debris to the brake shoe surface. In this experiment, it appears that a local instantaneous high temperature, due to the friction between iron and iron, causes the generation and growth of small iron fragments via the melting of the transferred iron fragments on the brake shoe. As a result of this process, damage and wear of the opposing brake drum are

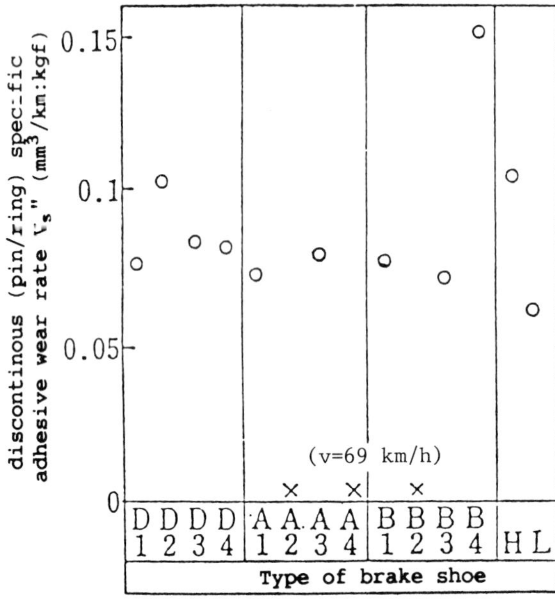

Fig. 4.79 Discontinuous (plate/ring) specific adhesive wear rate, V_s'', for various brake shoes using a brake-testing machine

TABLE 4.27
Specific adhesive wear rate V_s of various brake shoes

Brake Shoe Code	Specific Adhesive Wear Rate (mm^3/km·kgf) V_s' (Ogoshi type)	V_s'' (Brake tester)
D1	0.118	0.074
D2	0.090	0.101
D3	0.085	0.082
D4	0.076	0.082
A1	0.071	0.072
A3	0.168	0.076
B1	0.147	0.077
B3	0.085	0.071
B4	0.097	0.152
H	0.044	0.103
L	0.110	0.063
Sliding Velocity (km/h)	0.61	42

accelerated. Brake shoes made from DAP matrices and A1, B1 and D1 (made of the composites filled with soft metals, such as Pb or Cu) generate comparatively little iron debris.

Table 4.27 compares the specific adhesive wear rates, V_s' measured using an Ogoshi-type wear tester with those, V_s'', obtained using the brake-testing apparatus for each brake shoe. The table demonstrates that the V_s' ranges from 0.054 to 0.168 mm³/kgf·km, whilst V_s'' ranges from 0.07 to 0.152 mm³/kgf·km.

The relationship between V_s' and V_s'' is shown by Fig. 4.80, which demonstrates some correlation, although with appreciable scatter.

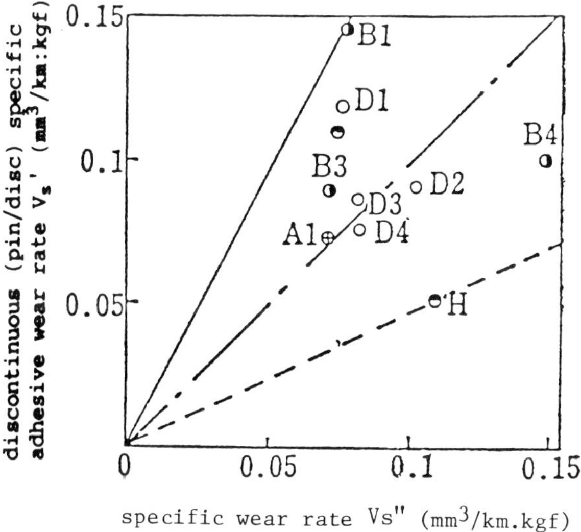

Fig. 4.80 Relationships between V_s' using an Ogoshi-type wear tester, and V_s'' using a brake testing machine

(ii) *Minimum effect of water absorption [50]*

The stability of a brake shoe's friction coefficient, μ, is an important and necessary characteristic in brake performance. However, the value of μ can change greatly in outdoor applications, particularly during rain, or as a result of the existence of water between the brake shoe and the drum. The resulting phenomenon is known as "waterfade" [43]. It is therefore essential that the difference in value of μ between wet and dry conditions should be minimized. In this section, experimental trials and results [54] to achieve this objective are described for various brake shoes made from plastic composites.

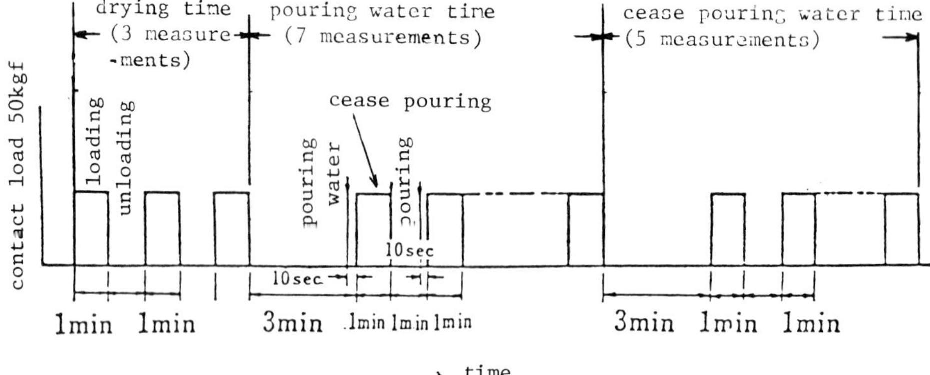

Fig. 4.81 Brake test sequence for dry / pouring water / cease pouring water conditions

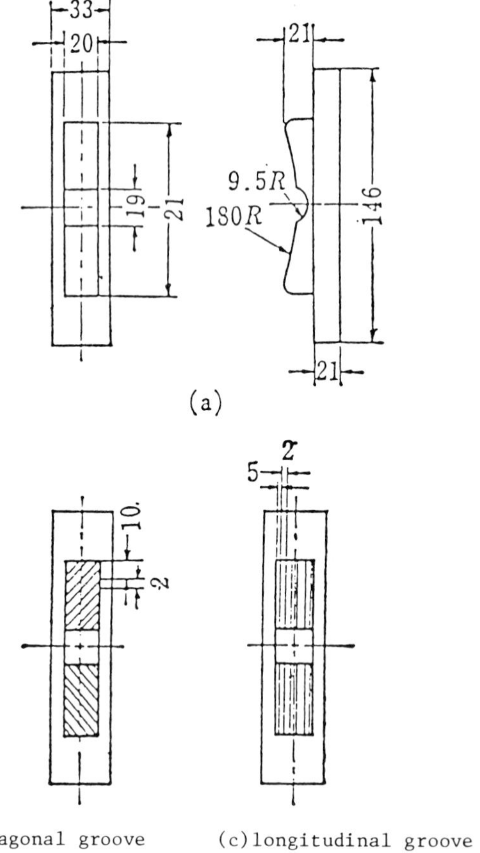

Fig.4.82 Surfaces and dimensions of brake shoes

TABLE 4.28
Materials of various brake shoes

Matrix	Code	Fillers (wt.%) Fe	Asbes- tos	Graph- ite	MoS_2	Al	Gum	Sil- ica	Glass Beads	ZnO	Mica	PPS	Pb	Cu	Carbon Fluor- ide	Clay	Carbon Fibres
DAP resin	D1	40	5	5			5						5	5			
	D2	40	5	5													
	D10	40	5	5		5											
	D12	40	5	5				10								5	
Phenolics only	A1	40	5	5			5						5	5			
	A107	30	10	10	10	5	5										
Phenolics modified with gum	B1	40	5	5			5						5	5			
	B10	40	5	5		5											
	B101	30	10	10		5	5	10			10						
	B102	30	10	10		5	5	10									
	B103	40	5						10								5
	B104	40	5														20
	B105	30	10	10		5	5				10						
	B106	30	10	20	5	5	5		10								
	B107	30	10	10	10	5	5										
	B108	30	10	10	5	5	5										
	B109	30	10	10	5	5	5			15							
	B110	40	5	10	5												
	B111	40	5	10								20					
	B112	40	5	10							15				4		
	B113	40	5							10							
Mix of MF and PF	MA107	30	10	10	10	5											

(1) *Experimental method.* The testing apparatus has already been described and shown in Fig. 4.65. For the present experiments water was poured between the brake drum (12) and the brake shoe (13) to maintain wet conditions. The friction coefficient of the brake μ is obtained from equation (4.7), i.e. $\mu = (2T/DP)$, where T is the braking torque (kgf-m), measured by a strain gauge attached to the shaft, D is the outside diameter of the brake drum (m), and P is the contact load (kgf).

Two types of dry-wet tests were carried out at a constant speed of 42 km/h: "completely dry/completely wet" and "dry/intermittent supply/supply stopped", as shown in Fig. 4.81. The friction coefficient μ was measured under each condition.

TABLE 4.29
Materials of brake shoes

Code	Asbestos	Filler (wt.%) $BaSO_4$	Other Fillers
P1	40	30	Red Ochre: 10
P2	40	30	Glass Powder: 10
P3	40	30	Fe: 10
P4	40	30	Fe: 20
P5	40	30	Alumina: 10
P6	40	30	Olivine Sand: 10
P7	40	30	Al Powder: 10
P8	40	30	Graphite: 10
P9	40	30	Porcelain Clay: 10
P10	40	30	$CaCo_3$: 10

Matrix: pure phenolics

TABLE 4.30
Filler details

Filler	Manufacturer
Asbestos	Nippon Asbestos Co., Crysotile W-30
Graphite	Hidachi Funmatsu Co., GP-304
MoS_2	Hidachi Funmatsu Co., MD-208 below 10μm
Polyphenylene sulfide (PPS)	Phillips Co., Ryton P-4
Carbon Fibre	Nippon Carbon Co., CLF
Graphite Fluoride	Nippon Carbon Co.
Porcelain Clay	Kugita Yakuhin Kogyo Co. Kolin
Fe	Yoneyama Yakuhin Kogyo Co. 50-200 μm size
Al	"
Pb	"
Cu	"
Mica	"
ZnO	"
Silica	Yoneyama Yakushin Kogyo Co. Silica anhydride
Gum	Nippon Goseigum Co., NBR, JSRN 203S
Glass Beads	Shinetsu Kagaku Co., MS-MH, 100 mesh

The shape and dimensions of the specimen used are shown in Fig. 4.82(a). Figure 4.82(b) shows the diagonal (b) and longitudinal (c) grooves on the shoe surface which are effective in preventing the generation of a water film. Table 4.28 shows details of the materials, i.e. plastic composites, based on four plastics; DAP, phenolics, modified phenolics and a mix of melamine and phenol resins, and sixteen types of fillers. In order to clarify the effects of the filler on the stability of the μ value, other "dry/ intermittent supply/supply-stopped" brake tests for various brake shoes (details in Table 4.29) were carried out. In this case the ten types of brake shoes, P1 to P10, were plastic composites composed of a phenolic matrix, two base fillers, asbestos and $BaSO_4$, and other fillers, as shown in Table 4.29. The details of the fillers are shown in Table 4.30.

(2) *Experimental results*

(a) *"Completely dry/completely wet" tests.* Graphs of the relationships between the frictional coefficient μ or braking torque T and contact load P during the "completely dry/completely wet" brake tests are given in Figs. 4.83 and 4.84. Figure 4.83 shows the relationships for each brake shoe, A1, B1 or D1. It is clear that the braking torque T is directly proportional to the contact load P, and that μ is not much affected by P, but does scatter somewhat. However, the values of μ or T, at the same contact load, are very different under dry and wet conditions. Figure 4.84 shows similar relationships for the B10 brake shoe both with longitudinal grooves on its surface and without grooves. It is clear that the general shapes of the μ-P and T-P relationships are not affected by the existence of grooves; however, the μ values for a shoe with grooves are smaller in dry conditions and greater in wet conditions than those for a shoe with no grooves.

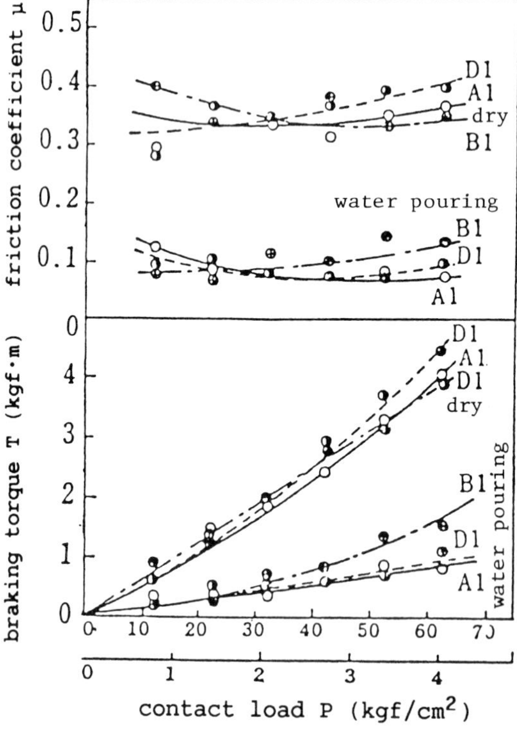

Fig. 4.83 Relationships between μ, braking torque, T, and contact load for brake shoes with various matrices

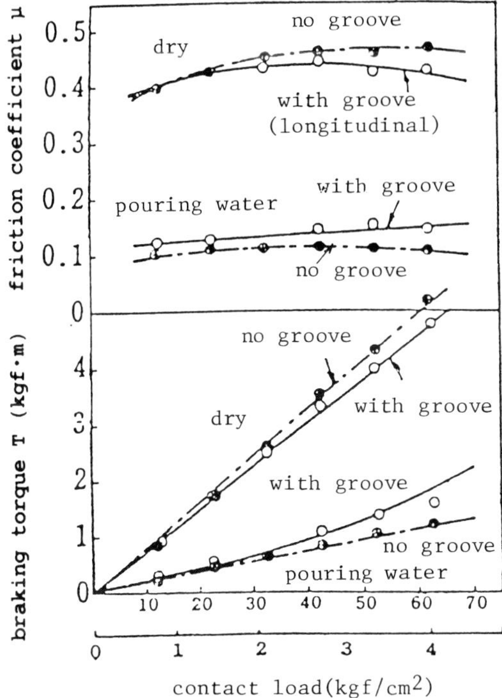

Fig. 4.84 Relationships between μ, braking torque, T, and contact load for brake shoes with or without grooves

Table 4.31 gives the average friction coefficient, μ_d (in dry conditions) and μ_w (in wet conditions), and the wet-dry ratio of friction coefficients $\beta(=\mu_w/\mu_d)$ for each brake shoe. The effects of matrix, filler and grooves on the value of β are as follows.

(I) The value of β for a brake shoe consisting of a DAP resin matrix and six types of fillers (Fe, asbestos, graphite, gum, Pb and Cu) is 0.24; that of a phenolic matrix with the same fillers is 0.26 and that for a modified phenolic matrix and the same fillers is 0.3. The largest value of β (i.e. the smallest change in μ between dry and wet conditions), is thus obtained with the modified phenolic matrix.

TABLE 4.31
Average μ value for dry or wet conditions and their ratio β of various brake shoes

Brake Shoe Code	Filler	Dry Condition Avg.μ_d	Wet Condition Avg.μ_w	$\beta = \frac{\mu_w}{\mu_d}$
D1	Gum, Pb, Cu	0.357	0.086	0.24
D2		0.376	0.148	0.39
D10	Al	0.350	0.123	0.35
D12	Silica, white clay	0.376	0.101	0.27
A1	Gum, Pb, Cu	0.333	0.088	0.26
B1	Gum, Pb, Cu	0.355	0.108	0.30
B10	Al, Pb, Cu	0.442	0.110	0.25
B10(longitude)	Al, Pb, Cu	0.423	0.134	0.32
B101	Al, gum, silica, mica	0.385	0.110	0.26
B101(diagonal)	Al, gum, silica, mica	0.385	0.118	0.31
B102	Al, gum, silica, glass beads	0.437	0.179	0.41
B103	Carbon fibres	0.432	0.05	0.13
B104	Carbon fibres	0.447	0.212	0.47
B105	Al, gum, glass beads, mica	0.362	0.090	0.25
B106	5% MoS_2, Al, gum	0.292	0.134	0.46
B107	10% MoS_2, Al, gum	0.342	0.218	0.37
A107	" " " "	0.339	0.118	0.35
MA107	" " " "	0.417	0.106	0.25
B108	5% MoS_2, Al, gum, ZnO_2	0.408	0.189	0.46
B109	5% MoS_2, Al, gum, PPS	0.364	0.164	0.45
B110	MoS_2	0.414	0.154	0.37
B111		0.40	0.125	0.313
B112	CF	0.375	0.109	0.29
B113	ZnO_2	0.412	0.214	0.52

(II) The value β for the D2 brake shoe, consisting of DAP resin and three base fillers (Fe, asbestos and graphite) is 0.39. This is greater than that of the D10 brake shoe, containing the base fillers and Al, and the D12 brake shoe, containing base fillers, silica and white clay, which are 0.35 and 0.27, respectively. The β value for the B10 shoe consisting of modified phenolic composites containing three base fillers and carbon fibres is 0.47; B101 containing the same base fillers and Al, gum, silica and glass beads is 0.41. The B102 shoe, having a hard and rough surface, due to the presence of spherical glass beads and hard silica, appears to break the water lubricant film. The highest values of β are those for B108 and B113, based on modified phenolics, 0.46 and 0.52, respectively, whilst that for B109 containing PPS is also high at 0.46.

(III) The β value for the B10 brake shoe with longitudinal grooves is 0.32, whilst that without grooves is 0.25; the value for the B101 brake shoe with diagonal grooves is 0.31, whilst that without grooves in 0.26. The presence of grooves therefore increases β by about 17 - 20%. It appears that the generation of a lubricating water film is prevented by leakage of the water through the grooves.

TABLE 4.32
μ values under dry/pouring/cease pouring conditions

Shoe Code	Dry μ_{d1}	Water Supplied μ_w	Water Supply Stopped Average μ_{d2}	$\beta' = \frac{2\mu_w}{\mu_{d1}+\mu_{d2}}$
P1	0.226	0.061	0.242	0.262
P2	0.240	0.052	0.236	0.220
P3	0.258	0.074	0.268	0.287
P4	0.329	0.119	0.343	0.356
P5	0.279	0.096	0.301	0.347
P6	0.297	0.092	0.294	0.310
P7	0.238	0.089	0.252	0.361
P8	0.276	0.076	0.278	0.281
P9	0.240	0.058	0.269	0.227
B113	0.380	0.301	0.419	0.744

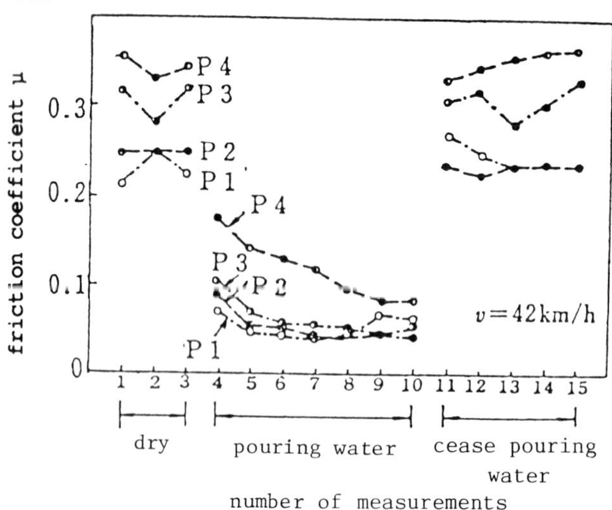

Fig. 4.85 Friction coefficients of various brake shoes during dry / pouring / cease pouring water conditions

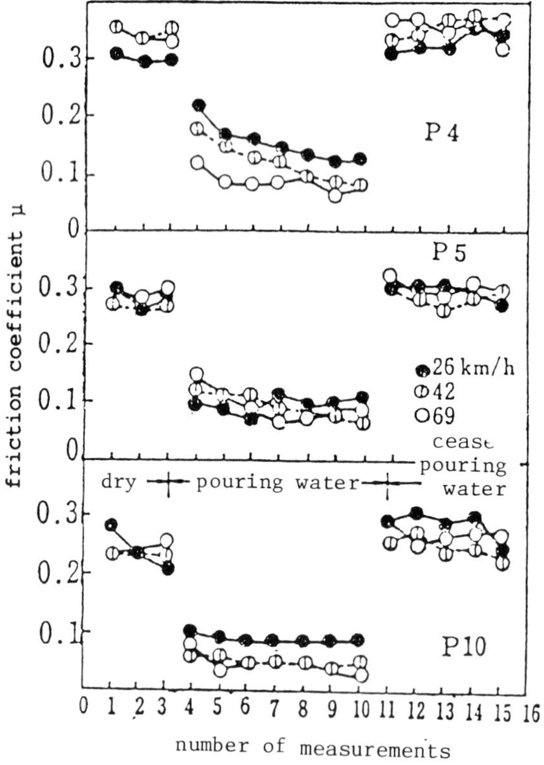

Fig. 4.86 Friction coefficients, μ, at various speeds during dry / pouring / cease pouring water conditions

(b) *"Dry/intermittent supply/supply stopped" tests.* Table 4.32 shows the friction coefficients, μ_{d1}, μ_w and μ_{d2}, under each of the three conditions; dry, intermittent supply of water and supply of water stopped. The average wet-dry ratio of the friction coefficients β' for each brake shoe, P1 to P9, composed of phenolics and two base fillers, asbestos, $BaSO_4$ and other fillers is shown in Table 4.32. Figure 4.85 shows the three frictional coefficients, μ_d, μ_w and μ_{d2} for each of the shoes, P1, P2, P3 or P4, composed of phenolics and two base fillers and red ochre, glass powder or Fe, as shown in Table 4.32. It is clear from this figure that the μ_d value decreases when water is supplied and recovers when the supply of water is stopped. Figure 4.86 shows the three friction coefficients, μ_{d1}, μ_w and μ_{d2} for each of the shoes, P4, P5 and P10, at speeds of 26, 42 or 69 km/h. It is clear that μ_w is largest at the lowest speed; the μ_d value, however, is hardly affected by the speed.

(c) *Performance improvements with composites of diallyl phthalate resin.* Previously phenolic resins have been generally used as the main material for resin brake shoes. In this section, some experimental trials and results are described concerning the usefulness of diallyl phthalate resin (DAP) as a matrix, having a higher heat resistance and friction properties than phenolic resin, and on the application of DAP composites to brake shoes.

(I) Materials. Table 4.33 shows the types of fillers and their combinations examined to improve various brake characteristics, such as μ, limiting pv value, heat resistance, wear resistance, prevention of countersurface wear, mechanical strength and thermal conductivity [57].

TABLE 4.33
Properties required from filler

Properties	Filler
μ and pv	Silica powder
Heat resistance	Barium sulfide, Mica, $CaCO_3$, ZnO, TiO
Wear resistance	Talc, silica powder
Decrease of counterface wear	Al, Sn, Zn, Cu
Mechanical strength	Asbestos, Fe
Heat conductivity	Al, Cu, Fe

The composition of the brake shoes used for the experiments are shown in Table 4.34, i.e. the matrix (DAP) and the various fillers: iron powders of 30 - 200 μm dia., asbestos fibres (0.7 - 3.3 μm dia.), $BaSO_4$ (0.2 - 1 μm), ZnO (0.1 - 0.8 mm), silica sand (8 - 200 μm), $CaCO_3$ (0.7 - 3.3 μm), Alumina

(9 - 50 μm), mica (30 - 150 μm) and Al (7 - 33 μm). They were compression moulded after mixing them with a catalyst (t-butyl benzene) and extracting acetone. The relevant types of mouldings were then machined to form specimens of the required dimensions.

TABLE 4.34
Composition of specimens

Code	Filler (%)				DAP (%)
A	-	-	-		100
B	Fe powder 20	Asbestos 10	-		70
1	Fe powder 20	Asbestos 10	Barium sulfide	20	50
2	"	"	ZnO	20	50
3	"	"	Silica sand	20	50
4	"	"	$CaCO_3$	20	50
5	"	"	Alumina	20	50
6	"	"	Mica	20	50
7	"	"	Talc	20	50
8	"	"	Cu powder	20	50
9	"	"	Al	20	50

(II) Performance test. In order to measure the friction coefficient of the brake shoes, a Suzuki-type friction and wear testing machine (shown in Fig. 2.1(a)) and an FT type (shown in Table 1.5 were used). The Suzuki-type machine measures adhesive wear during continuous sliding in conformity contact whilst the Ogoshi-type tester measures adhesive wear in non-conforming contact (plate-ring). An abrader (JIS-K 7205) was also used to provide a comparison with abrasive wear.

(III) Results and discussion. Figure 4.87 shows the flexural strength of each specimen. For the composites, the range is 3 - 7.2 kgf/mm^2, while that of DAP itself is 0.9 kgf/mm^2. Figure 4.88 shows the relationships between the coefficient of friction, μ, measured by an FT-type tester and the sliding distance, ℓ, for each of the five composites. The values of μ for the different composites range from 0.35 to 0.5 but for any one composite are almost constant and increase only slightly with increasing sliding distance. The average values of μ during a sliding distance of 1000 metres are shown in Table 4.35, and range from 0.42 to 0.51; they are generally larger than those of the phenolic composites shown at the bottom of the table. Of these values, specimen 6, i.e. a DAP composite filled with mica, has the smallest μ value, whilst that filled with $CaCO_3$ has the largest.

Table 4.35 shows the wear characteristics of the DAP composites expressed in terms of four specific rates: the adhesive wear rate during continuous sliding in conforming contact, V_s; the adhesive wear rate in non-conforming contact (plate-ring), V_s'; the adhesive wear rate of the counterface material, $\overline{V}_s$; and the abrasive wear rate, V_{sa}. Figure 4.89 shows the two specific wear rates V_s and $\overline{V}_s$ for each of these composites.

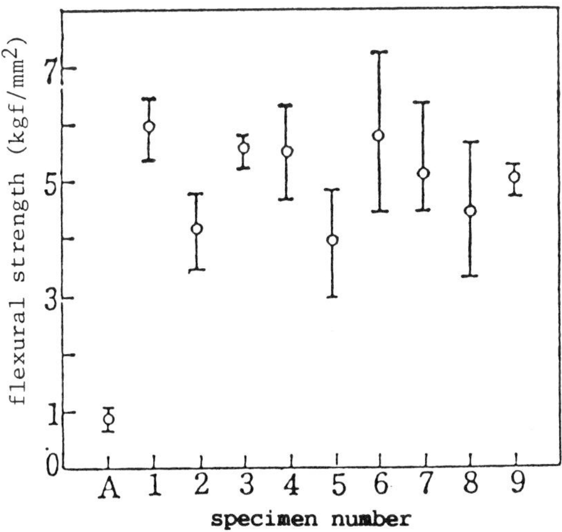

Fig. 4.87 Flexural strength of various specimens

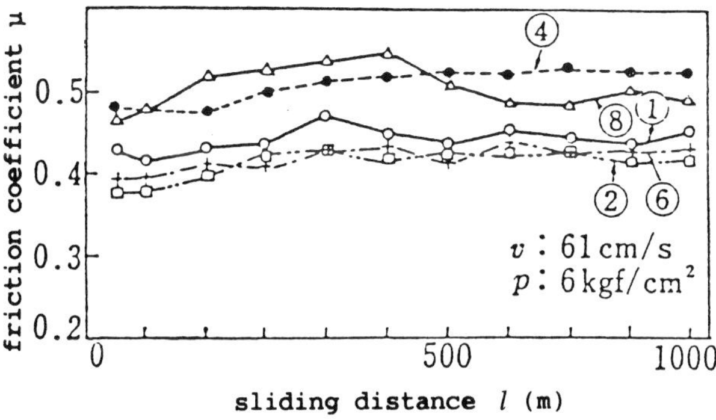

Fig. 4.88 Relationships between friction coefficient, μ, and sliding distance for various brake shoes

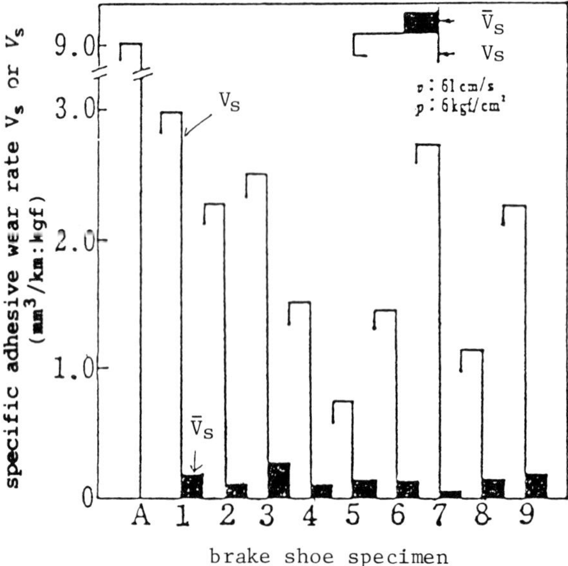

Fig. 4.89 Continuous specific adhesive wear rate, V_s, and that of the opposing material (S45C), $\bar{V}_s$, for each brake shoe

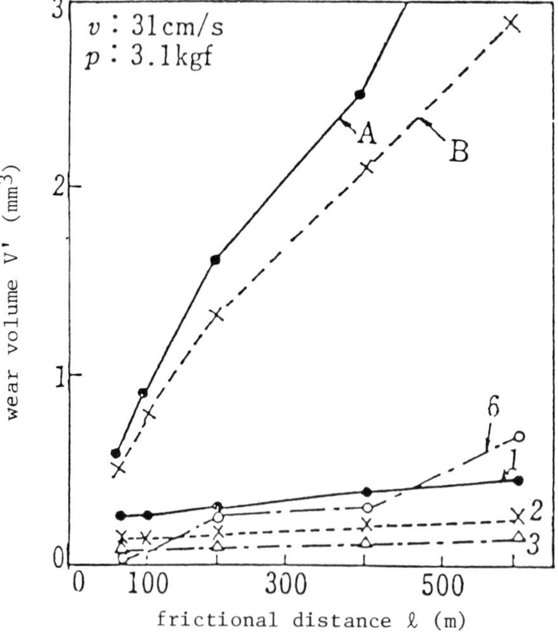

Fig. 4.90 Relationships between discontinuous sliding (pin/ring) wear volume, V' and sliding distance for each specimen

TABLE 4.35
Values of μ and specific adhesive wear rates V_s of various materials

Specimen: Matrix	Specimen: Filler	Code	Average μ v: 61 cm/s p:6kgf/cm^2	Conforming Contact V_s (mm^3/km·kgf)	Non-conforming Contact V_s' (mm^3/km·kgf)	Non-conforming Abrasive V_{sa}(mm^3/km·kgf)	Counterface Material $\bar{V}_s$ (mm^3/km·kgf)
	0	A			5.01		
	Fe, Asbestos	B			4.25		
DAP Resin	Fe, Asbestos, Barium Sulfide	1	0.442	2.97	1.19	396.6	0.16
	Fe, Asbestos, ZnO	2	0.426	2.41	0.59	344.4	0.09
	Fe, Asbestos, Silica	3	0.430	2.56	0.32	509.1	0.21
	Fe, Asbestos, $CaCO_3$	4	0.512	1.52	0.64	413.8	0.07
	Fe, Asbestos, Alumina	5	0.504	0.80	0.27	444.2	0.11
	Fe, Asbestos, Mica	6	0.424	1.45	0.56	814.9	0.11
	Fe, Asbestos, Talc	7	0.479	2.81	0.88	904.8	0.04
	Fe, Asbestos, Cu	8	0.506	1.20	0.64	331.9	0.08
	Fe, Asbestos, Al	9	0.493	2.17	0.98	679.4	0.10
Phenolics	Fe, Asbestos, Gum, Pd, Cu		0.333				
	Fe, Asbestos, Al, Pd, Cu		0.442				
	Fe, Asbestos, Al, Gum, Silica		0.385				
	Fe, Asbestos, MoS_2, Al, Gum		0.408				
	Fe, Asbestos, Graphite, Al		0.27-0.32		0.33		
	Fe, Asbestos, Graphite, Cu, Gum		0.28-0.34		0.3		
	Fe, Asbestos, Graphite		0.29-0.33		0.48		
	Cotton Cloth Laminate			2.4	0.24	344	
	60% Asbestos		0.2 -0.45		0.96		
				v=61cm/s	v=31cm/s	v=25.4cm/s	

Figure 4.90 shows the relationship between the wear volume in non-conforming contact, V', obtained used the Ogoshi-type wear tester, and the sliding distance, ℓ, for each composite, and Figure 4.91 shows the corresponding specific adhesive wear rates V_s'. The value of V_s for DAP itself is 9.0 mm^3/kgf·km, whilst those for the composites range from 0.8 to 2.97 mm^3/kgf·km, i.e. 1/3 to 1/11 that of DAP. It is clear from Fig. 4.91 and Table 4.35 that silica and alumina fillers lead to the lowest composite wear rates whilst talc and $CaCo_3$ fillers lead to the lowest counterface wear rates. It may also be noted that the range of wear rates for the DAP composites is not greatly different from that for the phenolic composites, shown in the lower part of Table 4.35.

(d) *Performance of melamine, polyimide, silicone and polyphenylene oxide composites [53].* The thermosetting plastics, melamine resin (MF), polyimide (PI) and silicone resin (SI), and the thermoplastic, polyphenylene oxide (PPO), are all highly heat resistant. This section describes the experimental brake characteristics for various composites of each of the above materials.

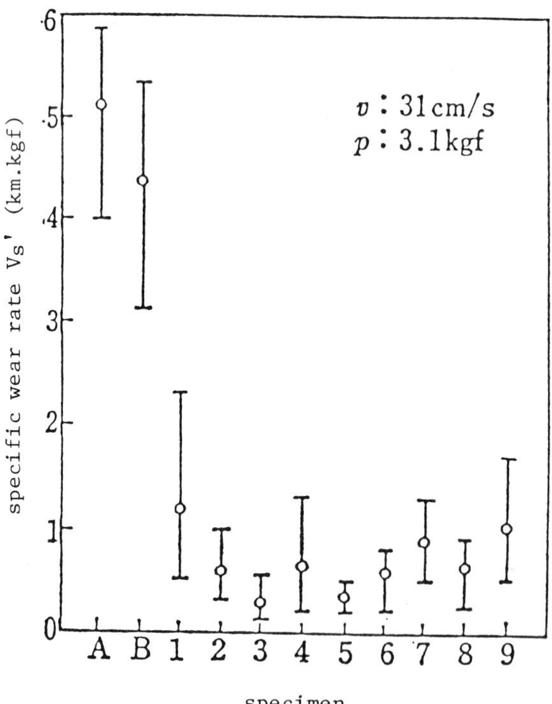

Fig. 4.91 Discontinuous (plate/ring) adhesive wear rate, V_s', for each specimen

Table 4.36 shows the details of the four matrix plastics and the eight types of fillers; iron powder, asbestos fibre, silica sand, ZnO, $CaCO_3$, Cu powder, mica and talc. Table 4.37 shows details of the combinations of the various plastics and fillers.

TABLE 4.36
Matrix and filler

Name		Details
Matrix Plastics:		
Melamine Resin		Fujikasei Co., ,MM-E
Polyimide		Loan Puran Co., Kermid 604
Silicone Resin		Shinetsu Kagaku Co., KMC 2103 G.F.30%
Polyphenylene oxide (PPO)		Engineering Plastics Co.
Filler:		
a	Fe	30-200 μm, Koso Kogaku Yakuhin Co.
b	Asbestos fibre	0.7-3.3 μm, Nippon Asbestos Co.
c	Silica	Cilicic anhydride, 8-200 .mm
d	ZnO	0.1-0.8 μm, Yoneyama Yakuhin Kogyo Co.
e	$CaCO_3$	Fine $CaCO_3$, 0.7-3.3 μm
f	Cu	30-150 μm, Yoneyama Yakuhin Kogyo Co.
g	Mica	30-150 μm
h	Talc	Kozakai Seiyaku Co.

TABLE 4.37
Composition of each specimen

Matrix	Code	Filler (%)			
Melamine Resin (MF)	A	-	-	-	
	B	Fe 20	Asbestos 10	-	
Polyimide (PI)	C	"	"	Silica Sand	20
	D	"	"	ZnO	20
Silicone Resin (SI)	E	"	"	$CaCO_3$	20
	F	"	"	Cu	20
Polyphenylene Oxide	G	"	"	Mica	20
(PPO)	H	"	"	Talc	20

TABLE 4.38(a)
Each characteristic value of various composites for brake shoe [53]

Matrix	Code	Specific adhesive wear rate (mm^3/km·kgf): Non-conforming contact V'_s	Conforming contact V_s	Counterface (S45C) $\bar{V}_s$	μ upper 4kgf bottom 7.2kgf load	Limiting pv value kgf/cm^2·cm/s	Specific gravity ρ	Hardness H_RM	Izod impact value kgf·cm/cm)	Flexural strength (kgf/mm^2)	Flex. modulus E_b (kgf/mm^2)
MELAMINE RESIN (MF)	A	0.170	2.639	0.049	0.873 0.640	900	1.48	122.2	3.03	10.4	679
	B	0.156	1.354	0.060	0.507 0.463	1300	1.91	118.5	2.97	9.0	720
	C	0.117	2.529	0.126	0.725 0.555	1300	2.08	113.0	3.34	8.7	914
	D	0.129	1.258	0.078	0.673 0.576	1300	2.29	116.3	2.52	9.7	917
	E	0.136	0.554	0.015	0.658 0.538	1200	2.10	115.3	3.26	9.7	864
	F	0.144	0.973	0.024	0.609 0.475	1300	2.39	114.1	3.14	8.9	927
	G	0.127	2.193	0.139	0.745 0.503	1600	2.09	95.0	3.96	8.0	938
	H	0.141	1.455	0.038	0.587 0.530	1600	2.10	108.8	3.43	7.3	845
POLYIMIDE (PI)	A	0.12	0.912	0.031	0.696 0.331	>3000	1.30	126.4	3.17	4.6	380
	B	0.105	0.357	0.033	0.603 0.485	>3000	1.70	121.8	3.57	6.8	623
	C	0.051	0.259	0.024	0.432 0.610	>3000	1.88	118.9	3.86	8.3	831
	D	0.076	0.328	0.020	0.520 0.406	2900	2.01	120.2	3.69	1.8	722
	E	0.085	0.329	0.066	0.502 0.417	2900	1.90	117.2	3.75	8.1	730
	F	0.091	0.166	0.022	0.566 0.425	2900	2.09	118.4	3.40	7.7	600
	G	0.069	0.616	0.022	0.475 0.573	>3000	1.86	100.1	4.12	7.7	811
	H	0.083	0.317	0.007	0.466 0.432	>3000	1.90	112.8	3.57	7.5	699

TABLE 4.38(b)
Each characteristic value of various composites for brake shoe [53]

Matrix	Code	Specific adhesive wear rate (mm^3/km·kgf): Non-conforming contact V'_s	Conforming contact V_s	Counterface (S45C) $\bar{V}_s$	μ upper 4kgf bottom 7.2kgf load	Limiting pv value $kgf/cm^2 \cdot cm/s$	Specific gravity ρ	Hardness H_RM	Izod impact value (kgf·cm/cm)	Flexural strength (kgf/mm^2)	Flex. modulus E_b (kgf/mm^2)
Glass Fibre Filled Silicon Resin (SI)	A	2.973	2.654	0.057	0.253 0.171	1700	1.93	91.6	15.71	6.3	769
	B	1.152	3.553	0.243	0.634 0.570	1100	2.35	92.8	6.84	5.3	892
	C	0.721	6.609	0.596	0.690 0.693	800	2.28	66.3	4.57	4.8	879
	D	1.127	3.217	0.229	0.566 0.612	1000	2.72	94.0	5.40	5.1	771
	E	0.716	3.944	0.255	0.570 0.544	1100	2.39	89.6	4.67	5.4	899
	F	0.992	5.029	0.556	0.669 0.610	1100	2.71	93.7	5.24	6.2	778
	G	2.185	8.055	0.413	0.624 0.579	800	2.22	20.3	5.24	3.4	517
	H	2.925	5.989	0.218	0.533 0.659	800	2.49	79.0	5.30	5.0	853
Poly-phenyl Oxide (PPO)	A	0.181	15.741	0.031	0.493 0.367	500	1.05	86.3	6.96	8.8	226
	B	0.145	0.313	0.090	0.399 0.429	900	1.33	84.8	5.59	6.2	297
	C	0.121	0.347	0.038	0.445 0.457	800	1.55	68.3	6.20	5.3	345
	D	0.145	0.199	0.015	0.376 0.507	800	1.66	59.1	5.22	3.4	247
	E	0.129	0.265	0.035	0.349 0.450	800	1.57	47.2	5.54	4.4	317
	F	0.134	0.644	0.115	0.439 0.443	1000	1.86	77.6	5.83	5.7	349
	G	0.136	0.437	0.038	0.503 0.512	800	1.55	28.4	5.64	4.0	369
	H	0.189	0.345	0.026	0.442 0.488	800	1.51	32.0	4.51	2.5	207

Table 4.38 shows the various brake characteristics of friction coefficient, composite wear rates under both conforming and non-conforming contact, counterface wear rates and the limiting pv value against steel obtained with a Suzuki-type testing machine. The table also gives the specific gravity, hardness, Izod impact strength flexural strength and flexural modulus of elasticity for all of the composites. The friction coefficients μ of each of the specimens are shown in Fig. 4.92.

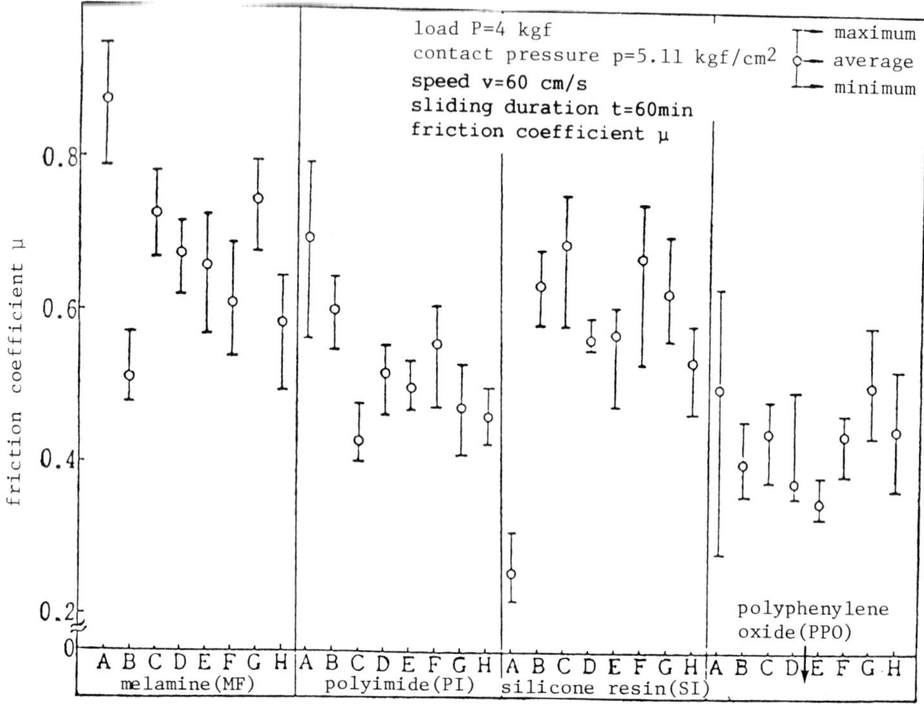

Fig. 4.92 Friction coefficients of each specimen

The values for melamine resin, polyimide and polyphenylene oxide (PPO) decrease following the addition of fillers, whilst those of silicone increase when fillers are added. The range of μ values for all the composites is 0.35 to 0.75, and generally decreases in the following order: melamine resin, silicone resin, polyimide, PPO composites.

Figure 4.93 shows the specific adhesive wear rates V_s' for these composites in non-conforming contact using an Ogoshi-type wear tester, and indicates that the value of V_s' decreases when fillers are added. For the silicone composites V_s' ranges from 0.71 to 2.97 mm^3/kgf·km and these values are the largest of all the four composite types. The values for the other three composites are generally smaller, ranging from 0.051 to 0.18 mm^3/kgf·km, and that for the polyimide composite C, filled with silica sand, is the lowest of all. The counterface wear rates, $\overline{V}_s$ against polyimide composites are generally small, especially that against H filled with talc which has the smallest value of 0.007 mm^3/kgf·km. The values of $\overline{V}_s$ against the glass fibre containing silicone composites are generally much larger.

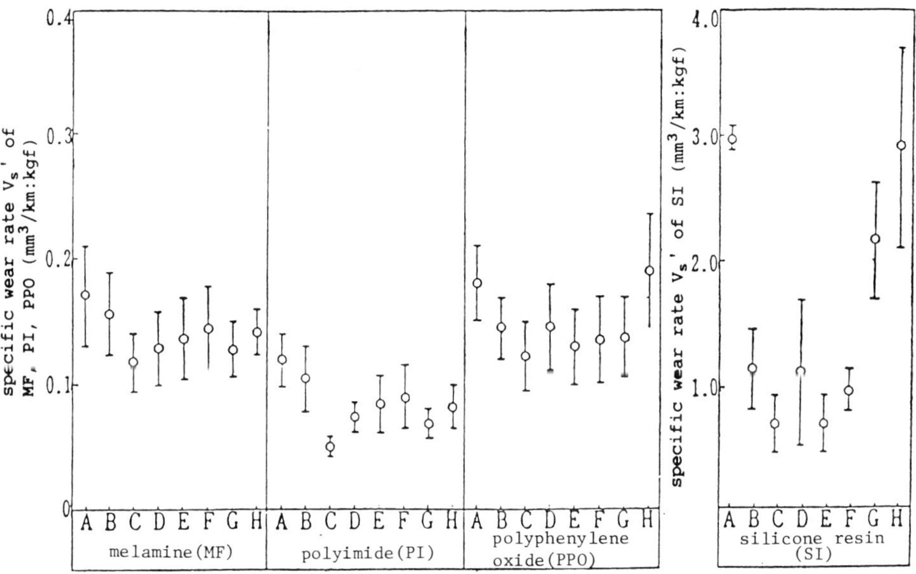

Fig. 4.93 Discontinuous (plate/ring) specific adhesive wear rate, V_s', of each specimen

4.3 GEARS

4.3.1 *Characteristics*

Although gears were first made of wood, metallic materials have been used almost exclusively until recently. Since the introduction of plastics, special gears constructed of laminated phenolics have been developed and the number of their applications to gears has been rapidly increasing. The advantages and disadvantages of plastic gears are now explained and compared with metallic gears.

(i) *Advantages*

(1) *Low specific gravity.* The specific gravity of plastics is, in general, between 0.83 and 2.1 (or about 1.2 on average) and equivalent to about 1/6 that of steel. Plastics are thus very useful as materials for instruments, transport, aeroplanes and automobiles.

(2) *No lubricant.* Plastic gears are able to transmit motion and power between themselves or metallic gears, without a lubricant. Metallic gears, however, normally require lubrication. Of course, oil or water lubricants could also be used, if desirable, in plastic gear transmissions.

(3) *Noiseless running.* In power transmission by metallic gears, high friction and contact noise occurs when the lubricant supply fails or the operating speed becomes especially high. These problems can be prevented, however, by the use of plastic gears for either one or both components.

(4) *Shock-absorbing properties.* In the power transmission of a metallic gear train, various problems can be caused either by the impact load generated at the starting-up of the motor or at the switch-over point from two contacts to one during operation. However, these problems are reduced by relaxing the impact load. A plastic gear can absorb the impact load due to instantaneous deformation as a consequence of its lower elastic modulus (about 1/100 that of a metallic gear).

(5) *High corrosion resistance.* Plastics generally have a higher resistance to air, water and chemical corrosion than steel; they are thus suitable for applications in hostile environments.

(6) *High wear resistance.* As described above, the wear resistance of plastic materials is generally as great as that of steel, although the hardness and tensile strength is much lower.

(7) *Excellent mouldability.* This is the greatest advantage of plastic gears. Metallic gears are produced using an automatic gear cutting machine, and this is a slow process. However, plastic gears can be produced quckly and easily by injection moulding, and therefore, when mass-produced, their cost is much lower than that of metallic gears.

(ii) *Disadvantages of plastic gears*

(1) *Low mechanical strength.* The tensile strength of most plastics, apart from those which are reinforced, lies between 1 - 9 kgf/mm^2. Moreover, the impact resistance of plastics is much lower, equivalent to 1/100 - 1/10 times that of steel. Plastic gears are thus useful for transmitting motion and minimal power, but are unsuitable for high power transmission because the size of the gear teeth would need to be extremely large.

(2) *Low heat resistance.* The melting temperature of plastic is generally between 120°C and 325°C (about 200°C, on average), which is far lower than that of steel, at 1500°C. The temperature range for which plastic gear teeth are durable over a long period of operation is generally limited to within 100°C. Their applications are therefore limited by the environmental temperature and the rising temperature due to the frictional heating.

(3) *Low durability.* Plastics generally have a lower durability to outdoor weathering, high temperatures and high humidity than metals. Plastic gears can also become weaker after extended application periods, even at average temperatures, due to their increased brittleness resulting from chemical or physical changes.

(4) *Low dimensional stability.* The linear thermal expansion of plastic is 4.5 - 18 x 10^{-5}°C^{-1}, which is about 3 - 6 times larger than that of steel, at 1.3 x 10^{-5}°C^{-1}. The water absorption of plastics is 0-8%, and leads to expansion. The dimensional changes of plastic gears, resulting from changes in the environment, temperature or humidity, are much greater than those of steel gears. Dimensional changes may also occur due to the residual stresses generated during the moulding process.

Taking into account the advantages and disadvantages of plastic gears, they have been used successfully in numerous applications. Studies of plastic gears have been published by Takanashi [59-61]; Kudo [62]; Arai [63]; Aota [64]; Ikegami [65]; Zumstein [66]; Lurie [67]; Kuhn [68]; Martin [69]; Pfuger [70]; Yamaguchi [71-74]; Oyanagi [75-76] and Tsukamoto [77-79].

4.3.2 *Dimensional accuracy of moulded plastic gears*

Plastic gears are generally manufactured from powder- or pellet-type moulding compounds in one step by compression or injection moulding processes and without machining. However, there is a problem as to how the dimensional accuracy of moulded gears can be maintained. Currently, fifth-class dimensional accuracy of plastic moulded gears is generally achievable by the most skillful technology. In this section, some experimental examples relating to the moulded gears are described.

(i) *Example of injection mouldings for a small gear [62]*

Figures 4.94 to 4.97 show illustrations of four small injection-moulded gear wheels: I, II, III, and IV, with between 24.5 and 38.55 mm addendum circle diameters, 0.6 module, and different insert and gate positions. Table 4.39 shows the metal mould dimensions for projecting injection-moulded gear wheels and types I and II from Nylon 6, polyacetal (POM) and polycarbonate (PC). The table also give the standard dimensions of the mouldings; errors from the standard dimensions in mm or accuracy in % of the addendum circle diameter; and outside diameter and length of the shaft on the mouldings. It is clear from this table that the largest shrinkage occurred in polyacetal, and the smallest in polycarbonate. Therefore, the greater the degree of crystallization in plastics, the larger the shrinkage.

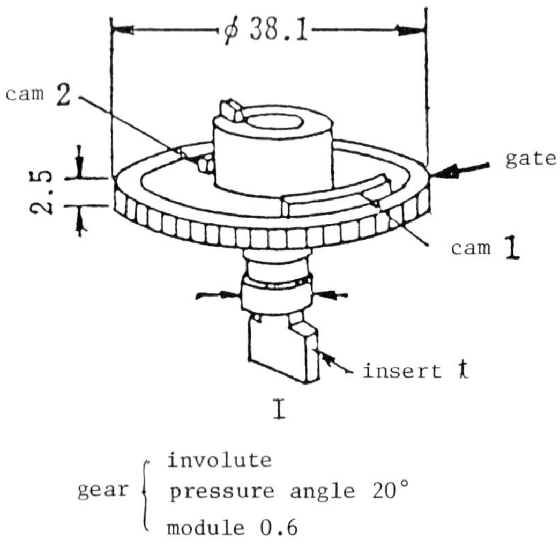

Fig. 4.94 Gear 1

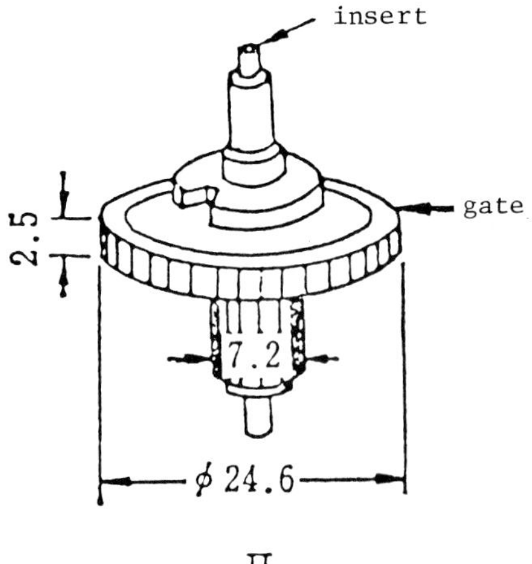

Fig. 4.95 Gear 2

TABLE 4.39
Dimensional accuracy of mouldings, mm. % accuracy in brackets ( )

		Aim Dimension		Nylon 6	Delrin (POM)	Polycarbonate
Specimen	Position	Mould Dimension and Tolerance	Standard Dimension	Error from Standard d.	Error from Standard d.	Error from Standard d.
I	dia. of addendum circle (a)	0 38.1 -0.05	38.02	-0.017 (-0.05)	-0.266 (-0.70)	+0.388 (+0.89)
I	outside dia. of shaft (c)	-0.02 8.0 -0.05	thick-ness 7.965 1.0	+0.036 (±0.45)	+0.053 (+0.66)	+0.070 (+0.88)
II	shaft length (j)	0 16.6 -0.1	16.55	-0.51 (-0.91)	-0.209 (-1.26)	-0.005 (-0.03)

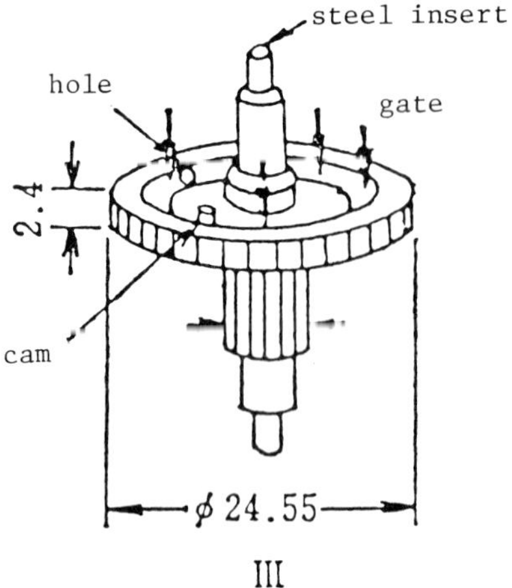

Fig. 4.96 Gear 3

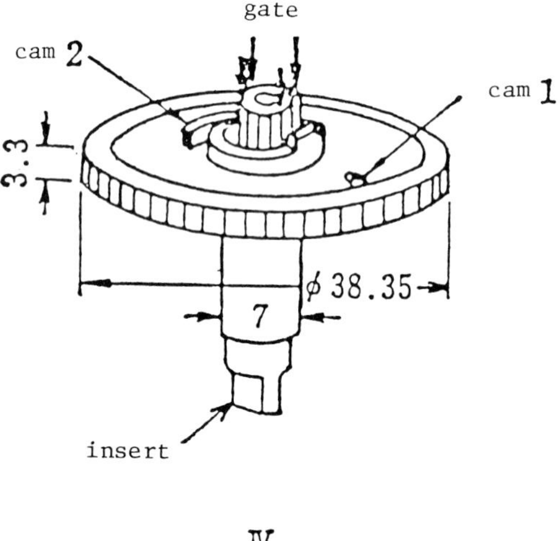

Fig. 4.97 Gear 4

Table 4.40 shows the errors from the true addendum circle diameters of Nylon 6 moulded gear wheels which were manufactured by injection moulding with three different types of moulds: single gate (I); four gates near the outer circle (III); and four gates near the shaft (IV). This table indicates that the dimensional accuracy of the moulding from the gate type (III) is the highest. Table 4.41 shows the various normal pitch errors in mm, or in the JGMG accuracy class, the Nylon gear wheels moulded with the same types of moulds, I, III and IV, mentioned above. Mouldings from the type III mould have a fifth-class accuracy, whilst those using a type I mould have, on average, a seventh-class accuracy.

TABLE 4.40
Error from true circle (mm, Nylon 6)

Specimen	Gate Type	Position	Error from True Circle
I	Single gate	Gear	< 0.015
III	4 gates near outer circle	Gear	< 0.005
IV	4 gates near shaft circle	Gear	< 0.015

TABLE 4.41

Specimen:	I		III		IV	
Error	Value (μm)	JGMG Class	Value (μm)	JGMG Class	Value (μm)	JGMA Class
Average normal pitch	-28	-	5	-	-9	-
Maximum normal pitch	-42	7th	18	5th	-26	6th
Maximum single pitch	23	7th	14	5th	17	5th
Maximum adjoint pitch	27	6th	17	5th	25	6th
Max. accumulated pitch	154	6th	17	5th	25	6th
Synthesis	-	7th	-	5th	-	6th

(ii) *Example of injection moulding for a medium-sized gear*

Three kinds of plastic spur gears (polyacetal (POM, Delrin), HI styrene and ABS resin) with the dimensions of module 3, 30 teeth and 90 mm pitch circle diameter (as shown in Fig. 4.98), were manufactured by injection moulding process. The maximum and minimum addendum circle diameters of the moulded gears are shown in Table 4.42. The dimensions of similar POM gears manufactured from a block by an automatic cutting machine are also shown in Table 4.42. The dimensions are illustrated in Fig. 4.99, which indicates that the addendum circle diameter of a moulded POM gear is smaller than the standard size, due to its high shrinkage. The addendum circle diameter of the machined POM gear is slightly larger than standard size, that of the ABS gear is the next largest, and that of HI styrene is the largest. This shows that the dimensions of a moulded gear are affected by the type of resin used.

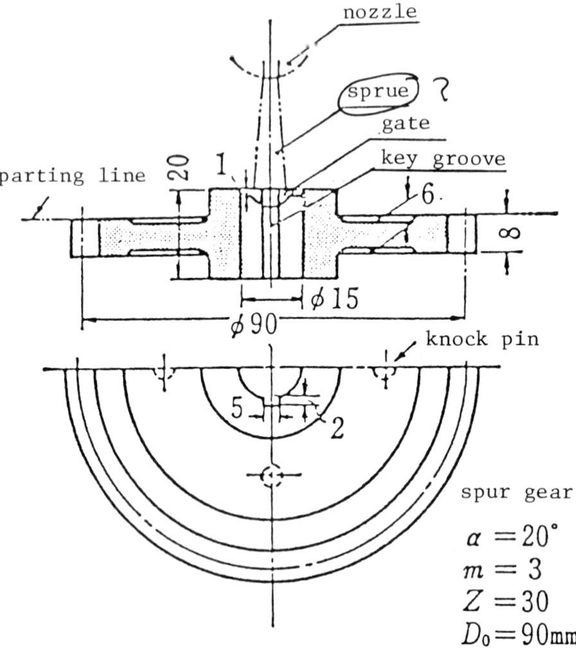

Fig. 4.98 Shape and dimensions of moulded gears

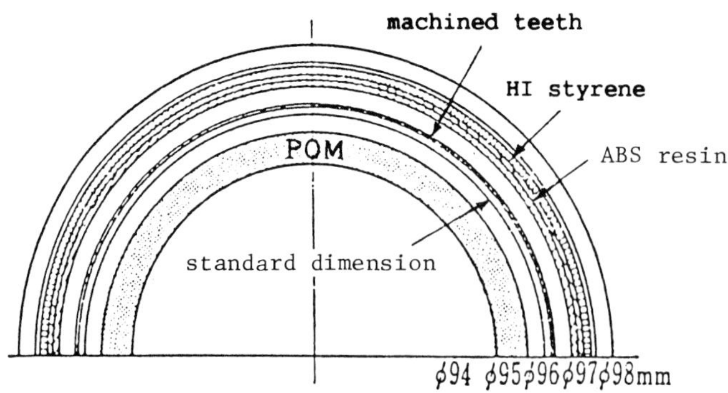

Fig. 4.99 Outside diameter of each medium spur gear

TABLE 4.42
Addendum circle diameter of each moulded gear (mm)

	Max. Value	Min. Value
POM (Delrin)	95.460	94.510
HI Styrene	97.329	97.003
ABS Resin	97.179	96,752
Machining (POM)	96.243	96.182
Standard spur gear	96.0	
Outside diameter of mould cavity	97.40 (average)	

Table 4.43 shows three pitch errors: single, adjoining and accumulated (in millimeters) and also the accuracy classes according to JIS (Japan Industrial Standard) of these four types of plastic gears. It can be seen that the three types of injection-moulded gears have a class of accuracy of between 7-8, while machined gears are between 2 - 5.

TABLE 4.43
Dimension accuracy of each moulded gear

	Single Pitch		Adjoining Pitch		Accumulated Pitch	
Mouldings	error μm	JIS class	error μm	JIS class	error μm	JIS class
POM (Delrin)	112	8	90	7	401	8
HI Styrene	108	8	95	7	345	8
ABS resin	98	8	98	7	292	8
POM (Delrin) by machining	15	4	9	2	99	5

Figure 4.100 shows a polypropylene injection-moulded gear and a melamine resin compression-moulded one manufactured using a similar mould, together with details of each tooth face profile; mould size (dotted line); standard size (chain line); and moulding size (full line). This figure indicates that, compared with the size of the mould, the addendum circle diameter of a moulded gear is smaller and the tooth thickness at the dedendum circle is larger.

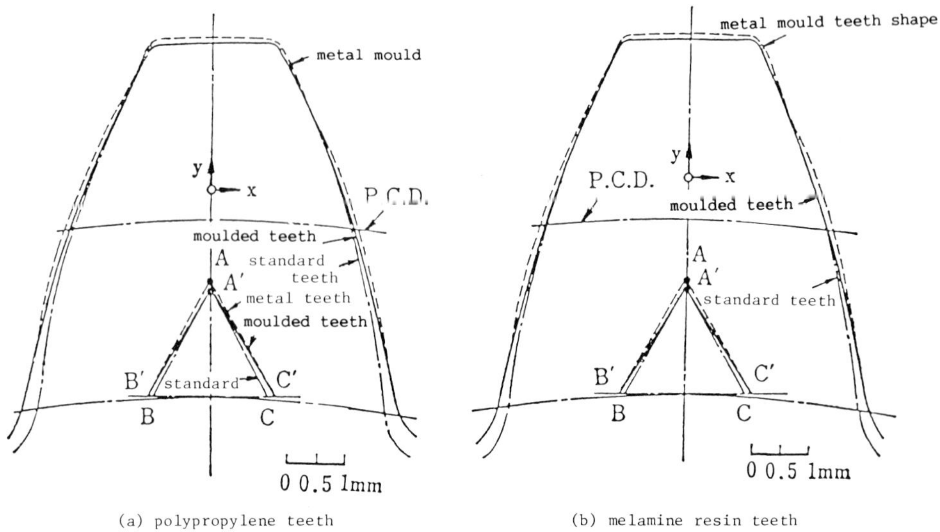

Fig. 4.100 Schematic comparison of moulded gear teeth

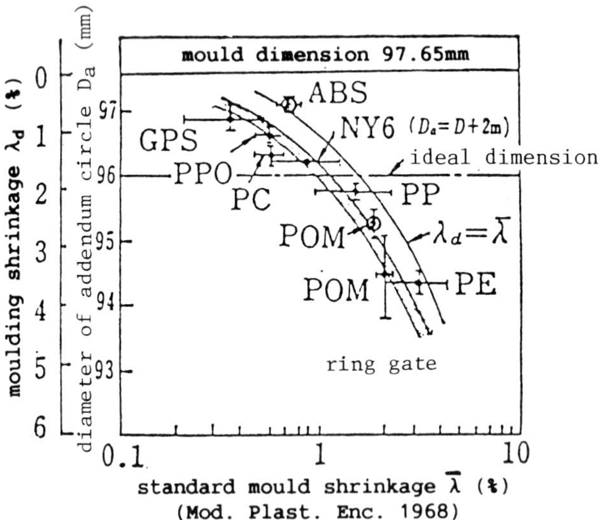

Fig. 4.101 Relationships between diameter of addendum circle and standard moulding shrinkage of moulded teeth for each material

Figure 4.101 shows the moulding shrinkage of the addendum circle diameter, λ_d, for each injection-moulded gear and the shrinkage of general standard injection mouldings, $\overline{\lambda}$ for each type of plastic. It also indicates that the relationships between them are proportional, but not directly proportional.

Injection-moulded plastic gears generally have residual stresses, especially at the teeth, and a suitable heat-treatment must be given to relieve these stresses. Figure 4.102 shows various dimensional changes (%) due to heat treatment of POM and ABS injection-moulded gears, at different temperatures, for: boss length $\Delta\ell_b$; inside and outside boss diameters Δd_1 and Δd_2; face width δ_b; normal pitch ΔE_n; and addendum circle diameter ΔD_a. It is clear from this figure that each dimension generally shrinks with heat treatment, with an especially large shrinkage near the transition temperature of 150°C for POM. The rate of change of the face width δ_b decreases when heat-treated at below 150°C.

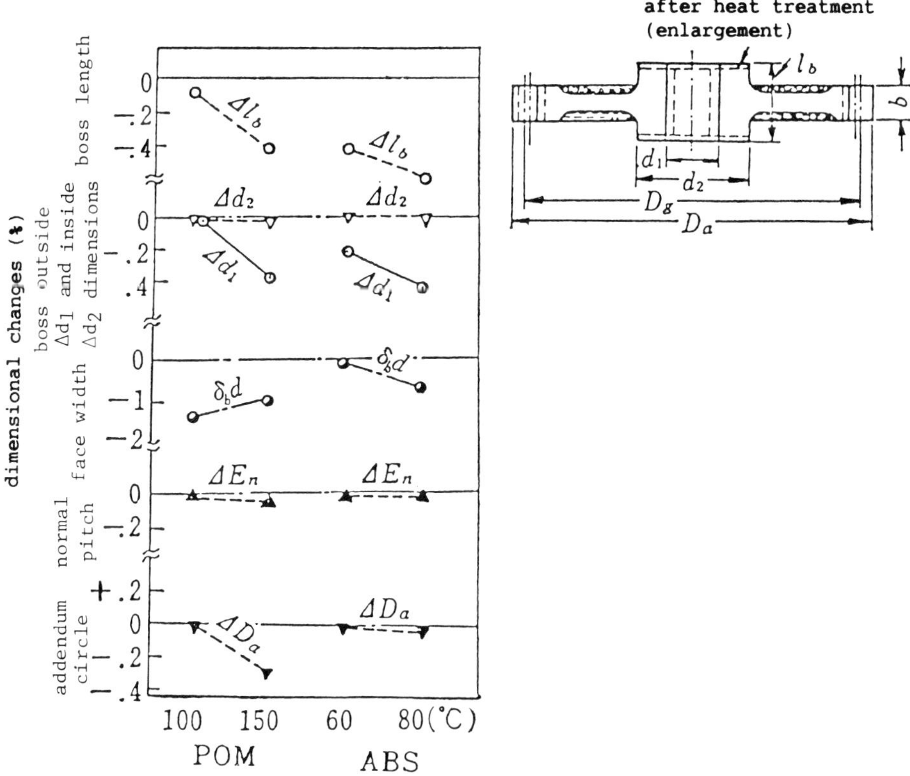

Fig. 4.102 Dimensional changes due to heat treatment for each moulded gear

(iii) *Example of a press-moulded gear*

Plastic gears are occasionally produced by press-moulding in which a solid plastic block is pressed into a gear mould under a suitable temperature and pressure. Figure 4.103 shows the useful moulding region using the coordinates of mould temperature T_m and pressure P_m for a plastic cylindrical block of dimensions 12 mm inner diameter, 25 mm outer diameter, and 45 mm length, pressed into a steel mould with gear cavity dimensions of 31.5 mm addendum circle diameter, 18 mm boss length, 16 teeth and 1.75 module. The upper figure shows the region for ABS resin preheated to a temperature of 130°C,and the bottom figure is for polypropylene (PP) preheated to a temperature of 170°C. Plastic gears produced by deformation processing or press-moulding generally have higher residual stress which, in order to retain dimensional stability and mechanical strength, must be removed by suitable heat treatment.

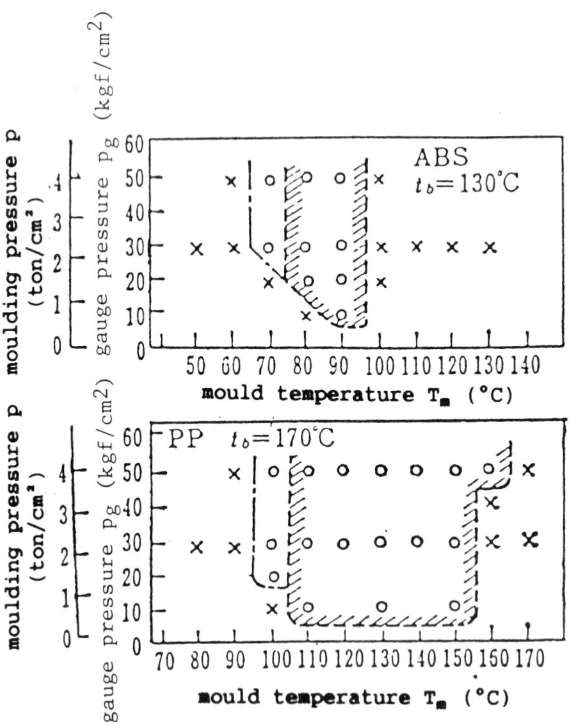

Fig. 4.103 Moulding range field of ABS and PP gears (effect of pressure)

Figure 4.104 shows the plastic gear tooth profiles before (full line) and after (dotted line) heat-treatment at each temperature, 60°C and 90°C, for ABS resin and polypropylene (PP). It is clear from this figure that the dimensions of the teeth do not change uniformly in all parts and the thickness increases at the dedendum circle diameter.

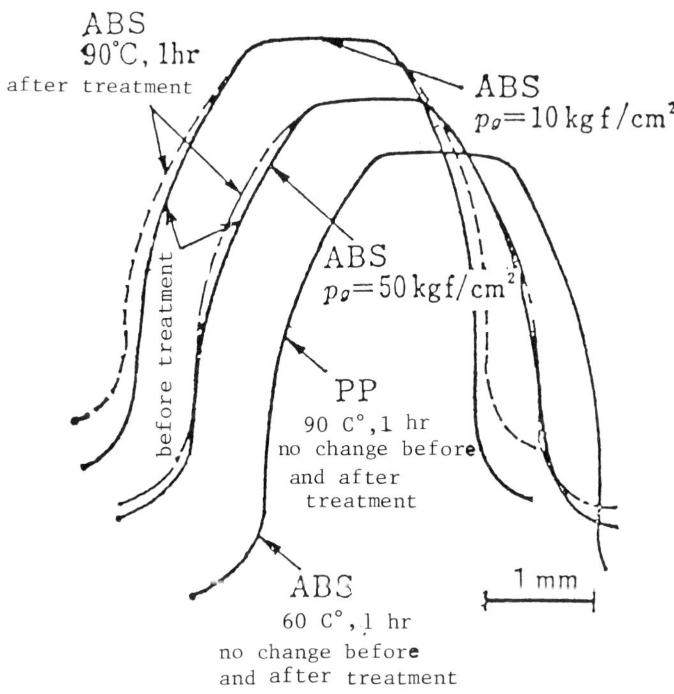

Fig. 4.104 Change of tooth shape due to heat treatment of ABS and PP gears

Figure 4.105 shows press-moulded gear tooth profiles for polyacetal (POM) before and after 1 hour of heat-treatment at 120°C, and indicates that there is little dimensional change in the profile due to heat-treatment at temperatures lower than the glass transition temperature. Table 4.44 shows the thickness errors; normal pitch error; single pitch error and the accuracy class in JIS, for a POM press-moulded gear produced and heat-treated under different conditions and indicates that the accuracy class is between 6 and 8.

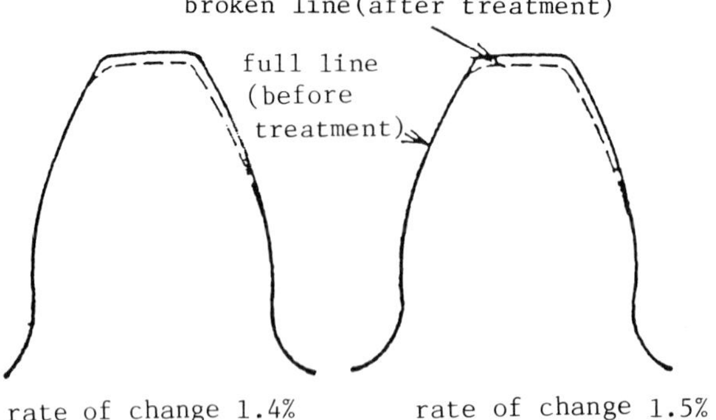

Fig. 4.105 Change of tooth shape due to heat treatment of POM gears (for addendum circle)

TABLE 4.44
Dimension accuracy and class (JIS) of POM pressed gears

Position	Manuf-acturing Condition	Pressed gear only: 7 min press time	5 min press time	3 min press time	2 min press time	100°C heat treat-ment (1 hr)	120°C heat treat-ment (1 hr)	Pressed gear with shaft
Error of thickness	μm max	175	273	296	258	290	209	183
	class	7	8	8	8	8	8	8
Error of normal pitch	μm max	47.4	33.6	75.0	99.2	86.0	113.2	43.7
	class	7	6	7	8	7	8	6
Error of single pitch	μm max	50.6	26.6	45.4	61.2	53.0	63.0	25.5
	class	7	7	7	7	7	8	6

4.3.3 *Experiments on the performance and efficiency of power transmissions*

Power transmitting efficiency, mechanical strength for transmitting load, durability and wear resistance are the main characteristics which affect gear performance. These characteristics have been obtained experimentally under various loading conditions using the gear test apparatus shown in Fig. 4.106.

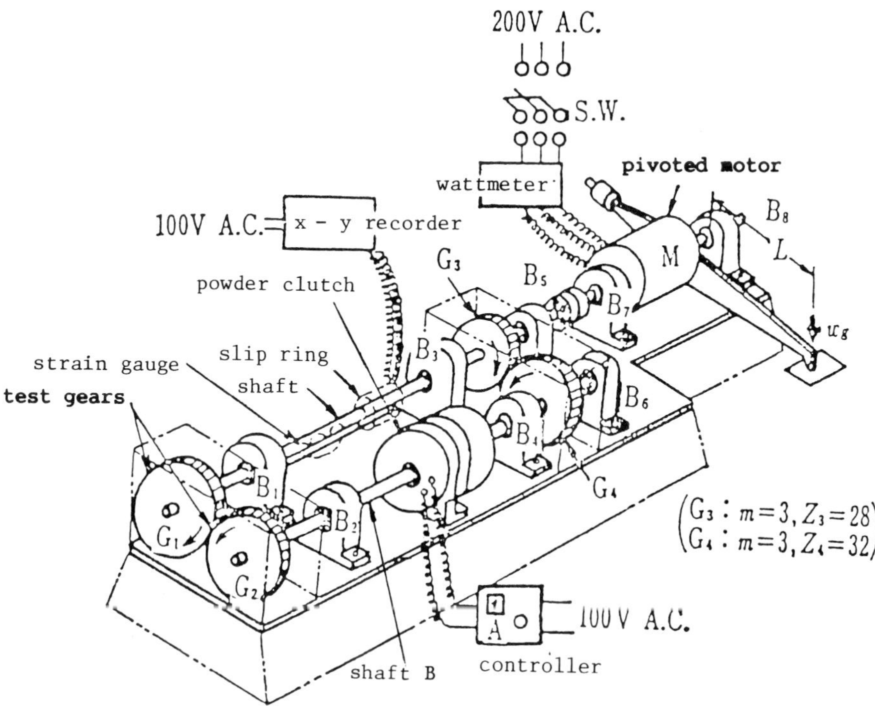

Fig. 4.106 Gear test apparatus

In this apparatus, if the actual power load on the gear teeth were applied by an external driving source and absorbed by an external energy absorber, a very large motor and brake would be needed. To reduce the size and power, however, the arrangement has been developed as a "power-circulating-type gear test machine", and its principles given in Fig. 4.107. In order to adjust the loading torque to about 0 - 2 kgf-m during running, a powder clutch was used as shown in Fig. 4.107 and Fig. 4.108(a). The characteristic relationship between the generated torque T_o and the magnetically-excited current I is shown in Fig. 4.108(b).

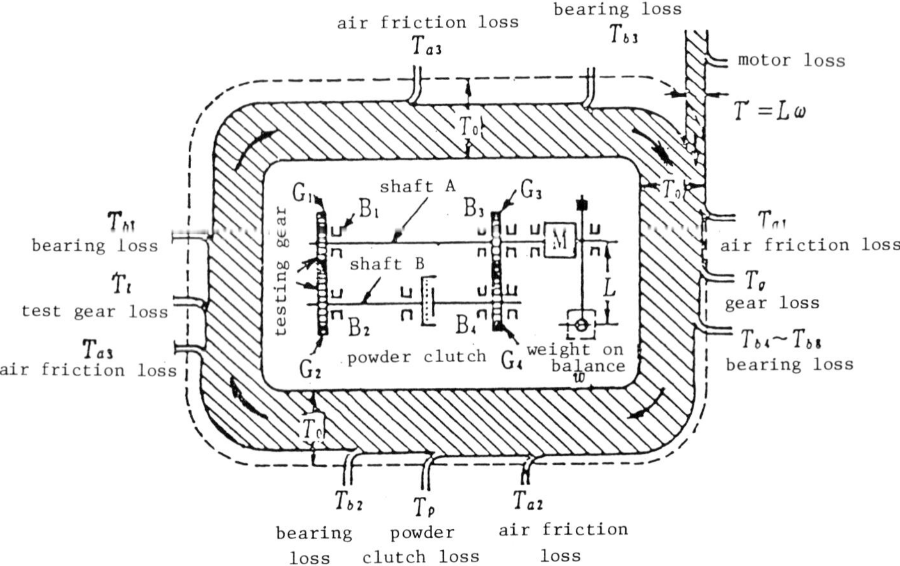

Fig. 4.107 Input and output of energy in power-circulating-type of gear test apparatus (in constant of rotating speed)

The loading torque T_o during running may be estimated from the value of I. Figure 4.107 shows each torque necessary to keep the machine running: bearing loss torque, T_b; friction loss torque at no load, T_a; sliding friction-loss torque in the powder clutch, T_p; friction loss torque at the intermediate gears, (G_3, G_4), T_g; and friction loss torque of the test gears, T_ℓ. If the running torque or total loss torque is T', the following relationship holds:

$$T_\ell = T' - (T_a + T_b + T_p + T_g) \tag{4.8}$$

and the power transmitting efficiency, η, of the test gears is

$$\eta = \frac{T_o - T_\ell}{T_o} = \left(1 - \frac{T_\ell}{T_o}\right) \times 100\% \tag{4.9}$$

Values for T_a, T_b, T_p and T_g were obtained experimentally or estimated, as shown in Fig. 4.109, for example; the value of η is then obtained from equation (4.9). Figure 4.109 shows the relationship between total loss torque, T', or power transmitting efficient, η, and the loading torque, T_o, for steel gears lubricated with oil.

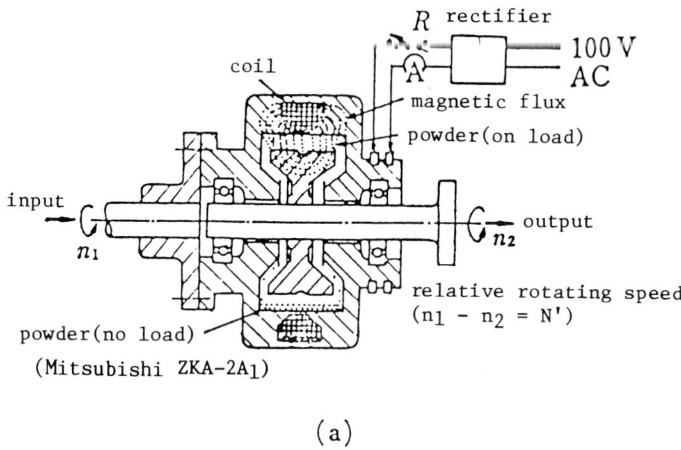

(a)

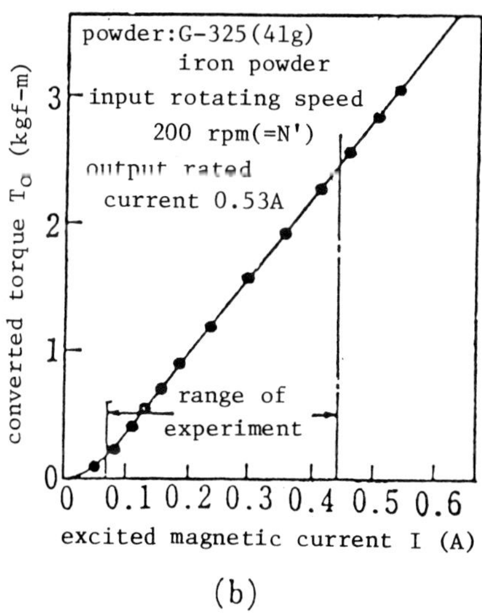

(b)

Fig. 4.108 Apparatus and characteristics of powder clutch

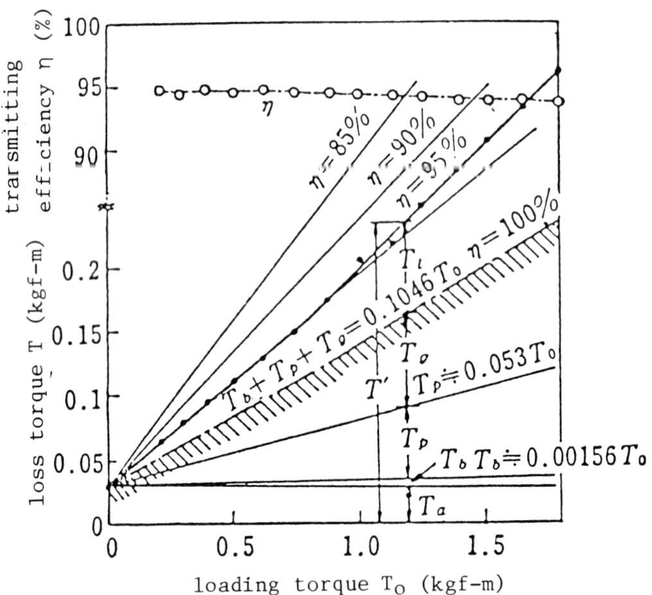

Fig. 4.109 Relationship between loss torque, or transmission efficiency, and loading torque for a steel gear (oil lubricant)

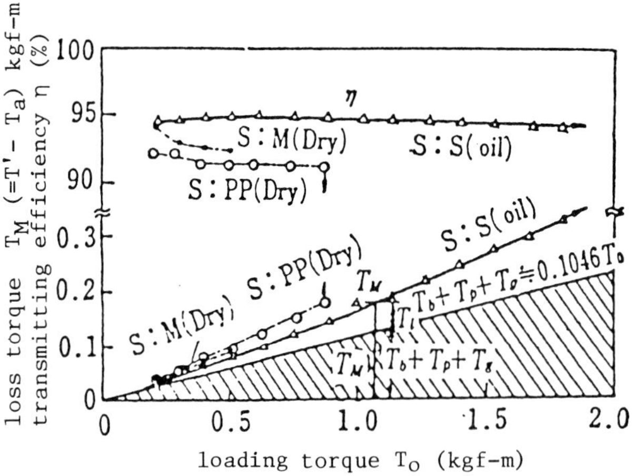

Fig. 4.110 Relationships between T_M or η and T_o for each combination of steel and plastic gears

Figure 4.110 shows similar relationships for steel gears lubricated with oil, and steel with polypropylene (PP) or melamine resin (M) gears, without lubricant (dry). Figure 4.111 shows similar relationships for PP gears without lubricant (dry), either unfilled PP) or with 2.5% graphite (2.5PP) or 5% graphite (5PP). It is clear from these figures that the value of η for oil-lubricated gears is about 95%. Without oil, the values are: steel:melamine resin gears, 92%-95%; steel:PP gears, 91%-92%; and PP:PP gears, 90%-93%.

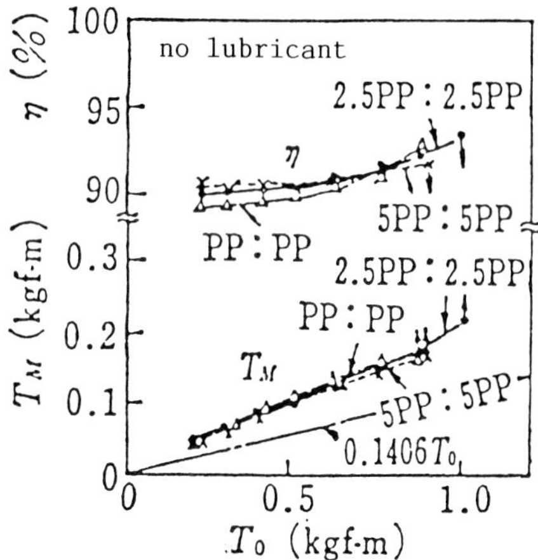

Fig. 4.111 Relationships between T_M or η and T_o for PP:PP gears (no lubricant)

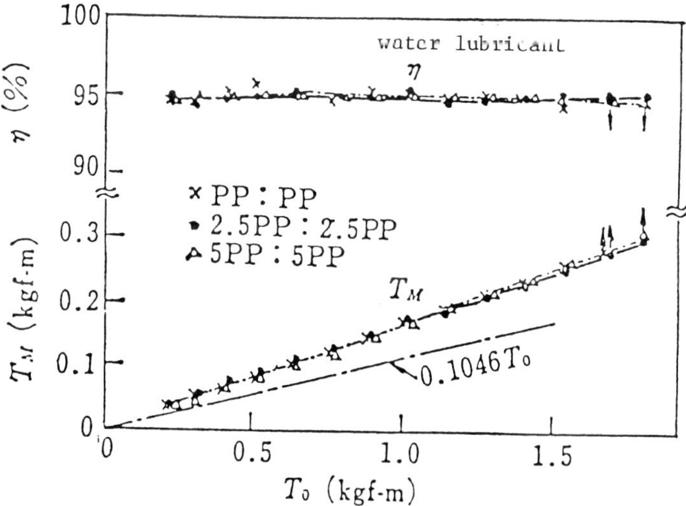

Fig. 4.112 Relationships between T_M or η and T_o for PP:PP gears lubricated with water

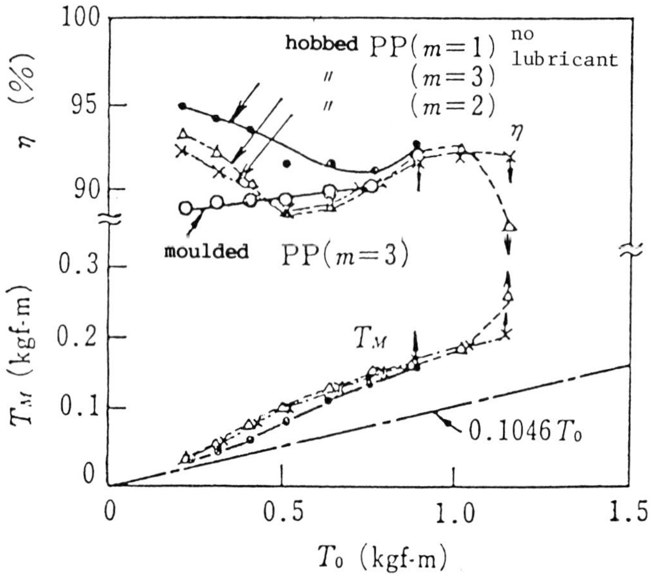

Fig. 4.117 Relationships between T_M or η and T_o for the combination of hobbed PP:moulded PP gears without lubricant

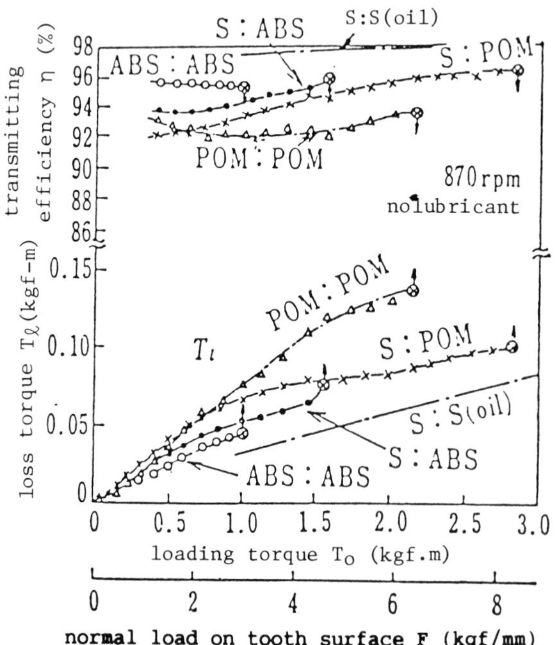

Fig. 4.118 Relationships between T_M or η and T_o or F for various combinations of plastic and steel gears either dry or oil-lubricated

Figure 4.118 shows the relationships between T_ℓ or η and T_o for the gear combinations POM:POM: S(Steel):POM; ABS:ABS; and S:ABS without lubricant and an S:S gear lubricated with oil. Figure 4.119 compares the similar relationships for two POM hobbed gears and a POM hobbed gear against a POM moulded gear without lubrication and indicates that the value of η is over 92% for either combination, but is less than 97.8%, which is the η value for S:S lubricated with oil.

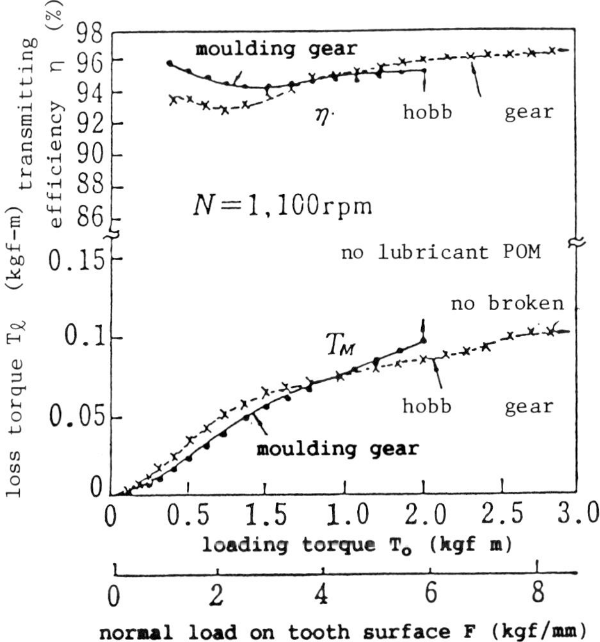

Fig. 4.119 Relationships between T_ℓ or η and T_o or F for hobbed and moulded gears without lubrication

Figure 4.120 shows the relationships between the maximum load torque Tm(= T_o - T'), F_m (maximum tooth face load per unit width) or $\bar{\eta}$ (η at 1.5 kgf/mm of F_m) and rotational speed N (rpm), for transmission between various combinations of gears without lubricant. The value of η is particularly high for the combination of ABS:ABS.

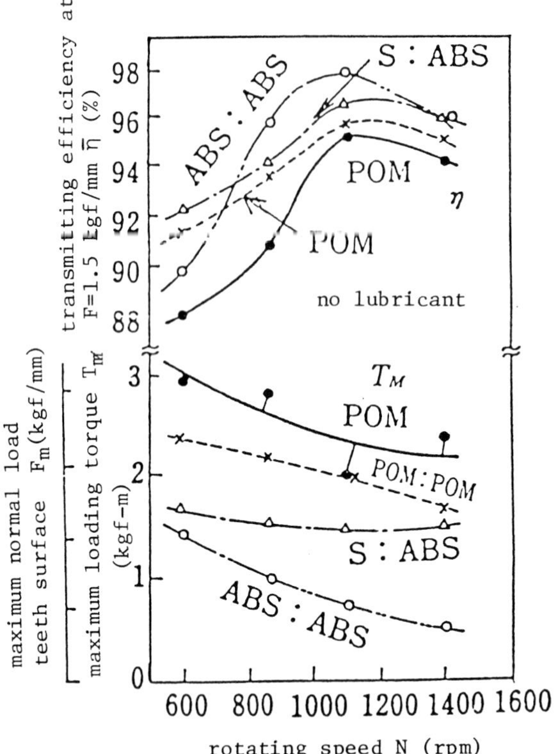

Fig. 4.120 Relationships between T_m, F_m, η and N for various combinations of gears without lubrication

4.3.4 *Mechanical strength and endurance*

(i) *Performance under stress*

A theoretical study of gear strength was developed by Wilfred Lewis (1893) and is still used today, as in the following.

In Fig. 4.121(a), when the transmitting force, P, is applied to the tooth tip, O, its component in the radial direction, N, generates a compressive stress, σ_c, and its component in the circumferential direction, F, (= P cosθ, θ:pressure angle 14.5° or 20°) transmits a rotating force to the opposing gear and also applies a bending moment, M or Fh (M_{max}), at the dedendum VF. The compressive stress σ_c due to N is negligibly small compared with the bending stress due to bending moment M_t. The distribution of bending stresses, σ_b, due to the bending moment M_t along the tooth height is represented by the horizontal length between the straight line BH and the parabolic curve BK'V, and the maximum bending stress σ_b generated at the dedendum is presented by the equation $\sigma_b = M_t/Z$ (where Z is the modulus of the section). Then, $\sigma_b = 6Fh/(bt^2)$ (where t is the thickness of the tooth at the dedendum and b is the face width). As △BGV ∞ △VHG, $x/(t/2) = 2t/h$, and $h = t^2/4x$, the following relation is obtained:

$$F \cdot \frac{t^2}{4x} = \sigma_b \cdot \left(\frac{bt^2}{6}\right) \tag{4.10}$$

Therefore,

$$F = \sigma_b \cdot b \cdot \left(\frac{2}{3}\right) \cdot x \tag{4.11}$$

Multiplying the right hand side of equation (4.11) by p_c/p_c, where p_c is the circular pitch,

$$F = \sigma_b \cdot b \cdot p_c \cdot \left(\frac{2x}{3p_c}\right) \tag{4.12}$$

and the value in brackets is determined by the shape of the tooth and expressed as $2x/3p_c = y$, which is Lewis' "gear factor". Therefore,

$$F = \sigma_b \cdot b \cdot p_c \cdot y \tag{4.13}$$

$p_c = \pi M$, where M is the module and so

$$F = \sigma_b \cdot b \cdot \pi M \cdot y \tag{4.14}$$

Therefore,

$$\sigma_b = \frac{F}{b \cdot \pi M \cdot y} \tag{4.15}$$

The value of y is presented as follows:

> $y = 0.124 - 0.684/n$ for an involute tooth with a pressure angle of 15°, where n is the number of teeth and
> $y = 0.154 - 0.916/n$ for an involute tooth with a pressure angle of 20°.

In the design of a gear, σ_b must be the allowable bending stress $\overline{\sigma}_b$, and is estimated, generally, for steel gears with a peripheral velocity, v (m/s), as:

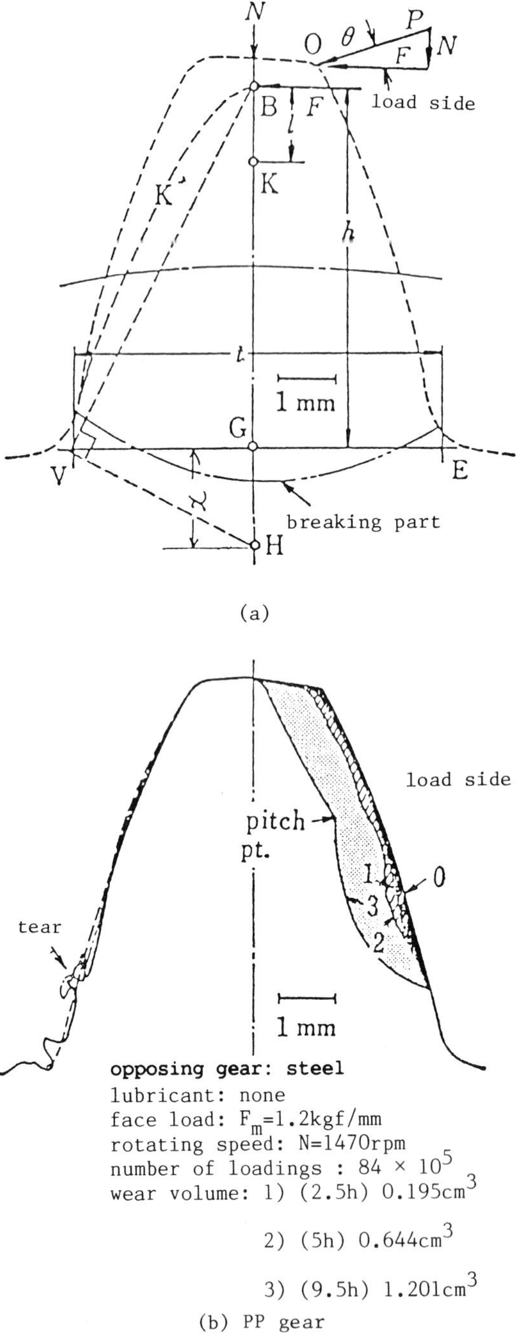

Fig. 4.121 Profile changes of PP gear teeth due to wear

$$\overline{\sigma}_b = \sigma_{bo} \cdot \left(\frac{a}{a + v} \right) \tag{4.16}$$

where a is the velocity factor and is equivalent to 6 for medium speeds, and σ_{bo} is the allowable bending stress for a static load, i.e. when v = 0.

(ii) *Failure as a result of transmitting power*

The relationship between the stress generated at the gear teeth and the static transmitting force is represented by equations (4.11) to (4.15). The actual failure process of a plastic gear can be explained from the experimental results as follows.

Figure 4.121(a) shows the breaking point of a melamine resin gear tooth which is comparatively hard and brittle, and indicates that the gear tooth failure is a result of reaching the limiting bending stress at the dedendum part (VE) where the maximum bending stress is generated. However, in a polypropylene gear, which is comparatively soft, the loading surface is worn by repeated contacts as shown in Fig. 4.121(b), and a crack due to fatigue is generated at the dedendum on the gear face and results in a large deformation. Due to this deformation, the power transmission ceases and, therefore, complete breakage of the tooth does not occur.

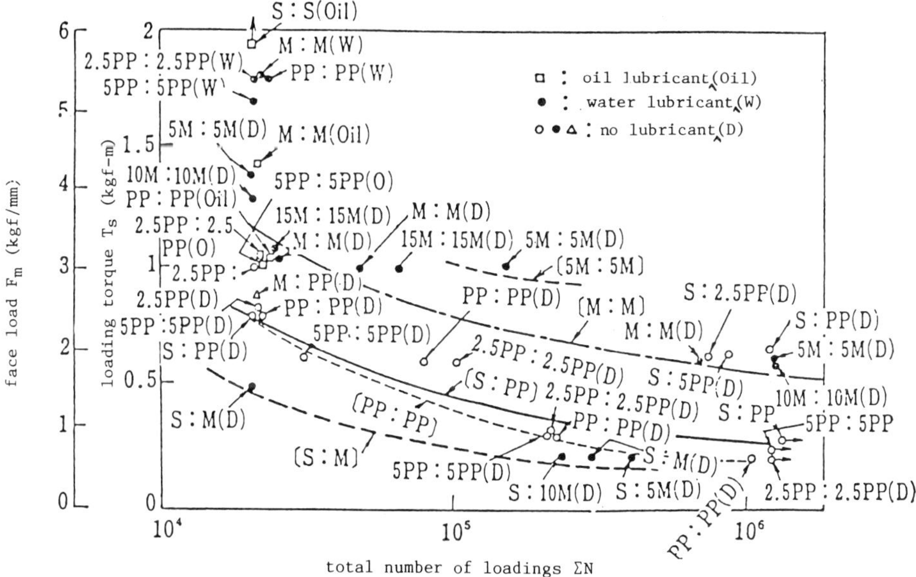

Fig. 4.122 Relationships between loading torque, T_s, or face load, F_m, and the total number of loadings for various combinations of gears

Figure 4.122 shows the relationships between the applied tooth loading torque, T_s (kgf-m), or tooth face-load per unit width, F_m (=F/b, kgf/mm), and the total number of load cycles, ΣN, necessary to reach the breaking point loadings for various combinations of PP gears; melamine resin gears (M); and steel gears (S) with either oil, water or no lubricant. The curves are similar to the S-N curves of fatigue characteristics. It is clear from this figure that for ΣN of about 2 x 10^6, the value of F_m for the combination of PP:PP gears without lubricant is about 0.5 kgf/mm; for S:PP gears about 0.8 kgf/mm; and for M:M gears about 1.7 kgf/mm. These values increase with the use of a solid lubricant. In addition the F_m value for the combination of S:S gears with oil and PP:PP gears lubricated with water should also increase considerably. Figure 4.123 shows the relationship between the face load F_m and the total number of loading cycles ΣN at a rotational speed of 1400 rpm for each combination of S:POM gears; POM:POM gears; S:ABS gears; and ABS:ABS gears without lubricant, and indicates that the value of F_m for a ΣN of 2 x 10^6 is 3.2 kgf/mm for S:POM gears; 1.3 for POM:POM gears; and 0.7 kgf/mm for ABS:ABS gears.

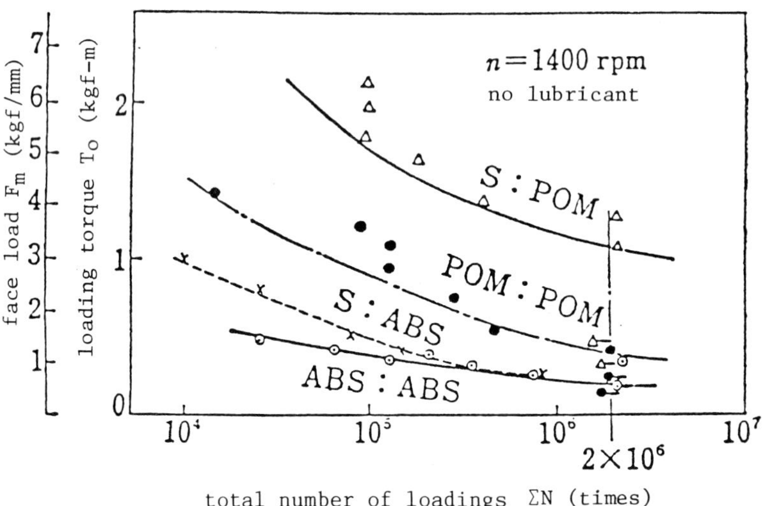

Fig. 4.123 Experimental results of endurance tests

Figure 4.124(a) shows the relationships between the maximum bending stress σ_b, or the maximum loading torque T_o, and ΣN at each rotational speed of 850, 1200 or 1450 rpm for the combination of ABS:ABS gears without lubricant, and indicates that the position of the σ_b-ΣN curves becomes lower with an increase in speed. Figure 4.124(b) shows corresponding relationships for POM_{co}:POM_{co} gears without lubricant, and demonstrates a similar effect of speed on the F_m-ΣN curves to that of ABS:ABS gears, but of smaller magnitude.

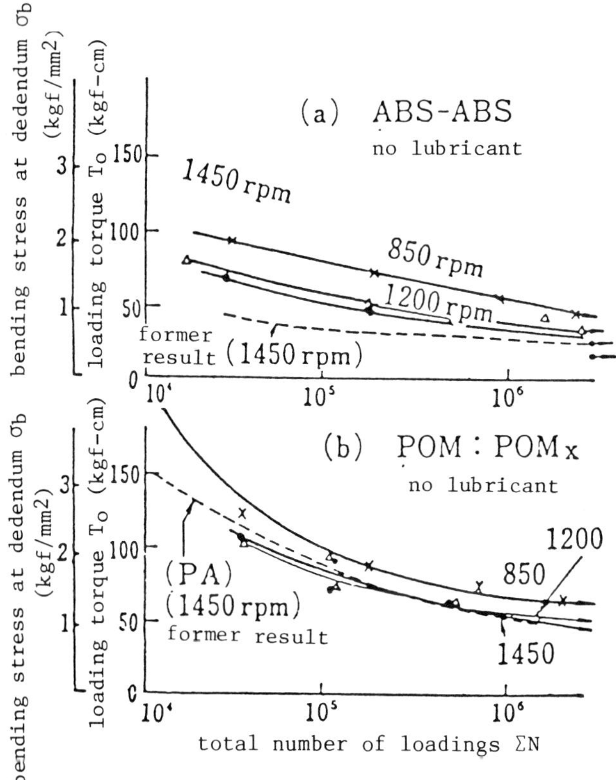

Fig. 4.124 Relationships between σ_b and ΣN for ABS:ABS or POM:POM gears

Figure 4.125 shows the relationships between bending stress at the dedendum σ_b and ΣN at different speeds for combinations of S:POM gears and S:POM_{co} gears without a lubricant. In these F_m-ΣN or T_o-ΣN curves, the breaking face load F_m for $\Sigma N=10^6$ is defined as a limiting face load F_{mf}(kgf/mm), and some typical values are shown in Table 4.45. The limiting transmitting horsepower L (HP) is obtained by the following equation: $L = F_a \cdot v/75$ ($F_a = F_{mf} \times b$ and v is the peripheral speed of the gear, m/s) and is shown in Table 4.45. The kinetic friction coefficients μ_k for these combinations of gears are also shown in the Table. It is clear that the limiting transmitting horsepower between plastic gears, without lubricant, is only between 1/3 to 1/10 that of steel gears lubricated with oil. The smaller the value of μ_k, the greater the value of F_m or L.

TABLE 4.45
Limiting load values for each gear

Application Condition	Limiting[2] face load F_{mf} (kgf/mm)	Limiting[1] transmitting horsepower L (HP)	Kinetic Friction Coefficient μ_k
Steel:steel, 1470 rpm, oil lubrication	5.52	5.9	
Steel:M, 1470 rpm, no lubricant	0	0.32	0.6
Steel:PP, 1470 rpm, no lubricant	0.8	0.51	0.3
Steel:PP (2.5% carbon filled)	(0.8)	(0.5)	
M:M (25% carbon filled) no lubricant	1.8	1.2	0.07
M:M (5% PTFE filled)	(3.0)	(2.0)	
PP:PP (5% PTFE filled), no lubricant	0.6	0.4	0.35

1) From the Lewis equation
2) Normal load on the gear surface per unit width (at the pitch point)

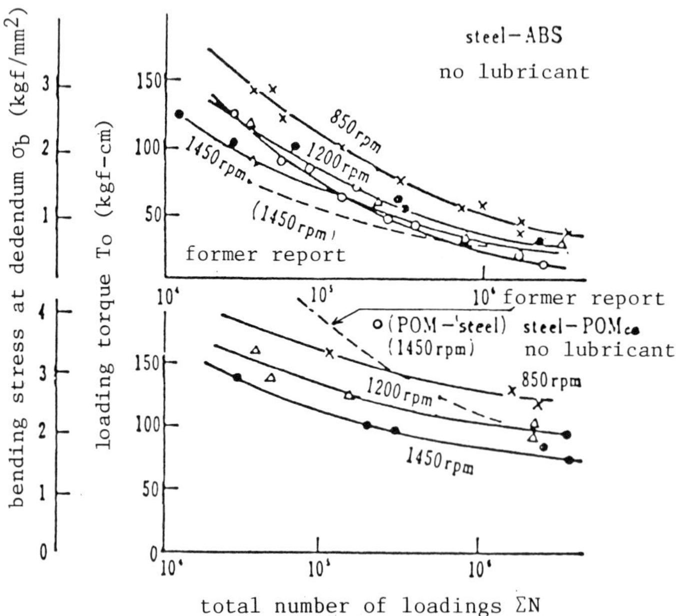

Fig. 4.125 Relationships between σ_b and ΣN for steel:ABS and steel:POM gears

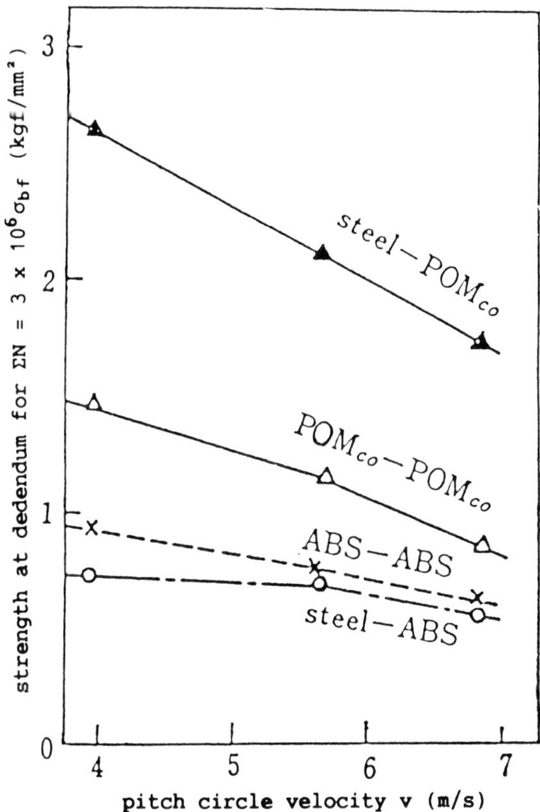

Fig. 4.126 Relationships between fatigue bending strength, σ_b, and pitch circle velocity for various combinations

According to equation (4.16), the allowable stress on a gear tooth $\bar{\sigma}_b$ depends on the pitch circle velocity, v; that is, it decreases with an increase of the constant a. Figure 4.126 shows the relationships between the limiting bending stress σ_{bf}, obtained from the limiting face load F_{mf} in equation (4.14), or σ_b in Figs. 4.124 or 4.125, and the pitch circle velocity, v. Figure 4.127 shows the relationships between the endurance limit ratio α (the ratio of the limiting bending stress σ_{bf} to the static bending strength σ_{bo}) for each plastic gear, and the pitch circle velocity, v. These relationships are represented by the following equation:

$$\alpha = \frac{\sigma_{bf}}{\sigma_{bo}} = \alpha_o - mv \qquad (4.17)$$

where α_o is the value of α at v = 0, and m is a constant depending on the type of material.

The values of σ_{bo}, α_o and m, obtained from the curves in Figs. 4.126 and 4.127 are presented in Table 4.46.

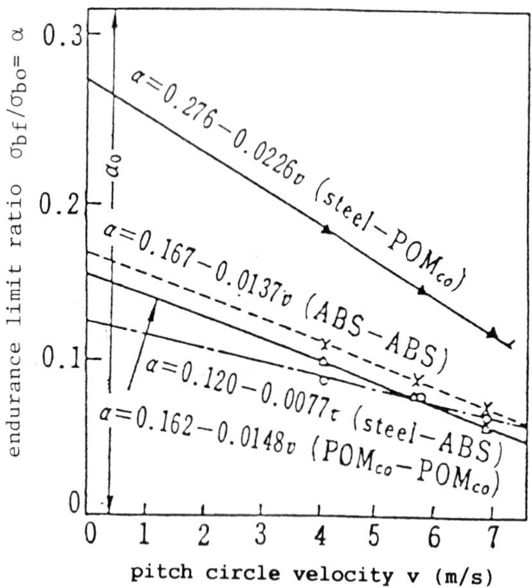

Fig. 4.127 Relationships between the endurance limit ratio, α, and pitch circle velocity for various combinations

TABLE 4.46
Running characteristics values

Gear Combination	σ_{bo} (kgf/mm^2)	α_o	m	a(m/s)	μ_k[82]
S-POMco	14.3	0.276	0.023	1.44	0.155+
POMco-POMco	14.3	0.163	0.015	1.88	0.177+
S-ABS	8.5	0.120	0.008	4.19	0.371
ABS-ABS	8.5	0.167	0.014	1.23	0.180

TABLE 4.47 Running characteristics [72]

Combination	Transmission Efficiency at F=1.5 kgf/mm, η	Average Effici- ency η_m	Max.Load at Tooth Face F_{mo}	Endurance Load for ΣN=2x106 F_{mf}	Static Flexural Strength σ_{bo}	Kinetic Friction Coeffic- ient μ_k	Estimated Critical Temper- ature T_m	α F_{mf}/F_{mo}	C $(=F_a/F_s)$
	%	%	kgf/mm	kgf/mm	kgf/mm²		°C		
S-POM	95.2	95.4	7.1	3.1		0.155		0.44	0.203
POM moulded-	94.9	95.3	5.5	1.0	14.3	0.177	150°C	0.18	0.064
POM hobbed	93.4	94.8	8	-		-		-	-
S-ABS	95.9	96.5	5.1	0.6		0.371		0.12	0.066
ABS-ABS	95.9	97.6	1.8	0.6	8.5	0.180	80-90	0.33	0.066
S-PU	86.2	87.4	2.4	-	-	-	80-100	-	-
PU-PU	93	92.1	2.0	-	-	-		-	-
S-PP	92	91.2	3.0	0.65		0.308		0.21	0.094
PP-PP	94	62.6	2.65	0.50	6.5-7.0	0.350	100-120	0.19	0.073
(2.5%C)PP-(")PP	90.5	-	3.0	0.55		-		0.18	-
S-PE	94	93.8	3.3	0.5		0.124		0.15	0.13
PE-PE	96	93.9	2.5	0.4	3.4	0.071	100-110	0.16	0.100
S-M	92	-	1.6	0.4		0.627		0.25	0.027
M-M	92.5	-	3.1	1.5	14.0	0.071	130	0.49	0.100
(10%PTFE)M-(")M	93	-	4.3	1.5	9.0	-		0.35	-

S: hobbed steel gear (S45C), POM: polyacetal, PU; polyurethane (Elastlan E1095), PP: polypropylene (Noblem JH-M), PE: polyethylene (Hizex,1100J), M: melamine resin (MM50), (10% PTFE)M: melamine composites filled with 10% PTFE

The values of the velocity factor "a" in equation (4.16) are between 1.23 - 4.19, as shown in Table 4.46, and are much lower than the steel gear transmission value of 6. This indicates that the rate at which the allowable bending stress decreases with speed for plastic gears is greater than that for the steel gears.

Various running characteristics for different gear combinations are shown in Table 4.47. They include: power transmission efficiency at 1.5 kgf/m of face load, η; average transmission efficiency, η_m; maximum static face load, F_{mo}; endurance face load at $\Sigma N=10^6$, F_{mf}; static bending strength, σ_{bo}; kinetic friction coefficient, μ_k; estimated critical temperature, T_m; endurance limit ratio, α; and strength ratio of plastic to steel gear, C; for each combination of gears.

4.3.5 *Wear*

The tooth face of a gear is worn away adhesively, depending on the relative sliding between it and the opposing gear tooth face under a load due to the transmitting power. However, it is very difficult to estimate the wear of a gear accurately, because the face load and relative sliding speed between two faces is variable, depending on the position of the contact area. In this section, summaries of some reports [72,73,74] are described.

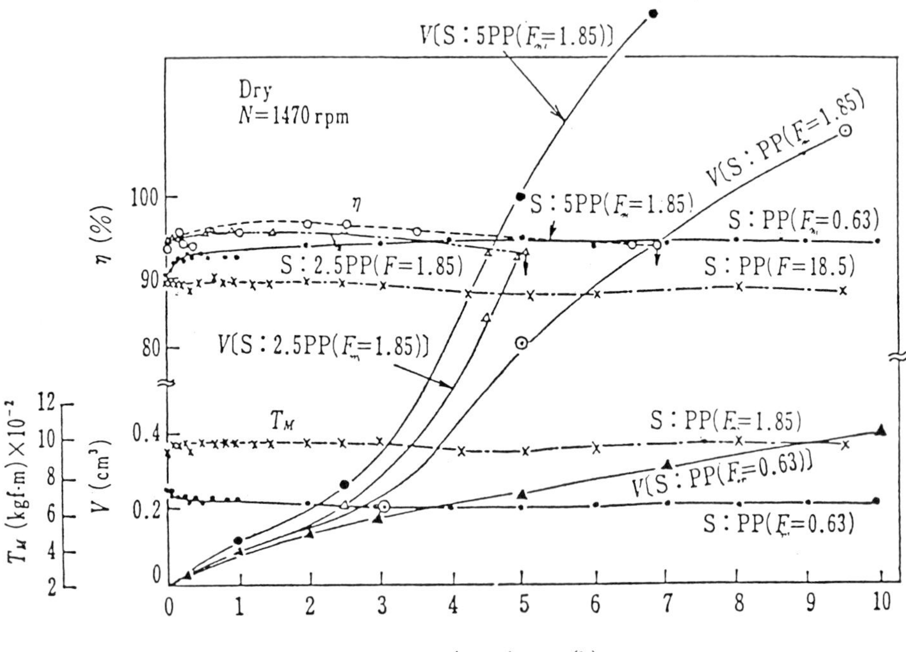

Fig. 4.128 Relationships between wear volume of the plastic gear, V, T_m or η and running time, t, for combinations of steel:PP gears

Figures 4.128 to 4.131 show the relationships between the power transmitting efficiency η (%); the mechanical loss torque T_M (kgf-m); the adhesive wear volume of a plastic gear V (cm^3), and running time t (h), for various combinations of gears.

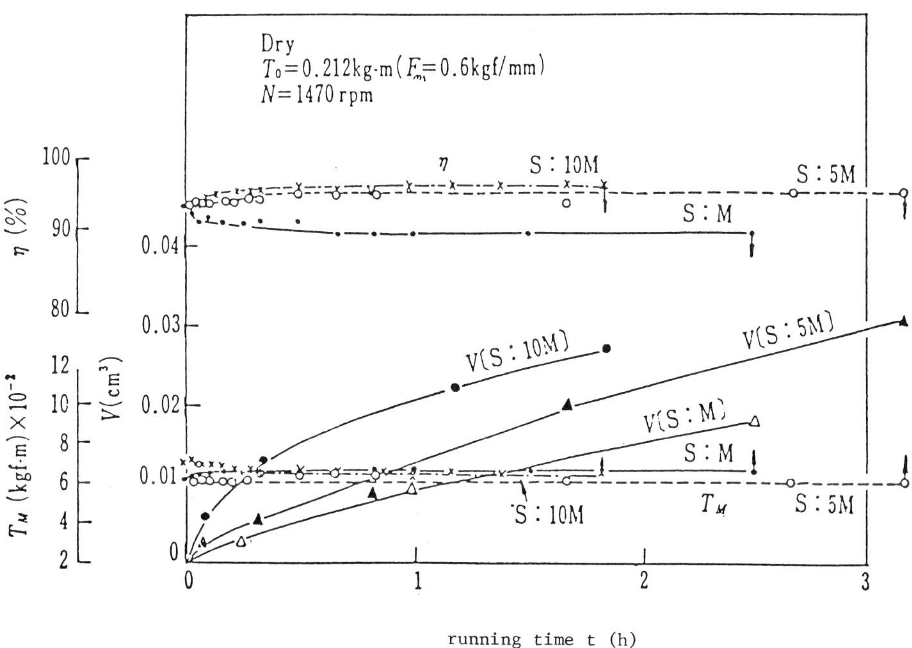

Fig. 4.129 Relationships between wear volume of the plastic gear, V, T_m or η and running time, t, for combinations of steel:MF gears

Figure 4.128 is for the combination of a PP gear and a steel one (S) at different face loads without lubricant (dry), and Fig. 4.129 is for the combination of each steel gear (S) with various melamine resin gears (M) at a face load of 0.6 kgf/mm and again without lubricant. Figure 4.130 shows similar relationships for the combinations of various PP gears at a face load of 0.63 kgf/mm without lubricant. Figure 4.131 shows the same relationships for combinations of various melamine resin gears (M, 5M, 10M) at 1.85 kgf/mm of F_m under similar conditions.

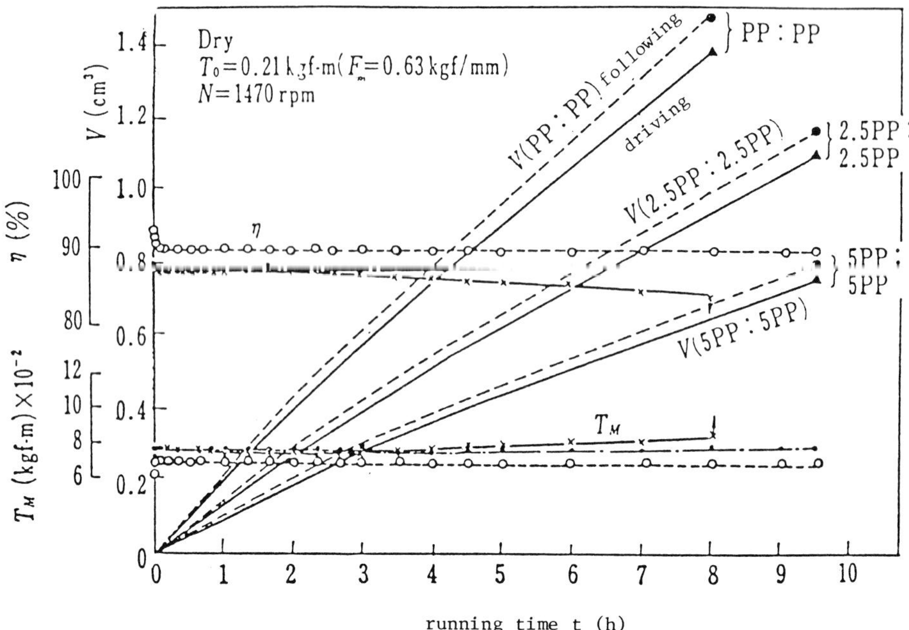

Fig. 4.130 Relationships between V, T_m or η and running time for combinations of PP:PP gears

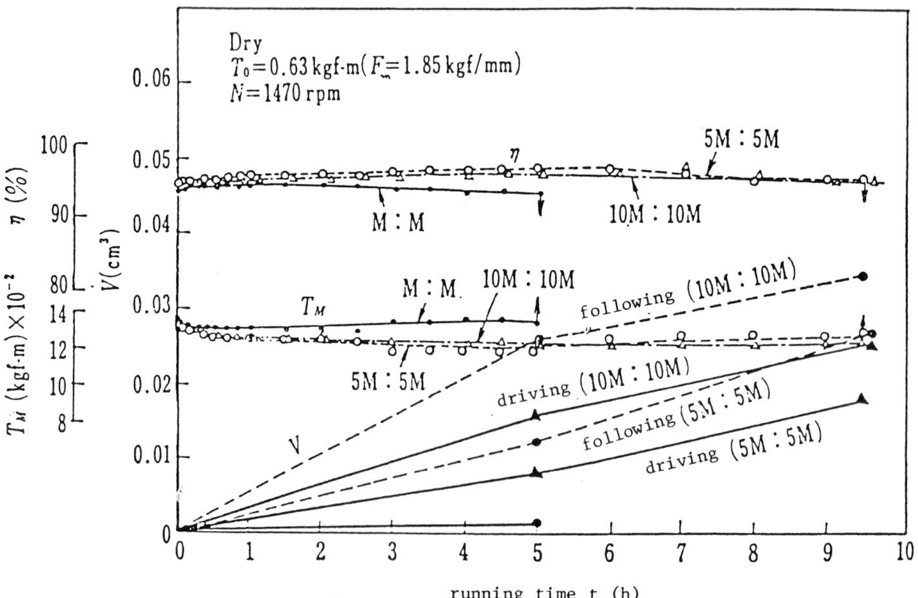

Fig. 4.131 Relationships between V, T_m or η and running time, t, for combinations of MF:MF gears

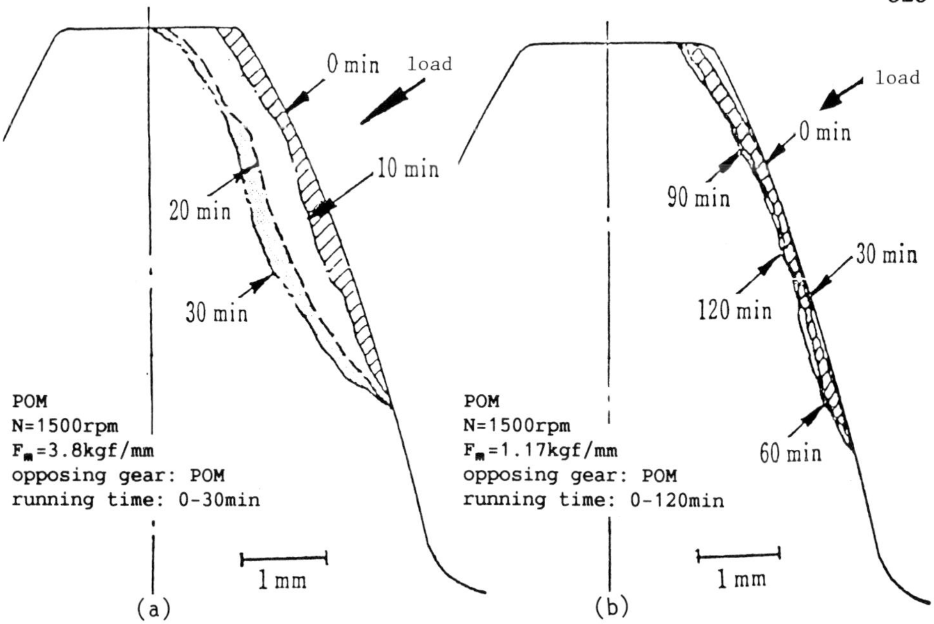

Fig. 4.132 Profile changes of POM gear teeth due to wear

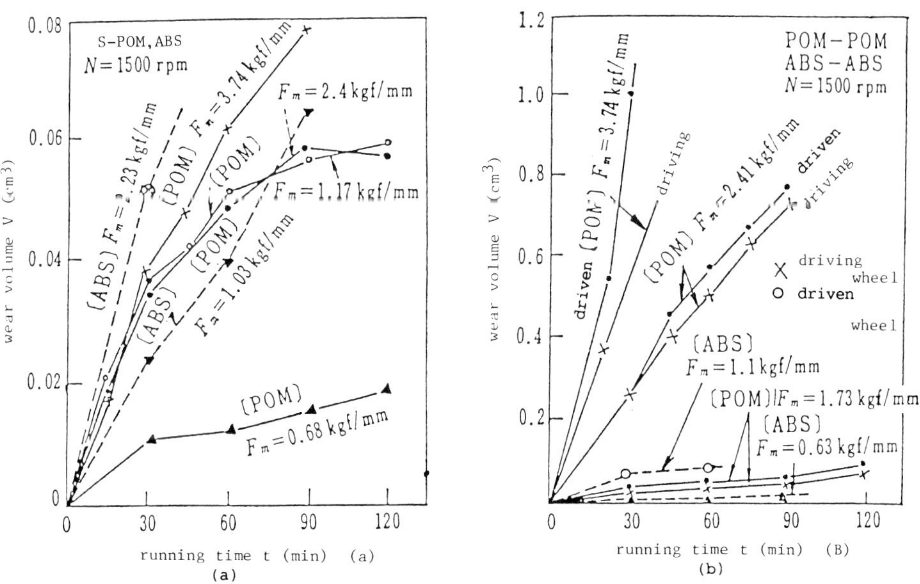

Fig. 4.133 Relationships between V, and running time at different face loads, F_m, and rotating speeds, N, for combinations of steel:ABS or steel:POM gears (a), and POM:POM or ABS:ABS gears (b)

Figure 4.132 shows the worn tooth profiles of a polyacetal (POM) gear from a POM/POM combination after various running times at 1500 rpm and face loads, F_m, of 3.8 kgf/mm and 1.17 kgf/mm. It can be seen that the part which experiences a large relative sliding velocity is severely worn. Figure 4.133(a) shows the variation of wear volume for POM (full line) and ABS (dotted line) gears V with running time t at various face loads F_m without lubricant for the combinations of a POM or an ABS gear with a steel (S) one. Figure 4.133(b) shows the similar relationships for a POM/POM combination and one for ABS/ABS depending on whether the gears are driving or being driven.

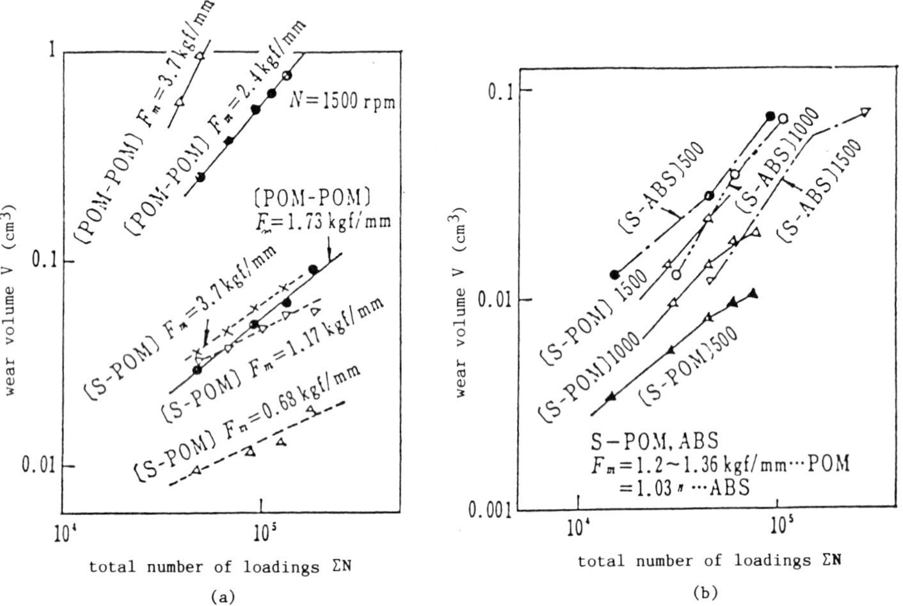

Fig. 4.134 Relationships between V and ΣN at (a) different face loads, F_m, or (b) different rotating speeds, N, for various combinations of gears

Figure 4.134(a) shows the relationship between the wear volume of POM gear V and the total number of loadings ΣN at 1500 rpm at different face loads F_m for combinations of POM/POM gear and POM/steel (S) gears. Figure 4.134(b) shows the similar relationships at various speeds for combinations of POM/steel (S) and ABS/steel (S) gears.

Figure 4.135(a) shows the relationships between the wear volume of plastic gear V and running time t at speeds of 500 rpm, 1000 rpm and 1500 rpm, and different face loads F_m for the combinations of steel (S)/POM and steel/ABS gears. Figure 4.135(b) shows similar relationships for POM/POM, and ABS/ABS gear combinations and for both the driving gear (full line) and the driven one (dotted line). It is clear from the figures that the wear volume, V, for both plastic gear combinations is generally larger than that of plastic/steel gear combinations, increases with increasing face load F_m and speed N (rpm), and is slightly larger for the driving gear than for the driven one.

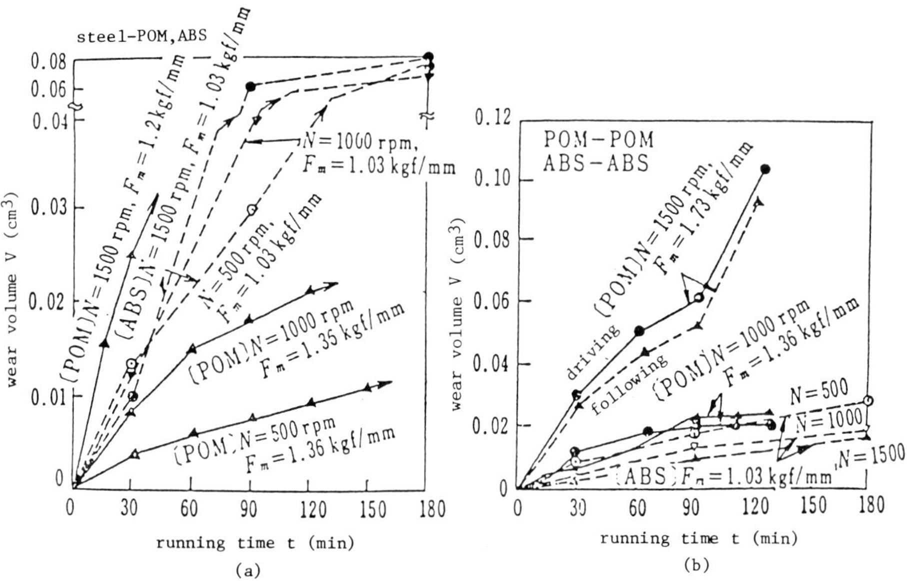

Fig. 4.135 Relationships between V and running time, t, at (a) different face loads, F_m, and (b) different rotating speeds, N, for various combinations of gears

Figure 4.136 (a) shows the relationships between the specific wear volume of the gear, V_s ((cm^3/contact/kgf/mm), wear volume per unit contact per unit face load) and graphite content for different combinations of PP/PP gears and PP/steel (S) gears. This figure indicates that the V_s is much greater for PP/PP gears than for PP/steel ones, increases greatly with increasing face load F_m and, for PP/PP at the lowest value of F_m, decreases with increasing graphite content. However, the filler has no effect on PP/steel gears. Figure 4.136(b) shows similar relationships between V_s and the PTFE content of combinations of PTFE-filled, melamine resin gears (M) and M/steel (S) gears. It indicates that the V_s values for the PTFE-filled melamine resin gears (M) are smaller than that of M/S gear combinations and increases with increasing PTFE concentration.

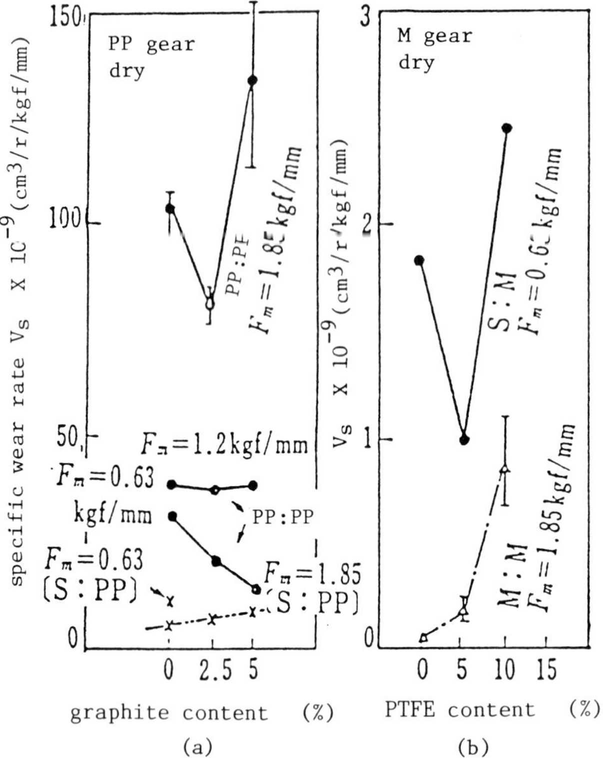

Fig. 4.136 Relationships between specific wear volume V_s and filler content (no lubricant)

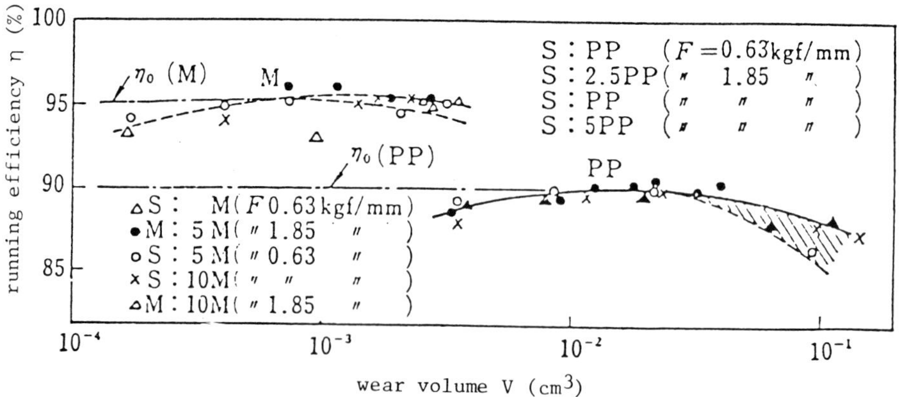

Fig. 4.137 Relationships between running efficiency η and the wear volume of gears for various combinations

Figure 4.137 shows the relationships between the relationships between the power transmission efficiency η and the wear volume for different combinations of melamine resin gears (M), PP gears and steel gears (S), where 2.5PP or 5PP refers to 2.5% or 5% graphite-filled polypropylene (PP) and 5M or 10M to 5% or 10% PTFE-filled melamine resin (M). It is clear from this figure that the value of η is only affected to a small extent by wear of the gear, as was also shown previously by Figs. 4.128 to 4.131.

4.4 PACKINGS

4.4.1 *Significance and classification*

A sealing device or a packing is used to prevent the leakage of substances such as oil, water, air and gas from containing vessels or from between moving surfaces. Recently, due to the development of fluid drives in various machines, higher performance sealing devices are required to withstand the higher temperatures and pressures, and improvements in packing materials are also necessary.

Sealing devices can be classified as gaskets, which seal water, for example, in static conditions, or as packings, which seal fluids at positions where there is relative motion, such as reciprocation, rotation or spiralling. Packings are also classified into the simple compression type, which seals a liquid by gland packing, and the self-sealing type, such as the V, L, U, Y and O-ring types, which seal automatically with an increase in pressure. The latter can be further classified into lip packings and compression packings. Mechanical seals are also used [83].

Packing materials can be classified into five types; leather, synthetic gum, cloth-filled synthetic gum, plastics and simple compression types of asbestos, cotton or hemp. Their applications are shown in Table 4.48. Of these materials, gum packings are the most widely used. However, plastic packings become more useful when the applied temperature and pressure are increased. This section details the results of some experiments with gum and plastic packings used under static and reciprocating sliding conditions.

4.4.2 *Experiments to determine the maintenance of side pressure [97]*

Many books on the study of packings have been published, such as: Iwanami [83]; Packing and Gasket Division Report [84]; Smoley [85]; Dunkle [86]; Watanabe [87]; there are also various research reports [88-96].

When a compressive force is applied to a static gland packing, the side force gradually decreases with time. This section examines this phenomenon and compares it with that of stress relaxation.

TABLE 4.48
Types of packing materials and their applicable ranges [83]

Types of Packing	Maximum Pressure Resistance (kgf/cm²) Air	Water	Mineral Oil	Range (°C)
Leather	20	500	1000	-70 - +100
Synthetic gum	50	500	500	-60 - +250
Cloth filled synthetic gum	50	500	850	-20 - +250
Plastic	-	500	500	-20 - +200
Simple compressed (asbestos, cotton, hemp)	-	-	-	-40 - +500

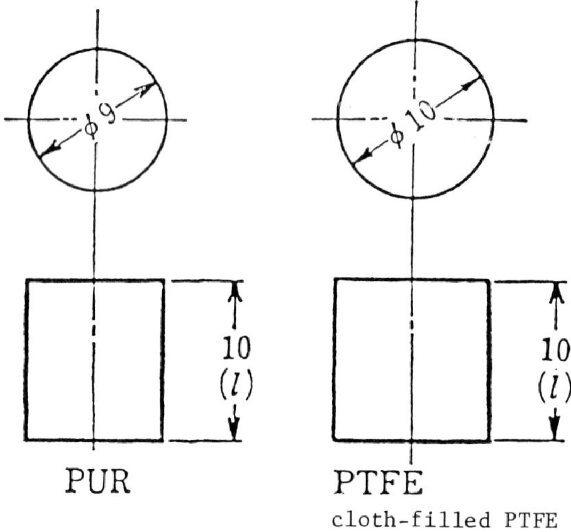

Fig. 4.138 Cloth-filled specimen

(i) *Experimental apparatus*

Figure 138 shows the shape and dimensions of the specimens used for compression, compressive fatigue and stress relaxation tests. Figure 4.139 shows the sections of the four kinds of packings, L, M, H and U types, used in the experiments. Details of the materials and manufacturing processes are shown in Table 4.49. Figure 4.140 shows an example of an adapter for a V packing. The test apparatus is shown in Figs. 4.141 and 4.142 and uses strain gauges to measure the pressure applied normal to the cylinder wall and the side pressure, p, which is caused by the generation of an outer axial compressive force F.

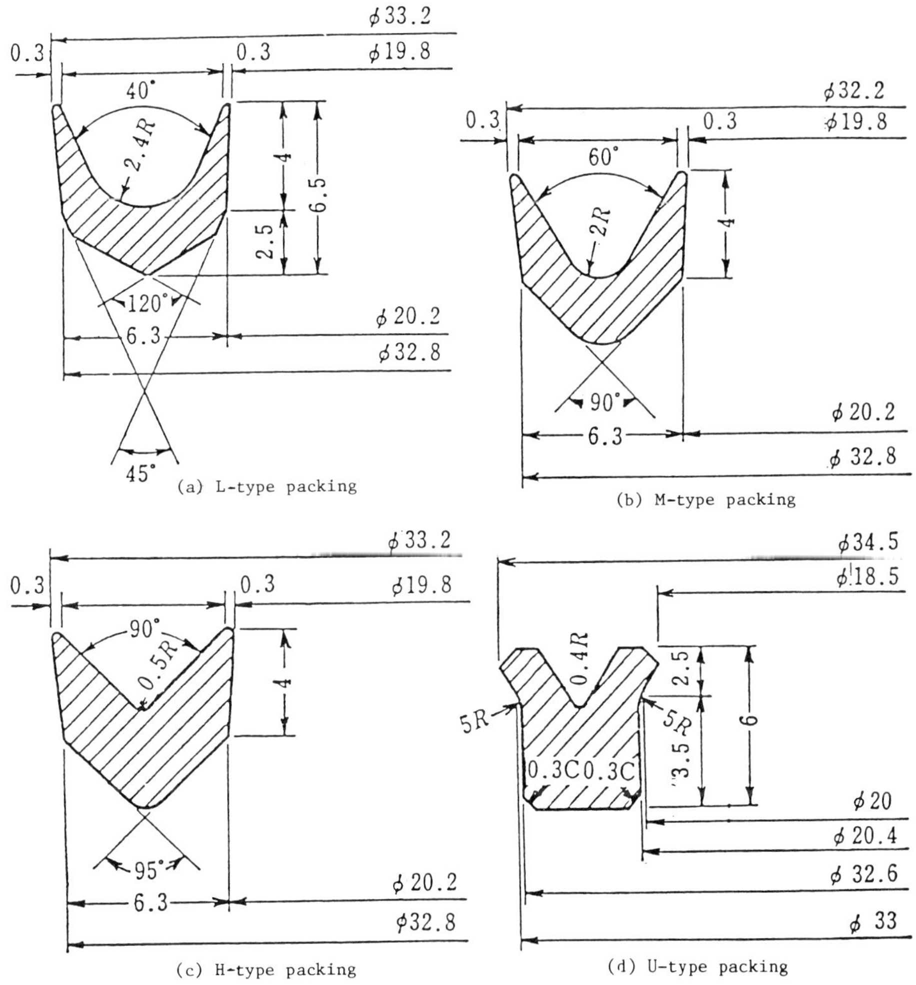

Fig. 4.139 Sections of different packing types

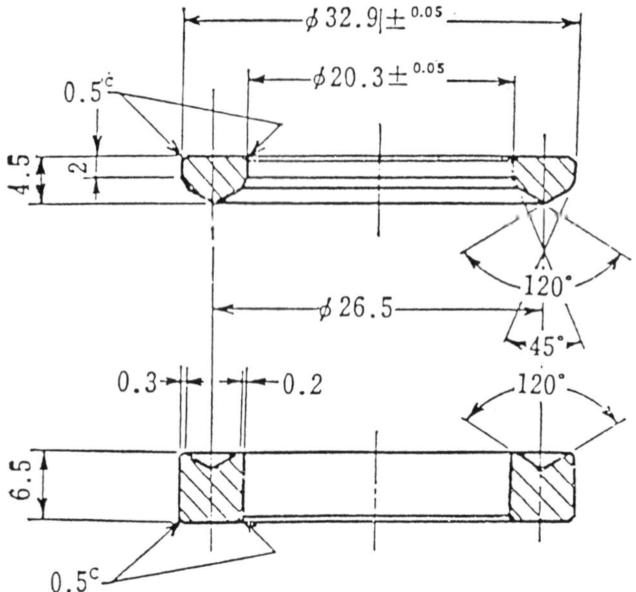

Fig. 4.140 V packing adapter (L type)

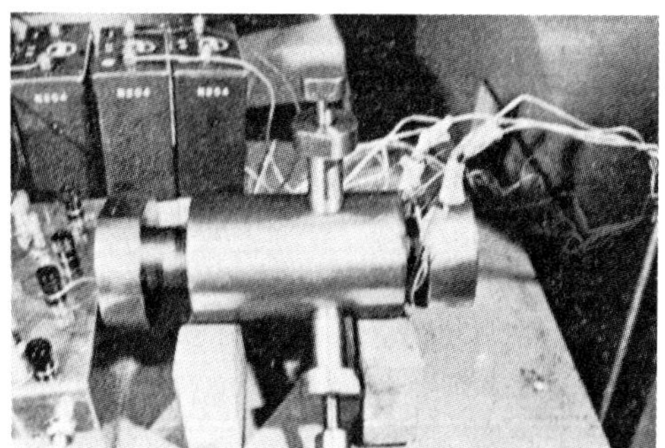

Fig. 4.141 Packing test apparatus

(ii) *Results*

(1) *Compressive characteristics.* Figure 4.143 shows the compressive stress-strain curves of cylindrical specimens (shown in Fig. 4.138) of polytetrafluoroethylene (PTFE), cloth-filled PTFE and polyurethane (PUR) at temperatures of 20°C, 60°C or 100°C. The compressive yield stresses of

these materials at each temperature, obtained from the above curves, are shown in Table 4.50. The relationships between yield stress, σ_y, and temperature, τ, are shown in Fig. 4.144.

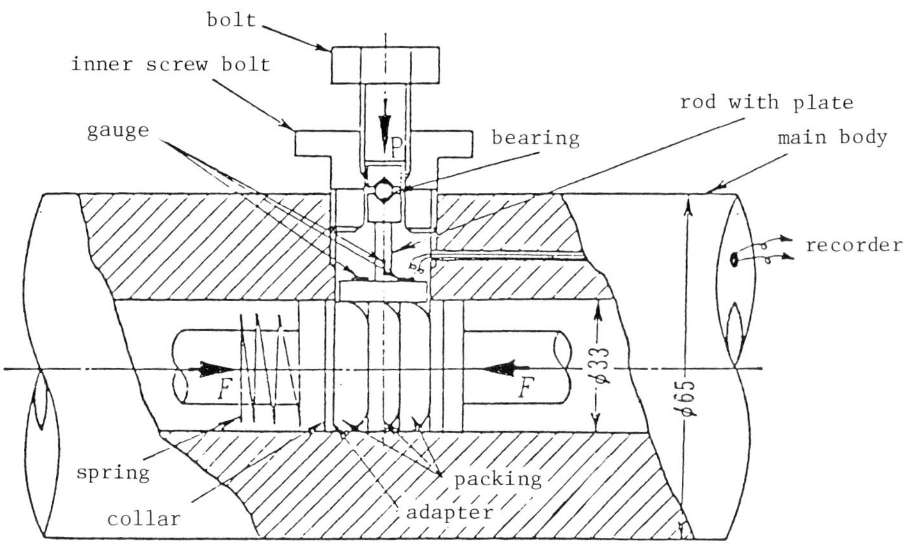

Fig. 4.142 Details of packing test apparatus

TABLE 4.49
Types of packing and moulding methods

Type	Material	Moulding Method	Inside Diameter (mm)	Outside Diameter (mm)	Production
U	PUR	Compression moulding	20	34.5	
V L,M,H type*	PTFE	Extrusion**	20	33	N.Co.
V	Cloth filled PTFE	Compression moulding	20	33	

*L type: low pressure; M type: medium pressure; H type: for high pressure
**Machine-finished after extrusion

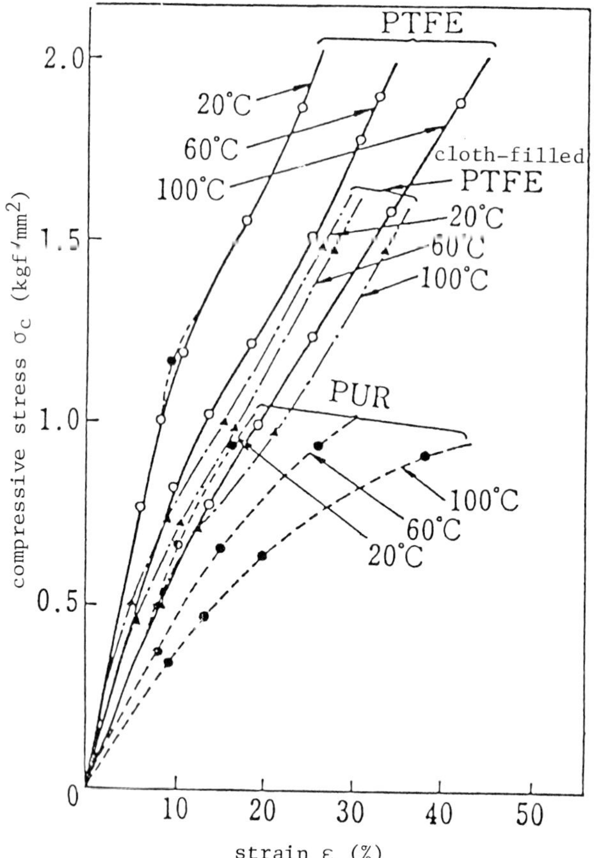

Fig. 4.143 Relationships between σ_c and strain ε for each plastic at different temperatures

TABLE 4.50
Compressive yield stresses of cylindrical specimens

Temperature (°C)	PUR	Material PTFE	Cloth-filled PTFE
20	0.85 kgf/mm^2	1.16 kgf/mm^2 (1.19 kgf/mm^2)	0.65 kgf/mm^2
60	0.69 "	0.88 "	0.58 "
100	0.52 "	0.63 "	0.49 "

( ): value obtained from another report

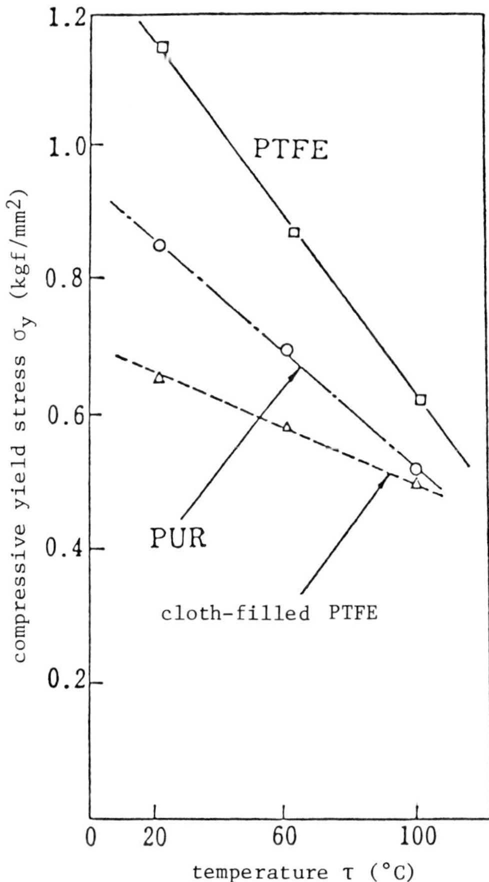

Fig. 4.144 Relationships between compressive yield stress, σ_y and temperature, τ, for each plastic

(2) *Compressive stress-relaxation characteristics.* Figure 4.145 shows the relationships between the compressive stress-retaining rate σ/σ_o (where σ_o is the initial stress, and σ is the stress at time t) and time t after loading, i.e. the compressive stress-relaxation curve. These results are for a PTFE cylindrical specimen at two initial compressive stresses σ_o and three temperature, 20, 60, or 100°C. Figure 4.146 shows similar stress-relaxation curves for a cloth-filled PTFE specimen, and Fig. 4.147 for polyurethane (PUR).

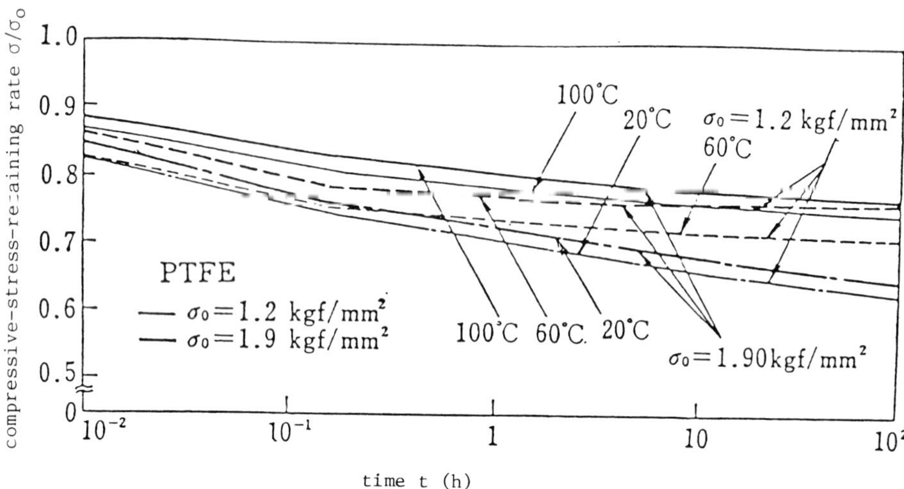

Fig. 4.145 Compressive stress relaxation curves of PTFE cylindrical specimens at two initial stress σ_o and 20, 60 or 100°C

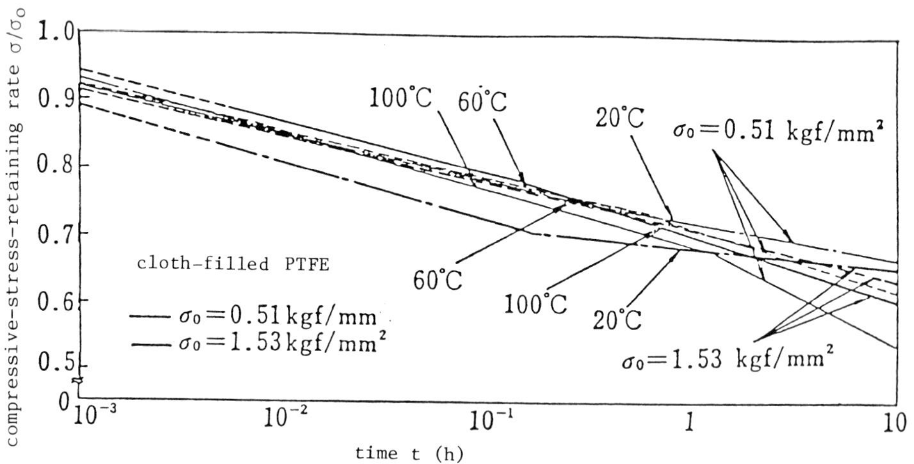

Fig. 4.146 Compressive stress relaxation curves of cloth-filled PTFE at two initial stresses σ_o and 20, 60 or 100°C

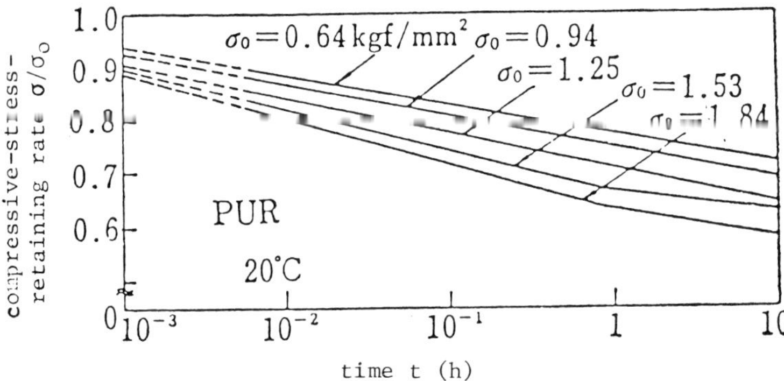

Fig. 4.147 Compressive stress relaxation curves for PUR specimens at 20°C and different initial stresses σ_o

From these figures, the following equations are obtained:

$$\frac{\sigma}{\sigma_o} = 1 - n \cdot \log t \tag{4.17}$$

or

$$\frac{\sigma}{\sigma_o} = \frac{\sigma_1}{\sigma_o} - n \cdot \log \frac{t}{t_1} \tag{4.18}$$

where n is defined as a "stress-decreasing index" which represents the oblique degree of the stress-relaxation curve or stress changing rate ($\Delta\sigma/\sigma_o$) per $\log_{10}t$, and σ_1 is the stress at t_1 (1 minute or 1 hour) after loading. This indicates that the smaller the n value, the higher will be the compressive-side-pressure-retaining rate. The n values for each material at each temperature is shown in Table 4.51.

(3) *Compressive fatigue characteristics.* Figure 4.148 shows the relationships between the dimension-retaining rate ℓ/ℓ_o (ℓ_o is the initial length and ℓ is the length after ΣN repeated loads) and the total number of repeated loads ΣN for a PTFE specimen at various compression stresses σ_c. This figure indicates that the smaller the compressive stress, σ_c, the higher is the dimension-retaining rate for the same number of repetitions. In these curves, there are deflection points which are marked by x in Fig. 4.148 or σ_c' in Fig. 1.49(a). These deflection points may be interpreted as points for fatigue. Similar relationships can be found for cloth-filled PTFE

and polyurethane (PUR) specimens. Figure 4.149 shows the relationships between the number of repeated loads at the yield point, N_y, and the repeated compressive stress σ_c for PTFE and cloth-filled PTFE specimens.

TABLE 4.51
Stress decreasing index value, n

Material	Temperature 20°C	60°C	100°C
Cylindrical Specimen:			
PTFE	0.069	0.073	0.090
Cloth Filled PTFE	0.082	0.097	0.117
PUR	0.071	0.078	0.161

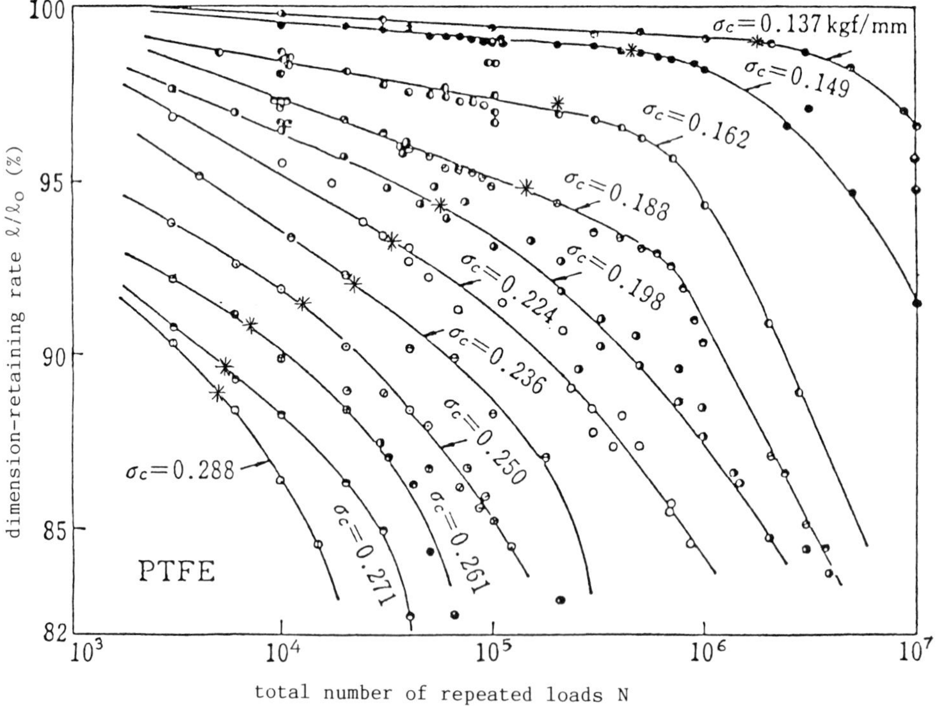

Fig. 4.148 Relationships between dimension-retaining rate ℓ/ℓ_o, and total number of repeated loads for PTFE specimens at various compressive stresses, σ_c

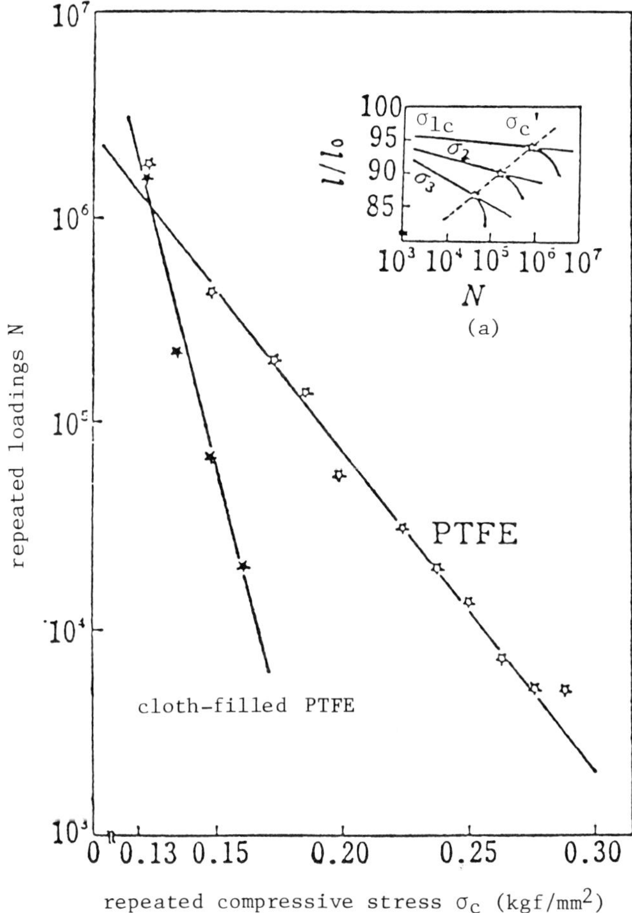

Fig. 4.149 Relationships between the total number of repeated loadings, N, and the repeated compressive stress at the deflection point

The following equation is obtained from this figure:

$$N_y = N_o \cdot e^{-m \cdot \sigma_c} \tag{4.19}$$

where N_o is the value of N_y at σ_c=0 (extrapolated) and m is the decrease index per unit change of repeated compressive stress σ_c. The value of log N_o is 8.18 for PTFE and 14.8 for cloth-filled PTFE, and the value of m is 16.38 for PTFE and 62.5 for cloth-filled PTFE.

(4) *Packing side-pressure relaxation characteristics.* Figure 4.150 shows the relationship between the compressive load F in the thickness or axial direction and the side pressure in the direction normal to the cylinder wall, expressed as the side expansion force P, which was measured experimentally using the packing test apparatus shown in Fig. 4.142. This relationship is represented by the following equation:

$$P = \alpha \cdot F \tag{4.20}$$

where α is a constant of approximately 1/10.

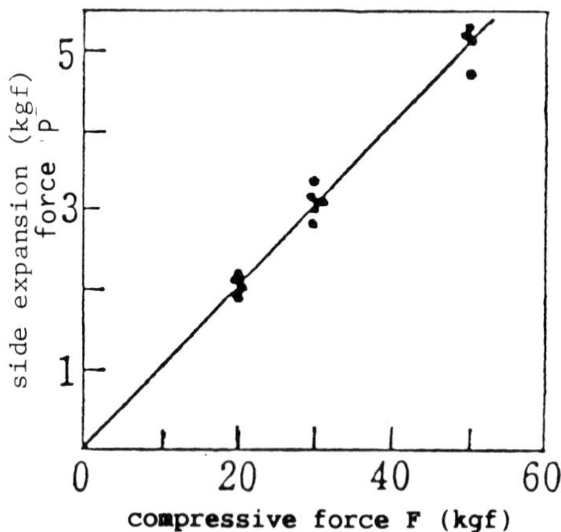

Fig. 4.150 Relationships between side expansion force, P, and compressive force, F

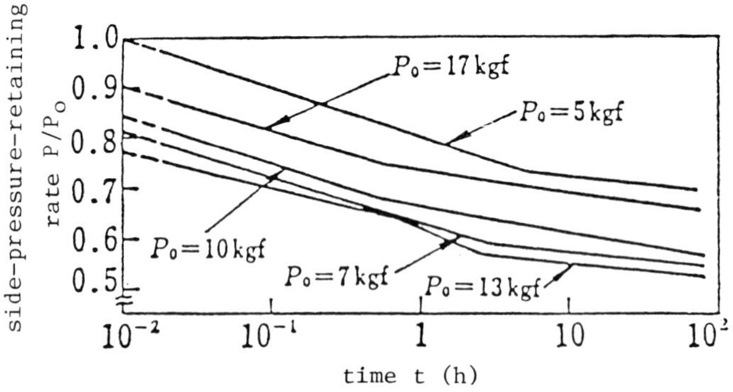

Fig. 4.151 Relationships between side-pressure-retaining rate, P/P_o, and time for different compressive forces, P_o for L type PTFE V packing

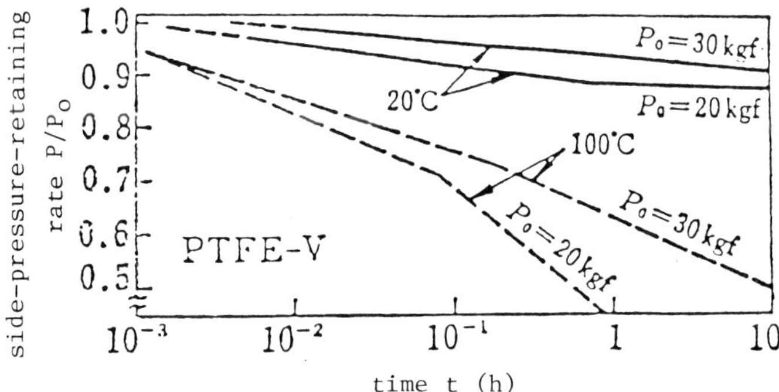

Fig. 4.152 Relationships between P/P_o and time for a PTFE M-type V packing at 20°C or 100°C and different initial compressive forces, P_o

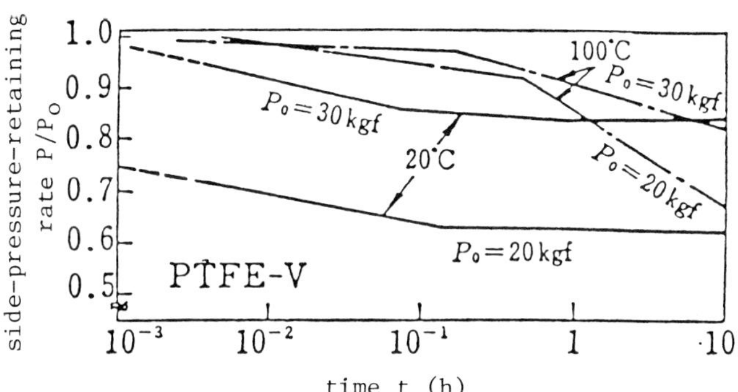

Fig. 4.153 Relationships between P/P_o and time, t, for a cloth filled PTFE V packing (3 sheets) at different initial compressive forces, P_o

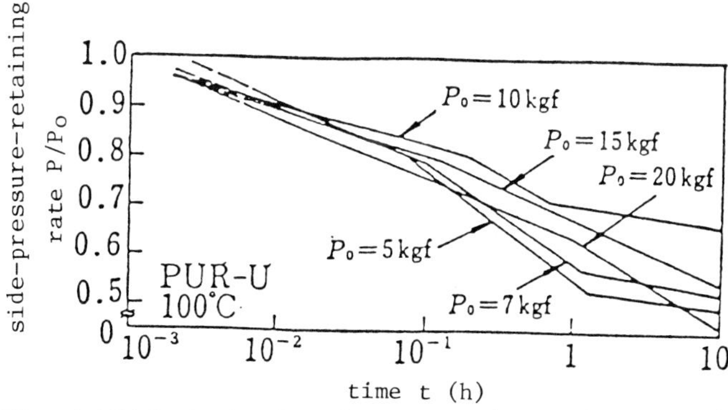

Fig. 4.154 Relationships between P/P_o and time, t, for a PUR U packing at 100°C and different initial compressive forces, P_o

Figures 4.151 to 4.154 show the relationships between the side-pressure-retaining rate P/P_o (P is the side pressure at time t and P_o is the initial side pressure) and duration of pressure t for different initial side pressures. Figure 4.151 is for a PTFE V packing (L type, 3 sheets) at 20°C; Fig. 4.152 is for a PTFE V packing (M type, 2 sheets) at 20°C and 100°C; Fig. 4.153 is for a cloth-filled PTFE V packing at 20°C and 100°C, and Fig. 4.154 is for PUR V packing at 100°C.

The following equations are obtained from these curves;

$$\frac{P}{P_o} = 1 - \bar{n}\cdot\log t \tag{4.21}$$

or

$$\frac{P}{P_o} = \frac{P_1}{P_o} - \bar{n}\cdot\log\frac{t}{t_1} \tag{4.22}$$

The values of $\bar{n}$ are obtained from the above figures and are given in Table 4.52.

TABLE 4.52
Values of the side-pressure-retaining rate index, $\bar{n}$

Material	Temperature		
	20°C	60°C	100°C
Packing:			
PTFE L Type (2 sheets)	0.031	0.021	0.069
PTFE M Type (2 sheets)	0.032	0.026	0.189
PTFE H Type (2 sheets)	0.023	0.015	0.131
Cloth-filled PTFE	0.010	0.098	0.126
PUR	0.090	0.105	0.180

Equations (4.21) and (4.22) are similar to the earlier equations (4.17) or (4.18), representing the compressive stress relaxation characteristics. The differences between n and $\bar{n}$, shown in Tables 4.51 and 4.52 respectively, thus indicate the degree of similarity between the compressive stress relaxation characteristics and the side-pressure-retaining ones. The relationships between n, $\bar{n}$ and temperature, τ, for each packing are shown in Fig. 4.155 (n-τ (thick lines) and $\bar{n}$-τ (thin lines)). For PTFE the values of n and $\bar{n}$ range from 0.03 - 0.075 and 0.025 - 1.8 respectively. In general, n is greater than $\bar{n}$ at 20°C and 60°C; however, both increase at high temperatures (100°C) and become about equal. The n and $\bar{n}$ values for PUR range from 0.07 - 0.16 and 0.08 - 0.18. Generally, n is slightly smaller than $\bar{n}$. In cloth-filled PTFE, both n and $\bar{n}$ values range from 0.075 - 0.13. In other words, the side-pressure-retaining characteristics are very similar to the compressive stress relaxation characteristics.

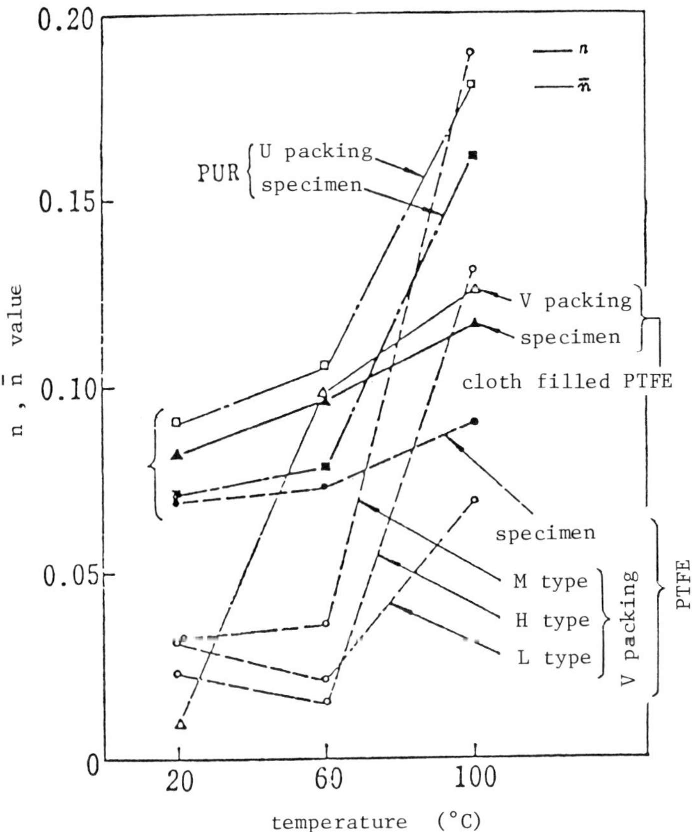

Fig. 4.155 Relationships between the values of n or $\bar{n}$ and temperature for PTFE or cloth-filled PTFE and PUR cyclindrical specimens or packing

4.4.3 *Results of experiments to determine the endurance of sliding packings [98]*

The main problems associated with packings in static conditions are the deterioration of packing material with time and a decrease in the side pressure as a result of stress relaxation. With sliding packings, however, in which there is relative motion between the packing and a counter-surface material, the problems of friction, frictional heat and wear must be taken into account. In this section, experimental results are described [98] from studies of a sliding packing between a piston and a cylinder.

(i) *Experiments*

The packings used in the experiments were: circular with a Y-type section and made from two types of material; Eskid rubber (GLY-30, S.Co., A type); and Looblan (PGY-40, S.Co., B type). Their shapes and dimensions are shown in Fig. 4.156. The composition and dimensions of the piston, cylinder and packings in the test apparatus are shown in Fig. 4.157, and its driving arrangement is shown in Fig. 4.158. The tensile properties of the packings were measured using a jig, as shown in Fig. 4.159. The tensile strength σ_t was obtained from the equation $\sigma_t = P_{max}/A$, where P_{max} is the maximum tensile load and A is a sectional area of the packing; the elongation was obtained from the distance of movement.

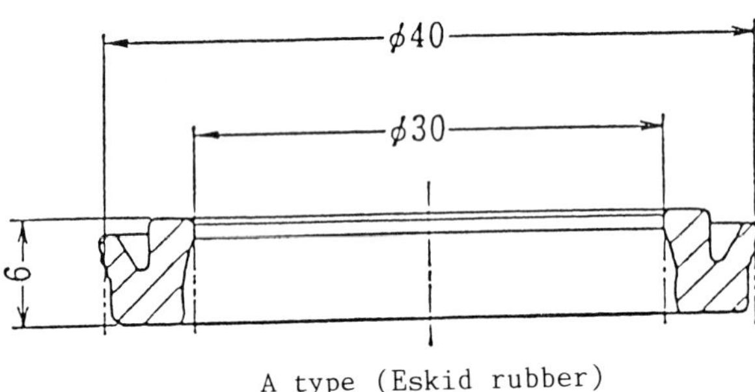

A type (Eskid rubber)

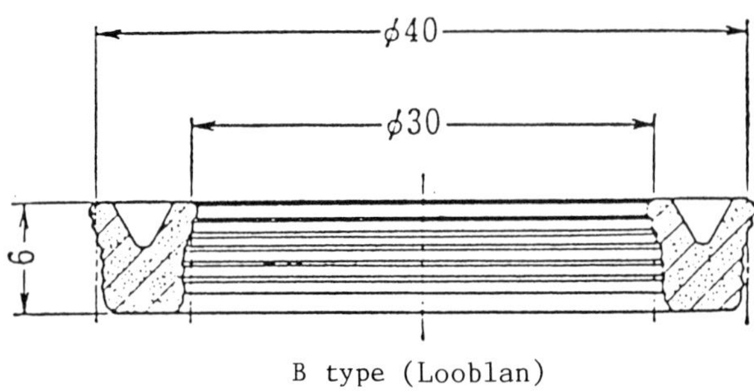

B type (Looblan)

Fig. 4.156 Shape and size of packing

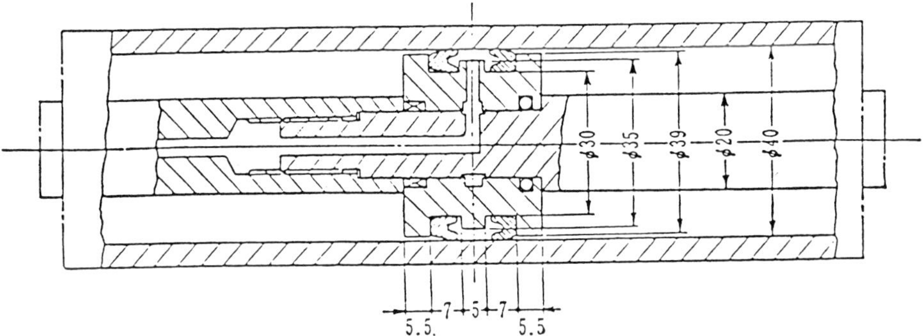

Fig. 4.157 Dimensions of piston and main part of the cylinder

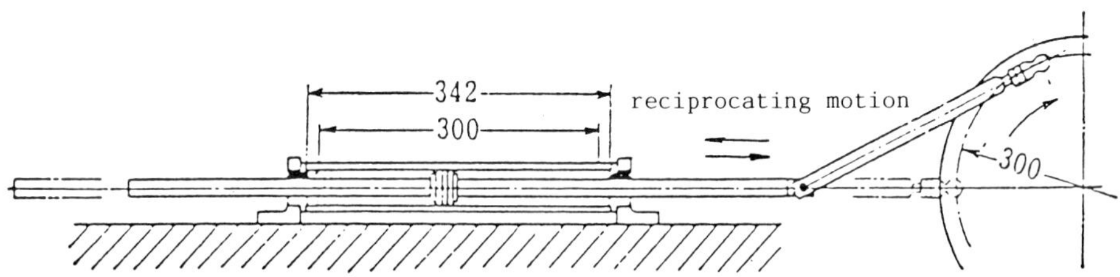

Fig. 4.158 Driving mechanism for sliding test

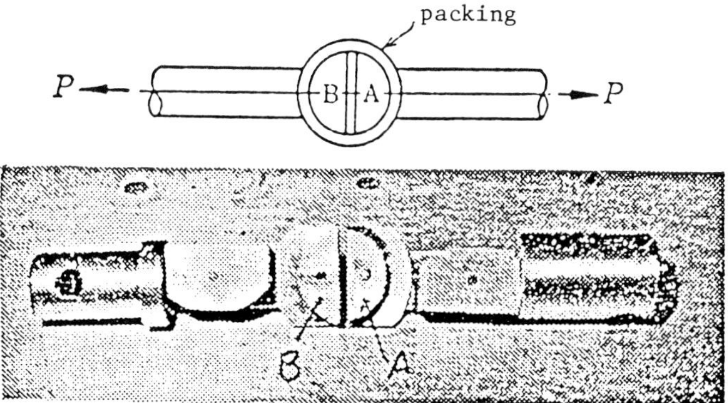

Fig. 4.159 Jig for tensile testing

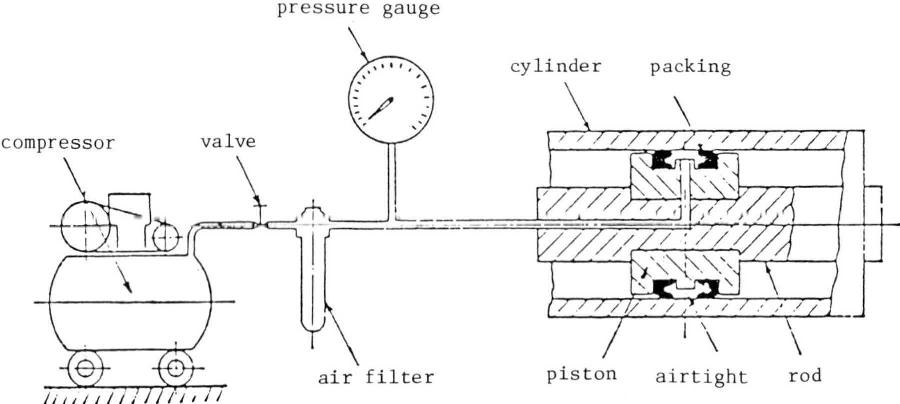

Fig. 4.160 Leakage measuring apparatus

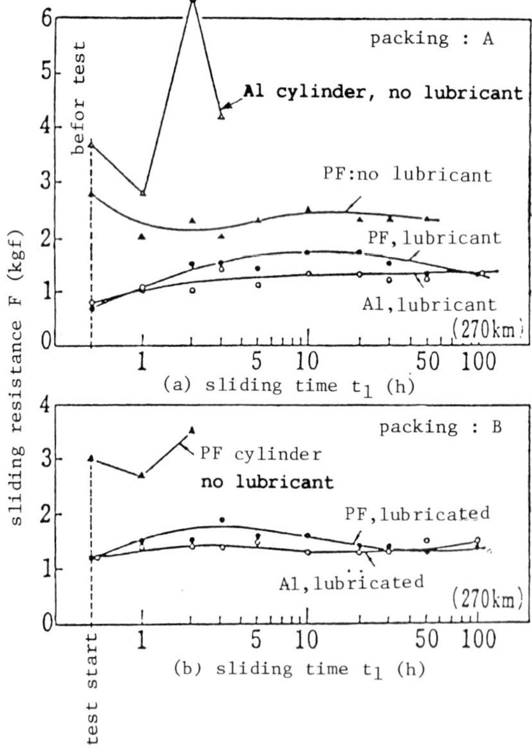

Fig. 4.161 Relationships between sliding resistance, F, and sliding time, t_1, with or without lubricant for each combination of an Al or PF cylinder and A or B packings

The measurement of leakage from the sealed part was obtained from the change of gauge pressure, using the apparatus shown in Fig. 4.160. The reciprocating speed of the piston was 75 times per minute, the stroke length was 300 mm and the average sliding velocity was 75 cm/s. Two kinds of material, phenolic (PF) and aluminum (Al) were used in the cylinder.

(ii) *Experimental results*

(1) *Sliding resistance, F.* For each combination of packing and cylinder, Fig. 4.161 shows the relationship between the sliding resistance, F(kg), of the piston moving reciprocating in the cylinder, and sliding time, t_1, for up to 100 hours, or equivalent to a sliding distance of up to 270 km, with and without lubricant. Figure 4.161(a) is for relationships for the A-type packing in both PF and Al cylinders and indicates that the values of F without a lubricant range from 2 - 6 kgf and are much greater than those with lubricant, 0.8 - 1.7. There is little difference between the F values for lubricated cylinders of Al and PF, and the only significant change during 100 hours of running occurs with the unlubricated Al cylinder. Similar relationships were found for the B-type packing, as shown in Fig. 4.161(b).

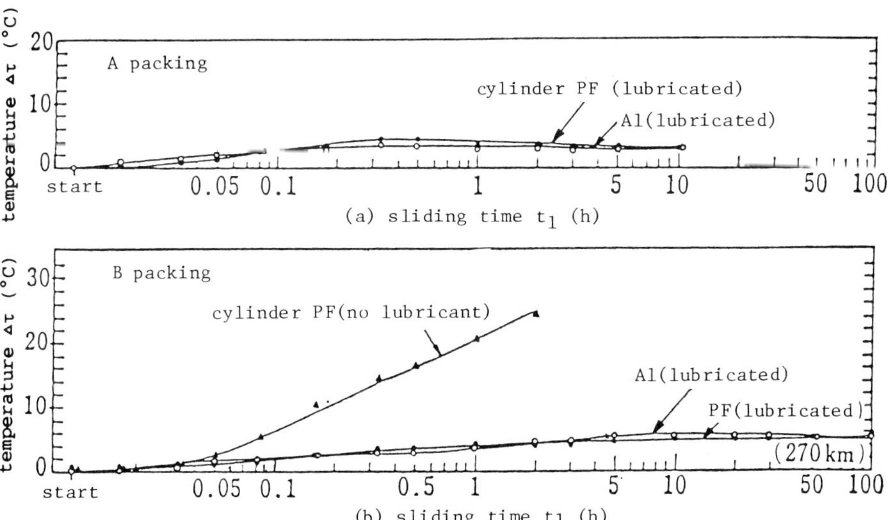

Fig. 4.162 Relationships between the change in temperature, $\Delta\tau$, on the outside surface of the cylinder and sliding time, t_1, using A or B packings with Al or PF cylinders (room temperature = 10°C)

(2) *Temperature of the cylinder wall.* The temperature of the cylinder was measured using a thermosensor attached to its outside surface. The relationships between the temperature rise, $\Delta\tau$, above air temperature and sliding time, t_1, for each combination of packing and cylinder, with or without lubricant, are shown in Fig. 4.162. It is clear that, in general, thermal equilibrium is reached after approximately 1 hour and the maximum temperature rise, $\Delta\tau$, for the combination of an A-type packing under lubricated conditions in both cylinders is only 5°C. However, for the combination of a B-type packing and an unlubricated PF cylinder, $\Delta\tau$ rises to 25°C after 70 minutes of running, and the difference in $\Delta\tau$ between the PF and Al cylinders is not recognized. The actual temperature rises in packing are likely to be far higher than these values.

(3) *Leakage from the sealing device.* Figure 4.160 shows the apparatus used to measure leakage. With the value initially set for an air pressure of 4 kgf/cm² between the two packings in the cylinder, the relationships between the gauge pressure, p, representing the sealing pressure between two packings, and time, t, for different sliding times t_1 were determined. These results for A- and B-type packings are shown in Figs. 4.163 and 4.164, respectively.

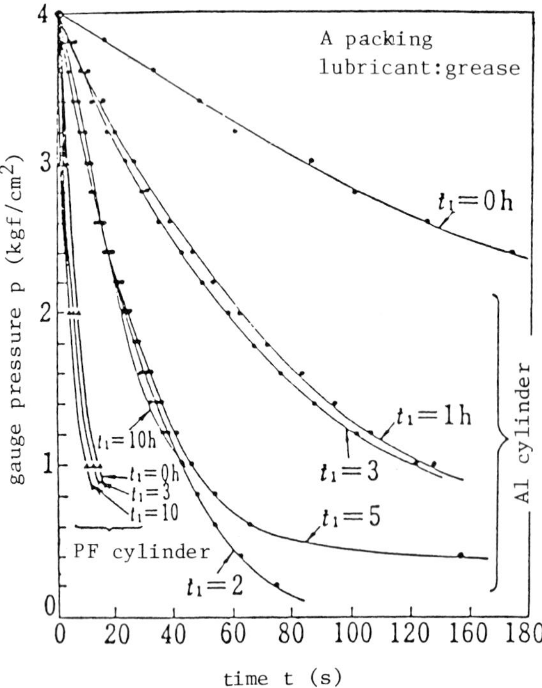

Fig. 4.163 Relationships between gauge pressure and time, t, for different sliding times, t_1, or an A packing with Al or PF cylinders

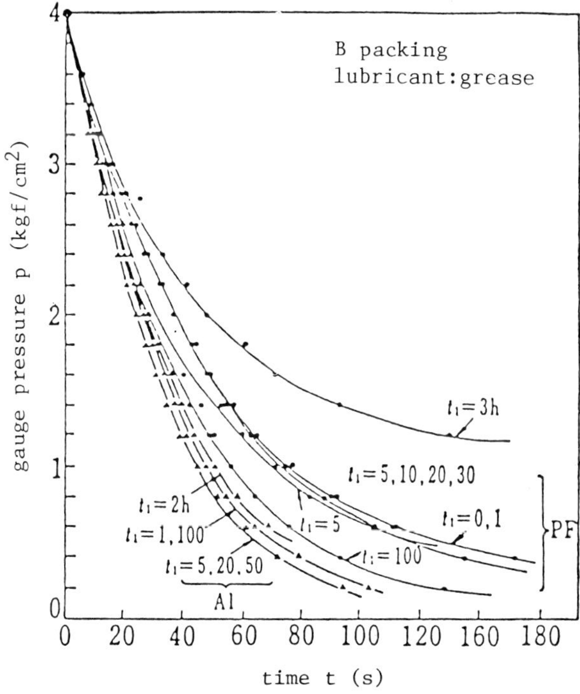

Fig. 4.164 Relationships between gauge pressure, p, and time, t, for different sliding times, t_1, of a B packing with Al or PF cylinders

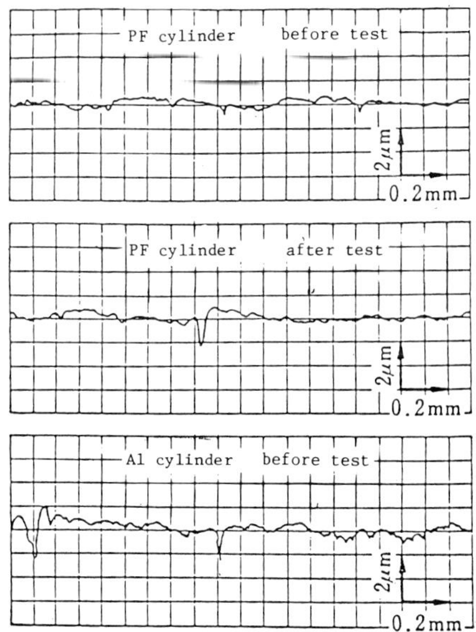

Fig. 4.165 Roughness of the inside surfaces of PF or Al cylinders

It is clear from these figures that the sealing pressure decreases with sliding time; that is, the amount of leakage increases with increasing sliding time, whilst the leakage in static conditions is minimal. The leakage rate for the combination of an A packing and an Al cylinder is smaller than that of an A packing and a PF cylinder. However, for a B packing and an Al cylinder, the leakage rate is greater than that of a B packing and a PF cylinder. The roughnesses of the cylinder wall surface before and after the experiments ranged from 1-2 μm, as shown in Fig. 4.165.

(4) *Radial pressure due to deformation of a packing.* In the first instance, in order to seek the radial pressure change due to sliding time the radial load P_g causing the diameter of the packing lip circle to shrink to a size equivalent to the inside diameter of the cylinder was measured using the apparatus shown in Fig. 4.166.

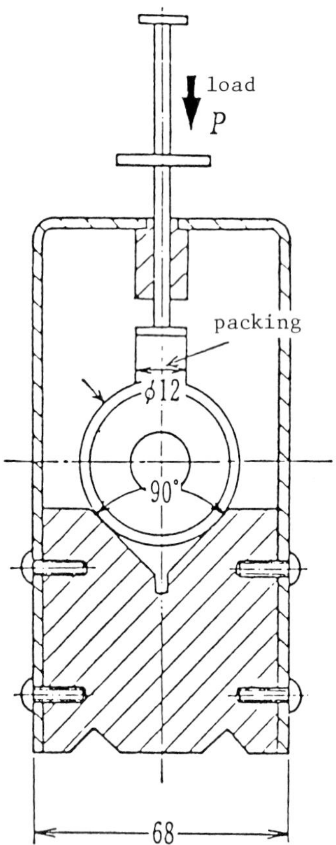

Fig. 4.166 Apparatus for measuring the load necessary to cause deformation of the packing

The relationships between P_g and sliding time, or distance, for each combination of packing and cylinder, with or without lubricant, are shown in Figs. 4.167 and 4.168.

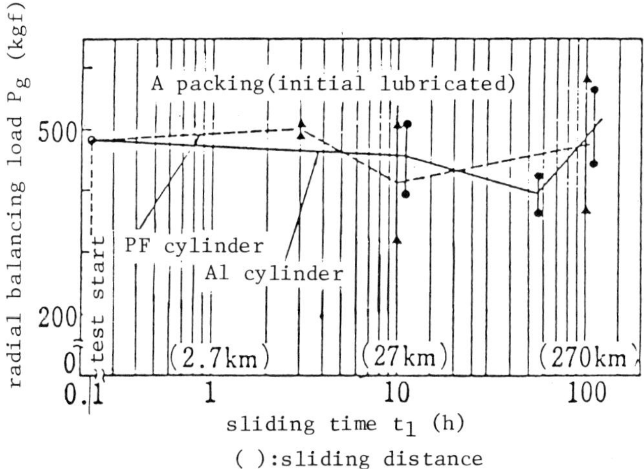

Fig. 4.167 Relationships between P_g and sliding time, t_1, for an A packing with Al or PF cylinders

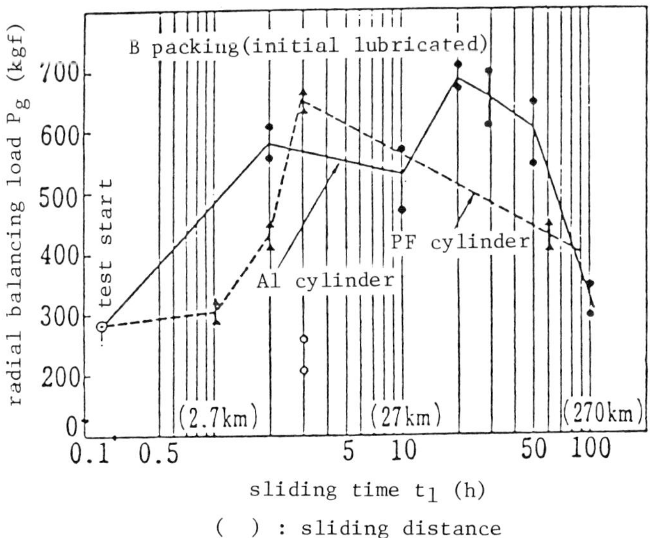

Fig. 4.168 Relationships between P_g and sliding time, t_1, for a B packing with Al or PF cylinders

Figure 4.167 gives the relationships for an A packing in both Al and PF cylinders with initial lubrication, and indicates that the P_g value decreases slightly over a sliding period of 10 hours. Figure 4.168 shows the same relationships for a B packing in both types of cylinder and again with initial lubrication. The P_g value increases during the initial 2 - 50 hours of sliding but then decreases gradually during the following 100 hours.

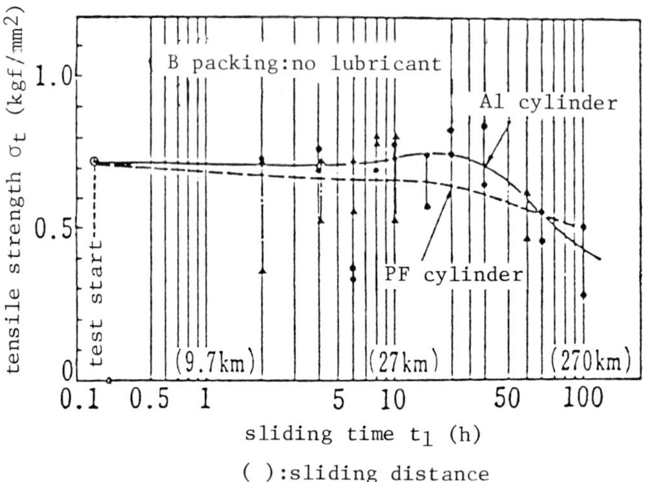

Fig. 4.169 Relationships between σ_t and sliding time, t, for a B packing with Al or PF cylinders

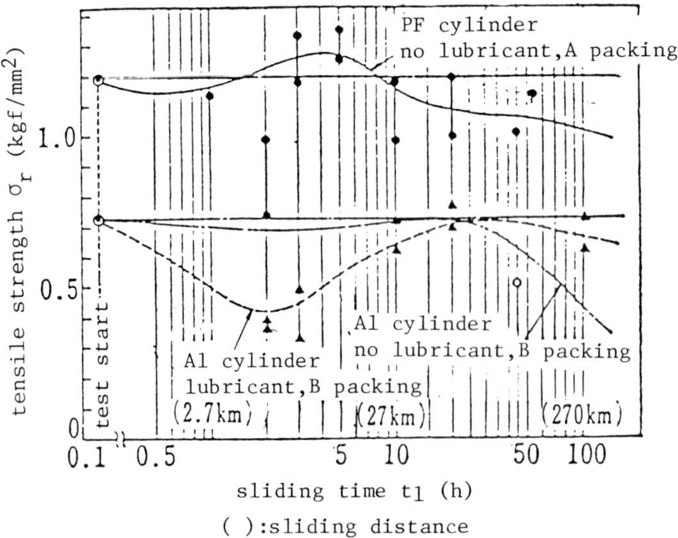

Fig. 4.170 Relationships between σ_t and sliding time, t, for A and B packings with Al or PF cylinders

(5) *Tensile strength of a packing material.* The change in tensile properties, such as tensile strength and elongation, were used as a scale for estimating the deterioration of packing materials during repeated sliding actions. Figures 4.169 and 4.170 show the relationships between the tensile strength of the packing material, σ_t, and sliding time, t_1, for various conditions. Figure 4.169 is for a B packing in both Al and PF cylinder without lubricant, and indicates that the tensile strength of packing, σ_t, decreases to 70% of its initial value after approximately 50 hours of sliding. The upper curves in Fig. 4.170 show similar relationships for the A packing and a PF cylinder without lubricant, and the characteristics are similar to those in Fig. 4.169. The lower curves in Fig. 4.170 show the relationships for a B packing and an Al cylinder with and without lubricant, and indicate that when a lubricant is present, σ_t only changes minimally during sliding up to 100 hours.

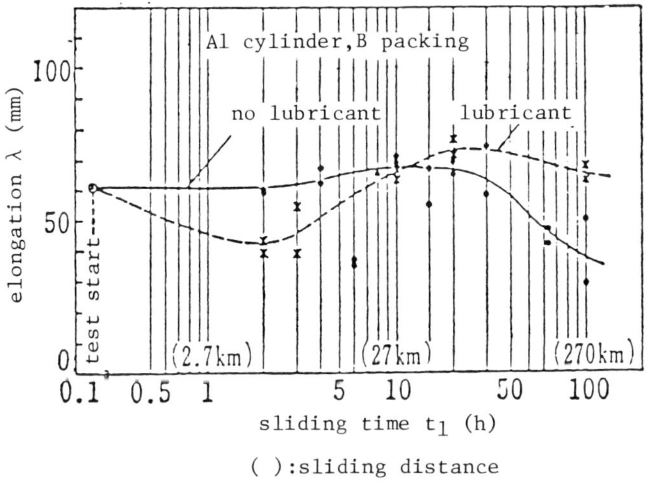

Fig. 4.171 Relationships between elongation, λ, and sliding time, t, for a B packing and an Al cylinder with or without lubricant

Figures 4.171 and 4.172 show the relationships between the elongation of the packing material, λ, and sliding time, t_1, for each combination. Fig. 4.171 is for B packing and an Al cylinder with or without lubricant, and indicates that without a lubricant λ decreases after 50 hours of sliding; the material has therefore deteriorated and increased in brittleness. With a lubricant, however, the λ value for this combination remains almost constant over this period of sliding.

Figure 4.172 shows the same relationships for A and B packings and a PF cylinder without lubricant. The λ value of the A packing is generally greater than that of B and decreases (corresponding to an increase in brittleness or hardening) after about 10 hours of sliding. The λ value for the B packing, however, decreases only slightly after 50 hours of sliding.

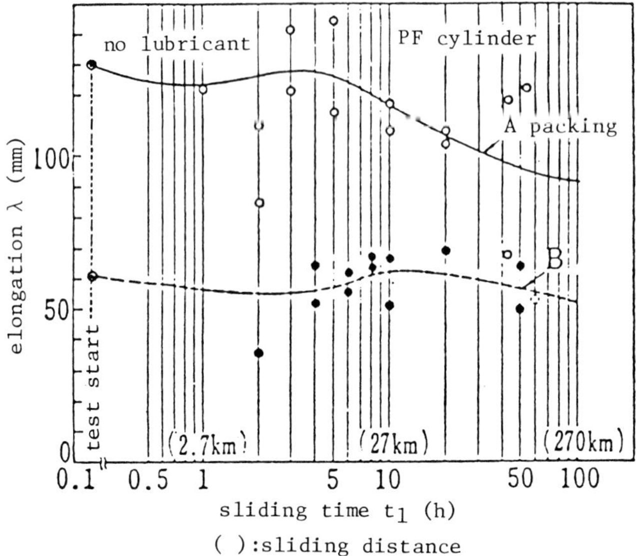

Fig. 4.172 Relationships between elongation, λ, and sliding time, t_1, for A and B packings and a PF cylinder without lubrication

(6) *Durometer hardness measurements.* The Durometer hardness can be used to characterize the deterioration of packing materials. The relationships between Durometer hardness, H_d, of a packing material, after sliding under various conditions, and sliding time, t_1 are shown in Fig. 4.173. For the combination of an A packing and a PF cylinder without lubricant, H_d value generally decreases slightly with an increase in sliding time.

(7) *Wear properties.* An example of the relationships between the wear volume of a packing, V, measured from the weight loss and sliding time are shown in Fig. 4.174. This figure is for the combination of a B packing in Al and PF cylinders with and without lubricant, and it indicates that the type of cylinder has little effect on V. The wear rate with a lubricant present is larger than that without lubricant for the first 5 hours, but this trend is then reversed.

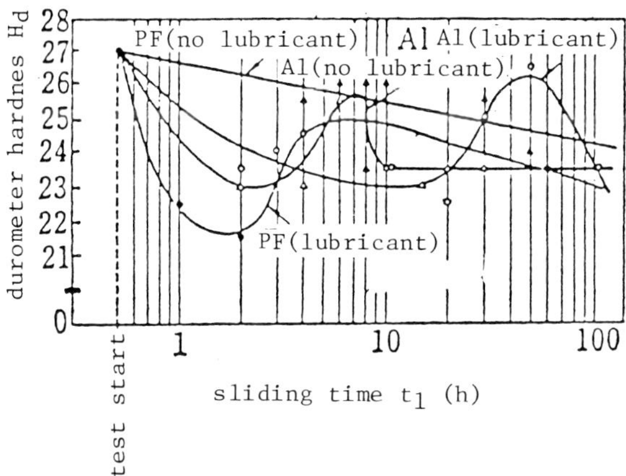

Fig. 4.173 Relationships between hardness H_d, and sliding time, t_1, for an A packing and Al or PF cylinders, with or without lubricant

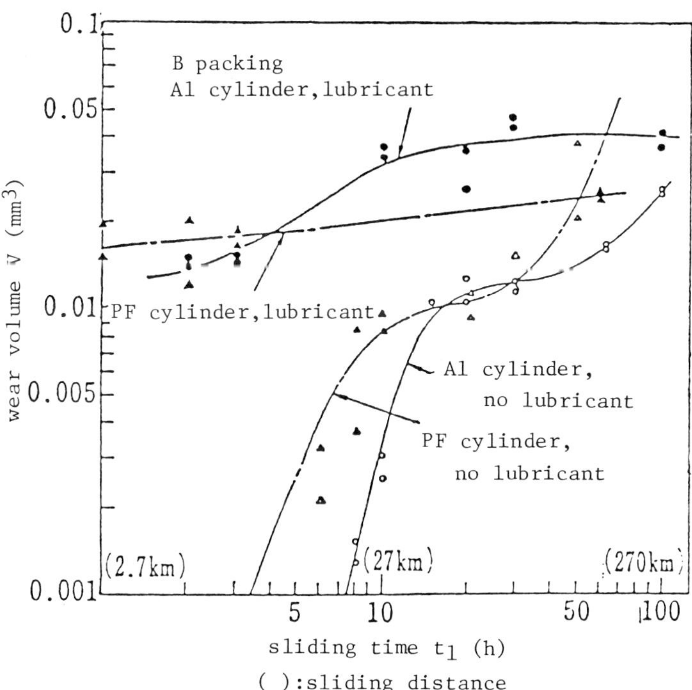

Fig. 4.174 Relationships between wear volume, V, and sliding time, t_1, for a B packing and Al or PF cylinders, with or without lubricant

REFERENCES

1 JSLE Lubrication Handbook, Yokendo, 1968, p. 621.
2 Standard Handbook of Lubrication Engineering, McGraw Hill Book Co., 1968, 18-6.
3 Sasaki, T. and Y. Sugimoto, Kyoto University Research Report, No. 1-No. 5; 1 (1952) 49; 2 (1952) 11; 3 (1952) 14 and 18; 4 (1953) 33; and 5 (1954) 92.
4 Craig, W.D., Lub. Eng., 20 (12) (1974) 456.
5 Pieper, H., Konstruktion, 22 (1970) 179.
6 Yamaguchi, Y., I. Sekiguchi et al., Kogakuin University Report, 26 (1971) 42.
7 Yamaguchi, Y., I. Sekiguchi et al., The Plastics (Goseijyushi) 17 (2) (1971) 42.
8 Yamaguchi, Y., I. Sekiguchi et al., Preprint of JSME, 218 (1969) 149.
9 Yamaguchi, Y., I. Sekiguchi et al., Preprint of JSME (1972) 2 or Kogakuin University Report, 33 (1973) 54.
10 Yamaguchi, Y., J. JSLE, 19 (1974) 718.
11 Sekiguchi, I., Y. Yamaguchi and T. Fukano, J. JSLE, 20 (1975) 513 or Proceedings of the JSLE-ASLE Int. Lub. Conf., 1975, p. 222.
12 Yamaguchi, Y., J. JSLE, 22 (1977) 215.
13 Sekiguchi, I., Y. Yamaguchi et al., Kogakuin Univ. Report, 45 (1978) 25.
14 Fuller, D.D., Theory and Practice of Lubrication for Engineers, John Wiley & Sons Inc., 1956, p. 235.
15 Braithwaite, E.R., Lubrication and Lubricant, Chapter 8; G.C. Pratt, Plastic-Based Bearing, Elsevier, Amsterdam, 1967, p. 377-426.
16 Clauss, F.J., Solid Lubricant and Self-Lubricating Solid, Academic Press, 1972, p. 146-163.
17 Kawasaki, KI., Oiless Bearing, Agne Co., 1973.
18 Willis, D.P., Machine Design, 27 (May, 1968) 130.
19 Twiss, S.B., Lub. Eng., 14 (1958) 255.
20 Pratt, G.C. and W.H. Wilson, Wear (1968) 73.
21 Yamaguchi, Y., I. Sekiguchi et al., Kogakuin Univ. Report, 33 (1973) 54.
22 Yamaguchi, Y., I. Sekiguchi, S. Takane and S. Shibata, J. JSLE, 25 (1980) 451 and International Edition No. 2 (1981) 169.
23 Yamaguchi, Y., I. Sekiguchi, M. Sakiyama and K. Mamiya, unpublished.
24 Wada, A., J. JSLE, 22 (1977) 589.
25 Matsubara, K. et al., Selection Point of Plastic Materials, JIS Pub. Co., 1976, p. 159 & 172.
26 Gremer, A., V.D.I., 97 (1955) 509.
27 Montalbano, J.F., Machine Design, 30 (1958) 96.
28 Boes, D.J., Lub. Eng., 19 (1963) 137.
29 Yamaguchi, Y., I. Sekiguchi et al., The Plastics, 15 (8) (1969) 54.
30 Yamaguchi, Y., I. Sekiguchi, K. Iwase and M. Sekine, J. JSLE, 18 (1973) 607.

31 Yamaguchi, Y., I. Sekiguchi et al., Kogakuin Univ. Report, 47 (1979) 41.
32 JSME Mechanical Engineering Handbook, 1968, 7-176.
33 Kanzaki, F., J. JSLE, 23 (1978) 241.
34 Wise, S. and G.R. Lewis, J. Inst. Mech. Eng., Railway Division 1, 4 (1970) 386.
35 Banard, J.H., Rail, International 1, 10 (1970) 694.
36 Tanaka, K., K. Nakata and S. Ueda, J. JSLE, 13 (1968) 662.
37 Fukuoka, K., The Plastics, 12 (8) (1966) 471.
38 Yamaguchi, Y., Engineering Materials, 22 (5) (1974) 31
39 Murata, K. and T. Takemoto, The Plastics, 16 (6) (1970) 337.
40 Nishimura, S. and S. Watanabe, The Plastics, 16 (6) (1970) 354.
41 Idemura, K., J. JSLE, 18 (1973) 805.
42 Idemura, K., J. JSLE, 21 (1976) 133.
43 Idemura, K., Engineering Materials, 25 (5) (1977) 38.
44 Hoshino, T., J. JSLE, 15 (1970) 425.
45 Hashikura, M. and T. Sugiki, Kikaishikensho Report, 8 (1954) 5, 30.
46 Hoshino, T., Kikaishikensho Report, 59 (1971) 11.
47 Fujii, K., J. JSLE, 18 (1973) 813.
48 Yoshikawa, Y., The Plastics, 16 (6) (1970) 349.
49 Yamaguchi, Y., I. Sekiguchi et al., Kogakuin Univ. Report, 24 (1973) 42.
50 Sekiguchi I., Y. Yamaguchi, H. Goto, W. Kono and K. Miyakoshi, J. JSLE, 21 (1976) 838.
51 Sekiguchi, I., Y. Yamaguchi et al., Kogakuin Univ. Report, 46 (1979) 39.
52 Sekiguchi, I., Y. Yamaguchi et al., The Plastics, 26 (10) (1980) 6.
53 Sekiguchi, I., Y. Yamaguchi, Y. Katsu, H. Kamoshida and T. Suzuki, J. JSLE, 27 (1982) 845.
54 Fukaya, T., JSME, 38th Meeting Text, 1973.
55 Sekiguchi, I., Y. Yamaguchi et al., Kogakuin Univ. Report, 40 (1979) 39.
56 Yamaguchi, Y., I. Sekiguchi et al., Kogakuin Univ. Report, 32 (1972) 65.
57 For example, Japan Railway Car Co.: Report on Special Railway Car Brake, 10 (1971).
58 Sekiguchi, I., Y. Yamaguchi et al., The Plastics, 26 (10) (1980) 6.
59 Takanashi, S. and H. Takahashi, J. JSPE, 32 (1966) 344.
60 Takanashi, S. and S. Shiraishi, Tohoku Univ. Report, 17 (1) (1969) 13.
61 Takanashi, S. and S. Shiraishi, Tohoku Univ. Report, 19 (1) (1970) 13.
62 Kudo, T. and S. Yamashiro, J. JSPE, 27 (1961) 142.
63 Arai, S., Kobunshi (High Polymers), 13 (1964) 18.
64 Aoto et al., JSME No. 46, Preprint No. 217 (1964) 73.
65 Ikegami, K., T. Sugibayashi and Y. Suzuki et al., Transaction of JSME, C47 (1981) 416, 466.
66 Zumstein, Am. Machinist, 95 (1951) 118.
67 Lurie, General Electric Review, 55 (1953) 34.
68 Kuhn, Mod. Plastics, 31 (1) (1953) 34.
69 Martin, Machinery, 62 (1955) 151 & 162.

70 Pfluger, British Plastics (1968) 522.
71 Yamaguchi, Y., Y. Oyanagi et al., Kogakuin Univ. Report, 26 (1969) 11.
72 Yamaguchi, Y., Y. Oyanagi et al., The Plastics, 15 (7) (1969) 1.
73 Yamaguchi, Y., Y. Oyanagi et al., The Plastics, 16 (10) (1970) 54.
74 Yamaguchi, Y., Machinist, 7 (1968) 48.
75 Oyanagi, Y., Y. Yamaguchi, S. Yasuda, T. Matsui and S. Mochizuki, J. JSTP, 17 (1976) 296.
76 Oyanagi, Y., Engineering Materials, 27 (5) (1979) 85.
77 Tsukamoto, N., Papers of JSME, 780 (8) (1978) 103.
78 Tsukamoto, N., Papers of JSME, 790 (75) (1979) 246.
79 Tsukamoto, N., Papers of JSME, 800 (15) (1980) 55.
80 Senba, M., Gear Wheel, Nikkankogyo Press, 1 (1953) 154.
81 Nakano, Y., Rolling Bearing, Nikkankogyo Press, 29 (1969).
82 Yamaguchi, Y. and I. Sekiguchi, J.JSLE, 11 (1966) 485.
83 Iwanami, S. and T. Kondo, Packing Handbook, Sangyo Pub. Co., 5, 1962.
84 J. JSME: Packing and Gasket Div. Report, 66 (1963) 1497 and 67 (1964) 1212.
85 Smoley, M. and E.C. Frazier, Mach. Design, 36 (15) (1964) 75.
86 Dunkle, H.H., Mach. Design, 34 (15) (1964) 90.
87 Watanabe, T., Selection of Packing, Nikkankogyo Press, 1963.
88 Hoda, T., Engineering Materials, 14 (2) (1966) 49.
89 Pool, E.B., J. 3rd England Seal. So., A3-29 (1967)
90 Yokoi, M., Engineering Materials, 11 (9) (1963) 94.
91 Nişhida, T., Valker Review, 24 (10) (1980) 1.
92 Hirano, F. et al., J. JSLE, 18 (1973) 587.
93 Hirawa, H.., J. JSLE, 24 (1979) 276.
94 Tagami, J. et al., J. JSLE, 13 (1968) 178
95 Ishiwatari, H. et al., J. JSLE, 14 (1969) 211.
96 Hirano, F. et al., J. JSLE, 24 (1979) 261.
97 Yamaguchi, Y. and H. Shinnabe, Kogakuin Univ. Report, 35 (1973) 40.
98 Amano, S., Y. Yamaguchi et al., 26th JSPT Preprint (1980) 86.